SYMBOLS

English Letters (*continued*)

MS	mean square	**17.5**
n	sample size	**4.3**
Pr	probability	**9.7**
p	probability of some outcome, given that the null hypothesis is true	**14.6**
r	sample Pearson correlation coefficient	**7.3**
r_s	sample Spearman correlation coefficient	**7.8**
S	sample standard deviation (descriptive statistics)	**5.6**
$S_{y \cdot x}$	standard error of prediction	**8.6**
SS	sum of squares	**17.6**
s	sample standard deviation (inferential statistics)	**13.2**
$s_{\bar{D}}$	sample standard deviation of difference scores	**15.3**
$s_{\bar{X}}$	estimated standard error of the mean	**13.2**
$s_{X_1 - X_2}$	$\{$ estimated standard error of the difference between two sample means **14.8**	
$s_{\bar{D}}$	estimated standard error of the mean of difference scores	**15.3**
s^2	sample variance	**14.7**
s_p^2	pooled sample variance	**14.7**
T	Wilcoxon **T** test for ranked data	**20.10**
t	t ratio	**13.3**
U	Mann-Whitney **U** test for ranked data	**20.3**
X	any observation or score	**4.3**
$\bar{X}$	sample mean	**4.3**
$x_1 - x_2$	difference between two sample means	**14.1**
Y	a score paired with **X**	**7.2**
Y'	predicted score	**8.4**
z	$\{$ standard score (descriptive statistics) **6.2** z ratio (inferential statistics) **11.3**	

STATISTICS

ROBERT S. WITTE

San Jose State University

HOLT, RINEHART AND WINSTON, PUBLISHERS

New York • Toronto • London • Sydney

© 1980 by Holt, Rinehart and Winston
All rights reserved
Library of Congress Catalog Card Number: 78-57922
ISBN: 0-03-055231-1
Printed in the United States of America
0123   039   987654321

TO DORIS

Preface

You could go through life quite successfully without ever learning statistics. Having learned some statistics, however, you will be less likely to flinch and change the topic when numbers enter a discussion; you will be more skeptical of conclusions based on loose or erroneous interpretations of sets of numbers; you will even be more prone to initiate a statistical analysis of some problem within your special area of interest.

People often approach statistics with great apprehension. For many, it is a required course to be taken only under the most favorable circumstances, such as during a quarter or semester when a light course load may be carried; for others, it is as distasteful as a visit to the dentist—to be postponed as long as possible, with the vague hope that the problem might miraculously disappear. Much of this apprehension doubtless rests on the widespread fear of mathematics and mathematically related areas.

You need not fear statistics. All unnecessary quantitative considerations have been eliminated from this book. When not obscured by mathematical treatments better reserved for more advanced books, some of the beauty of statistics, as well as its everyday usefulness, reveals itself to virtually everybody. The chances are excellent that you too will have this experience.

TO THE INSTRUCTOR

This book is designed for the introductory course offered by a variety of departments, including psychology, sociology, education, and health education. The level of presentation is geared to the non-mathematical student commonly found in this course.

The book contains more material than is covered in most one-quarter or one-semester courses. Various chapters can be omitted without interrupting the main development. An instructor who wishes to emphasize inferential statistics could omit Chapter 7 (Measures of Relationship: Correlation) and Chapter 8 (Prediction), while an instructor who desires a more applied emphasis could omit Chapter 12 (Two Types of Error and Sample Size Selection). After Chapter 14 (t Test for Two Independent Samples), you can proceed in cafeteria style, selecting only those chapters that are most relevant to the needs of your students.

ACKNOWLEDGEMENTS

I wish to acknowledge my children: Steve, Mike, Andrea, and John; brothers and sisters: Lila, J. Stuart, the late A. Gerhart, Henrietta, and Henry; deceased parents: Henry and Emma; and all friends and relatives, past and present, including Arthur, Betty, Bob, Brian, Cal, David, Floyd, George, Helen, Jim, Joyce, Kayo, Ralph, and Suzanne.

Numerous helpful comments were made by those who reviewed all or part of the book: Robert Clarke (San Jose State University), Mervin Dissinger (Rider College), Robert Fried (Hunter College, City University of New York), Joseph Porter (Virginia Commonwealth University), Calvin Thomson (San Jose State University), and Thomas Wickens (University of California, Los Angeles).

Excellent editorial and production support was supplied by Lloyd Black, Tom O'Connor, Amy Shapiro and Paul Streeter—all of W.B. Saunders Company. A special thanks to Baxter Venable, Psychology Editor at W.B. Saunders Company during most of the development of this book.

I am grateful to the Literary Executor of the late Sir Ronald A. Fisher, F.R.S., to Dr. Frank Yates, F.R.S., and to the Longman Group Ltd., London, for permission to reprint Tables IIi and IV from their book *Statistical Tables for Biological, Agricultural and Medical Research*, (6th ed., 1974).

Palo Alto, California
September, 1979

Robert S. Witte

Contents

Chapter 1

INTRODUCTION 1
    1.1   How to Use This Book 1
    1.2   What is Statistics? 1
THREE TYPES OF DATA 2
    1.3   Quantitative Data 2
    1.4   Interval-Ratio Measurement 2
    1.5   Measurement of Nonphysical Characteristics:
           A Complication 3
    1.6   Ranked Data 4
    1.7   Ordinal Measurement 4
    1.8   Qualitative Data 4
    1.9   Nominal Measurement 5
    1.10   Ordered Qualitative Data 5
    1.11   An Overview 5
    1.12   Summary 7

PART I DESCRIPTIVE STATISTICS

Chapter 2

ORGANIZING AND SUMMARIZING DATA WITH TABLES ... 11
    FREQUENCY DISTRIBUTIONS FOR QUANTITATIVE DATA   11
    2.1   Ungrouped Data 11
    2.2   Stem and Leaf Arrangements 12
    2.3   Grouped Data 13
    2.4   Minimizing Irregularities and Ambiguities
           in Format 13
    2.5   Selection of an Interval Width 15
    2.6   Gaps Between Boundaries 15

OTHER TYPES OF FREQUENCY DISTRIBUTIONS 16
   2.7   Relative Frequency Distributions 16
   2.8   Cumulative Frequency Distributions 17
   2.9   Frequency Distributions for Qualitative Data 17
   2.10 Frequency Distributions for Ranked Data 18
   2.11 Summary 18

Chapter 3

DESCRIBING DATA WITH GRAPHS 23
GRAPHS FOR QUANTITATIVE DATA 23
   3.1   Histograms 23
   3.2   Frequency Polygons 24
GRAPHS FOR QUALITATIVE DATA 26
   3.3   Bar Graphs 26
   3.4   Pie Charts 26
MISLEADING GRAPHS 26
   3.5   Some Tricks 26
   3.6   An Overview 28
   3.7   Summary 28

Chapter 4

DESCRIBING DATA WITH AVERAGES 35
AVERAGES FOR QUANTITATIVE DATA 35
   4.1   Mode 35
   4.2   Median 36
   4.3   Mean 36
   4.4   Which Average? 37
   4.5   Positively and Negatively Skewed Distributions . 39
   4.6   A Special Property of the Mean 39
   4.7   Averages for Grouped Data 41
AVERAGES FOR QUALITATIVE AND RANKED DATA 41
   4.8   Averages for Qualitative Data 41
   4.9   Averages for Ranked Data 41
   4.10 Using the Word "Average" 42
   4.11 Summary 43

Chapter 5

DESCRIBING VARIABILITY 46
MEASURES OF VARIABILITY FOR QUANTITATIVE DATA 46
   5.1   Range 46
   5.2   Standard Deviation 47
   5.3   Deviation Formula 48
   5.4   A Measure of Distance 49
   5.5   Raw Score Formula 50
   5.6   Variance 51

MEASURES OF VARIABILITY FOR QUALITATIVE AND
RANKED DATA 51
    5.7   General Comments 51
    5.8   Summary 52

Chapter 6

NORMAL DISTRIBUTIONS AND STANDARD SCORES 54
  NORMAL DISTRIBUTIONS 54
    6.1   The Normal Curve 54
    6.2   z Scores 56
    6.3   Standard Normal Curve 57
    6.4   Standard Normal Table 57
    6.5   Finding Areas 59
    6.6   Finding Scores 63
  STANDARD SCORES 65
    6.7   More About z Scores 65
    6.8   Other Standard Scores 66
    6.9   Percentile Ranks 67
    6.10 Summary 68

Chapter 7

MEASURES OF RELATIONSHIP: CORRELATION 71
    7.1   An Intuitive Approach 71
    7.2   Scatterplots 73
    7.3   Correlation Coefficient for Quantitative Data: r . . 75
    7.4   z Score Formula for r 75
    7.5   Raw Score Formula for r 77
    7.6   Interpretation of r 78
    7.7   Correlation Not Necessarily Cause-Effect 79
    7.8   A Correlation Coefficient for Ranked Data: r_s 79
    7.9   Summary 81

Chapter 8

PREDICTION 84
    8.1   A "Rough" Prediction 84
    8.2   A Prediction Line 84
    8.3   Least Squares Prediction Line 86
    8.4   Least Squares Equation 86
    8.5   Prediction: An Overview 88
    8.6   Standard Error of Prediction, $S_{Y \cdot X}$ 88
    8.7   Other Sources of Error 89
    8.8   Regression 89
    8.9   Assumptions 91
    8.10 Summary 91

PART II INFERENTIAL STATISTICS

Chapter 9

POPULATIONS, SAMPLES, AND PROBABILITIES 97
   9.1   Why Samples? 97
   9.2   Populations 98
   9.3   Samples 98
   9.4   Random Samples 99
   9.5   Tables of Random Numbers 100
   9.6   Some Complications 101
   9.7   Random Assignment of Subjects 101
   9.8   Probability 102
   9.9   Summary 104

Chapter 10

SAMPLING DISTRIBUTIONS 107
   10.1   An Example 107
   10.2   A Closer Look at Sampling Distributions 108
   10.3   Some Important Symbols 109
   10.4   Mean of All Sample Means 111
   10.5   Standard Error of the Mean 111
   10.6   Shape of the Sampling Distribution 112
   10.7   Why the Central Limit Theorem Works 114
   10.8   Types of Sampling Distributions 114
   10.9   Summary 114

Chapter 11

INTRODUCTION TO HYPOTHESIS TESTING: THE z TEST 116
   11.1   An Example 116
   11.2   Hypothesis Testing: An Overview 118
   11.3   The z Test 119
   11.4   Statement of Problem 121
   11.5   Null Hypothesis (H_0) 121
   11.6   Alternative Hypothesis (H_1) 122
   11.7   Statistical Test 122
   11.8   Decision Rule 122
   11.9   Calculations 123
   11.10   Decision 123
   11.11   Interpretation 123
   11.12   One-tailed and Two-tailed Tests 124
   11.13   Choosing a Level of Significance (α) 126
   11.14   Summary 127

Chapter 12

TWO TYPES OF ERROR AND SAMPLE SIZE SELECTION 131
    12.1   Vitamin C Experiment 131
    12.2   Four Possible Outcomes 133
    12.3 H_0 is True (& the Type I Error) 133
    12.4 H_0 is False (& the Type II Error) 134
    12.5   Influence of Sample Size 136
    12.6   Selection of Sample Size 137
    12.7   Small, Medium, and Large Effects 139
    12.8   Additional Comments about the Sample Size Table 139
    12.9   Summary 140

Chapter 13

t **TEST FOR ONE SAMPLE** 143
  t TEST FOR A POPULATION MEAN 143
    13.1   A Gas Mileage Investigation 143
    13.2   Estimating the Population Standard Deviation .. 144
    13.3   *t* Ratio 144
    13.4   *t* Distribution 145
    13.5   *t* Tables 146
    13.6   *t* Test for Gas Mileage Investigation 146
    13.7   Assumptions 147
    13.8   Degrees of Freedom 147
    13.9   Hypothesis Tests: An Overview 150
  t TEST FOR A POPULATION CORRELATION COEFFICIENT 150
    13.10 *t* Test for the Greeting Card Exchange 150
    13.11 Assumptions and a Limitation 151
    13.12 Summary 152

Chapter 14

t **TEST FOR TWO INDEPENDENT SAMPLES** 154
    14.1   Blood-doping Experiment 154
    14.2   Two Populations 155
    14.3   Sampling Distribution of $\overline{X}_1 - \overline{X}_2$ 155
    14.4   Mean of the Sampling Distribution 155
    14.5   Standard Error of the Sampling Distribution 156
    14.6   z Test 156
    14.7   Estimating the Population Variance 156
    14.8   Estimating the Standard Error 157
    14.9   *t* Ratio 157
    14.10 *t* Test for the Blood-doping Experiment 158
    14.11 Selection of Sample Size 159
    14.12 Assumptions 161
  MORE GENERAL COMMENTS ABOUT HYPOTHESIS TESTS 161
    14.13 Secondary Status of the Null Hypothesis 161

14.14 Statistical Significance vs. Practical Importance . 162
14.15 A Note on Usage 162
14.16 Suspending Judgment about the Null Hypothesis 163
14.17 Summary 163

Chapter 15

t TEST FOR TWO DEPENDENT SAMPLES 167
15.1   Matching Pairs of Athletes in the Blood-doping
       Experiment 167
15.2   Sampling Distribution of $\bar{X}_1$ - $\bar{X}_2$ for
       Two Dependent Samples 168
15.3   *t* Ratio 168
15.4   *t* Test for the Blood-doping Experiment
       (Two Dependent Samples) 169
15.5   Selection of Sample Size 170
15.6   Assumptions 172
15.7   To Match or Not to Match? 172
15.8   Using the Same Subject in Both Groups 172
15.9   Hypothesis Tests for Population Means:
       An Overview 173
15.10 Summary 173

Chapter 16

ESTIMATION 177
ESTIMATING ONE POPULATION MEAN 177
16.1   Investigation of SAT Verbal Scores, Revisited ... 177
16.2   Point Estimates for μ 178
16.3   Confidence Intervals for μ 178
16.4   Why Confidence Intervals Work 178
16.5   Confidence Intervals for μ Based on z 179
16.6   Interpretation of a Confidence Interval 181
16.7   Level of Confidence 181
16.8   Effect of Sample Size 182
16.9   Confidence Intervals for μ Based on *t* 182
ESTIMATING THE DIFFERENCE BETWEEN
POPULATION MEANS 184
16.10 Blood-doping Experiment, Revisited 184
16.11 Point Estimates for μ_1 - μ_2 184
16.12 Confidence Intervals for μ_1 - μ_2 184
16.13 Confidence Intervals for μ_1 - μ_2 Based on *t*
       (Two Independent Samples) 184
16.14 Confidence Intervals for μ_1 - μ_2 Based on *t*
       (Two Dependent Samples) 186
GENERAL COMMENTS ABOUT CONFIDENCE INTERVALS 186
16.15 Hypothesis Tests or Confidence Intervals? 186
16.16 Other Types of Confidence Intervals 187
16.17 Summary 187

Chapter 17

ANALYSIS OF VARIANCE (ONE FACTOR) · · · · · · · · · · · · · · · · · · 190

17.1 Introduction 190
17.2 Two Sources of Variability 191
17.3 F Ratio 192
17.4 F Test 193
17.5 Variance Estimates 194
17.6 Sum of Squares (SS) 195
17.7 Degrees of Freedom (df) 196
17.8 Mean Squares (MS) and the F Ratio 197
17.9 F Tables 199
17.10 Notes on Usage 199
17.11 Multiple Comparisons 200
17.12 Scheffé's Test 201
17.13 F Test is Nondirectional 202
17.14 Assumptions 202
17.15 Summary 203

Chapter 18

ANALYSIS OF VARIANCE (TWO FACTORS) · · · · · · · · · · · · · · · · 207

18.1 Data Interpretation with Graphs 208
18.2 Three F Ratios 210
18.3 Variance Estimates 212
18.4 Sum of Squares (SS) 212
18.5 Degrees of Freedom (df) 214
18.6 Mean Squares (MS) and F Ratios 214
18.7 F Tables 214
18.8 Interaction 215
18.9 Multiple Comparisons 217
18.10 Assumptions and a Final Caution 217
18.11 Other Types of ANOVA 217
18.12 Summary 217

Chapter 19

CHI-SQUARE (χ^2) TEST FOR QUALITATIVE DATA · · · · · · · · · · · 221

ONE VARIABLE 221
19.1 Observed and Expected Frequencies 221
19.2 Calculation of χ^2 222
19.3 χ^2 Tables and Degrees of Freedom 222
19.4 χ^2 Test 223
19.5 χ^2 Test is Nondirectional 225
19.6 Special Case: $df = 1$ 225
TWO VARIABLES 227
19.7 Observed and Expected Proportions 228
19.8 Expected Frequencies 229
19.9 χ^2 Tables and Degrees of Freedom 231

19.10 χ^2 Test 231
19.11 Special Case: $df = 1$ 232
19.12 Some Precautions 233
19.13 Summary 233

Chapter 20

TESTS FOR RANKED DATA 236
20.1 Three Tests for Ranked Data 236
MANN-WHITNEY U TEST (TWO INDEPENDENT SAMPLES) 237
20.2 Why Not a t Test? 237
20.3 Calculation of U 237
20.4 U Tables 239
20.5 Decision Rule 239
20.6 Null Hypothesis 240
20.7 Directional Tests 240
20.8 Large Sample Approximation for U 241
WILCOXON T TEST (TWO DEPENDENT SAMPLES) 241
20.9 Why Not a t Test? 242
20.10 Calculation of T 242
20.11 T Tables 243
20.12 Decision Rule 243
20.13 Null Hypothesis 244
20.14 Directional Tests 244
20.15 Large Sample Approximation for T 244
KRUSKAL-WALLIS H TEST
(THREE OR MORE INDEPENDENT SAMPLES) 245
20.16 Why Not an F Test? 246
20.17 Calculation of H 246
20.18 χ^2 Tables 246
20.19 Decision Rule 246
20.20 Null Hypothesis 247
20.21 H Test is Nondirectional 247
GENERAL COMMENTS 247
20.22 Ties 247
20.23 A Note on Terminology 248
20.24 Final Caution 248
20.25 Summary 248

APPENDIX A

Math Review 253
A.1 Pretest 253
A.2 Common Symbols 254
A.3 Order of Operations 254
A.4 Positive and Negative Numbers 255
A.5 Fractions 255
A.6 Square Root Radicals ($\sqrt{}$) 256

A.7   Rounding Numbers 257
A.8   Posttest 257
A.9   Answers (with Relevant Review Sections) 257

APPENDIX B

Answers to Exercises 259

APPENDIX C

Tables 297
  A.   Areas Under Standard Normal Curve for
       Values of z 297
  B.   Critical Values of t 300
  C.   Critical Values of F 301
  D.   Critical Values of χ^2 305
  E.   Critical Values of Mann-Whitney U 306
  F.   Critical Values of Wilcoxon T 308
  G.   Random Numbers 309
  H.   Approximate Sample Size for Hypothesis Test
       About Population Means (z and t Tests) 310

INDEX 311

1

Introduction

1.1 –How to Use This Book
1.2 –What Is Statistics?

THREE TYPES OF DATA

1.3 –Quantitative Data
1.4 –Interval-Ratio
    Measurement
1.5 –Measurement of
    Nonphysical
    Characteristics: A
    Complication
1.6 –Ranked Data
1.7 –Ordinal Measurement
1.8 –Qualitative Data
1.9 –Nominal Measurement
1.10–Ordered Qualitative Data
1.11–An Overview
1.12–Summary
1.13–Exercises

If you can add, subtract, multiply, and divide, your math background is adequate for this book.

Statistics consists of two major areas: an older area, descriptive statistics, and a newer area, inferential statistics.

Statistical tools are designed for use with words and labels, as well as with numbers.

Your zip code, your birth-order position in your family, and your current age reflect three different kinds of measurement and must be interpreted accordingly.

1.1 How To Use This Book

This book contains a number of features that will aid your study of statistics. The math review in Appendix A summarizes the basic math symbols and operations used throughout this book. If you're anxious about your math background—almost everyone is—check out Appendix A as soon as possible. Be assured that no special math background is required of you. If you can add, subtract, multiply, and divide, you can learn (or relearn) the simple math described in Appendix A. If this material looks unfamiliar, it would be a good idea to study Appendix A within the next few days (and to review this material throughout the term, as necessary).

Topic headings and important formulas are highlighted with color. Each chapter begins with both an outline and a preview and ends with a summary, including a list of important terms. Use these aids to gain an orientation before reading a new chapter and to facilitate your review of old chapters. Frequent reviews are desirable since statistics is cumulative, with earlier topics forming the basis for later topics.

Exercises are listed at the end of each chapter. Don't shy away from exercises; they will clarify and expand your understanding, as well as improve your ability to work with statistics. Solutions to all exercises are given in Appendix B.

As a reference aid, important statistical symbols are defined on the back of the front book cover, and a guide to important statistical tests appears on the corresponding part of the back book cover.

1.2 What Is Statistics?

Statistics consists of two main subdivisions: descriptive statistics and inferential statistics. The historically older area, **descriptive statistics,** supplies a number of tools such as tables, graphs, averages, and so on for *organizing and summarizing information about a collection of observations*. Automobile advertisements, stock market reports, baseball record books, environmental impact studies—all use statistical tools to organize and summarize information. Think of these everyday items as your headstart exposure to descriptive statistics.

1

The other area, **inferential statistics,** which is primarily a product of the twentieth century, supplies a number of tools for *generalizing information from a relatively small collection of observations, a "sample," to a relatively large collection of potential observations, a "population."* Tools from inferential statistics permit us to use samples to check a manufacturer's performance claim for a population of cars, a researcher's hypothesis that a population of meditators averages fewer headaches than a comparable population of nonmeditators, and a pollster's projection of a presidential winner on the basis of preliminary election returns. Most likely, you have little or no prior acquaintance with inferential statistics. With the proper mix of exposure and effort, however, you will eventually claim the area as your own.

In this book you will encounter the most essential tools and concepts of descriptive statistics (Part I) beginning with Chapter 2 and those of inferential statistics (Part II) beginning with Chapter 9.

THREE TYPES OF DATA

A statistical analysis is performed on data, that is, on a collection of observations from a survey or an experiment. As will be seen in later chapters, the precise form of a statistical analysis often depends on whether data are quantitative, ranked, or qualitative. It's important, therefore, to distinguish among these three types of data. These distinctions are based, in part, on the kind of measurement—or level of measurement—that produces a given set of data, and this fact is reflected in subsequent sections.

1.3 Quantitative Data

When observations are described by numbers that indicate an amount or a count, the data are **quantitative.** The weights reported by 53 female statistics students in Table 1.1 are quantitative data since each number indicates an amount of weight. Other examples of quantitative data include sets of observations based on IQ scores (where, for instance, 113, 97, 136, 104 represent amounts of intellectual aptitude), family size (where 2, 7, 4, 3 represent counts of people), and daily cigarette consumption (where 10, 42, 12, 36 represent counts of cigarettes smoked each day).

1.4 Interval-Ratio Measurement

You don't need a statistics book to tell you that the difference between students who weigh 120 and 130 pounds represents the same amount of weight as the difference between students who weigh 165 and 175 pounds. Or that someone who weighs 200 pounds possesses twice the amount of weight as someone who weighs 100 pounds. It might be news to you, however, that these interpretations are valid only because the assignment of numbers to weight—that is, the measurement of weight—complies with two special properties known as "equal intervals" and a "true zero."

When measuring weight with a bathroom scale, the achievement

TABLE 1.1
QUANTITATIVE DATA: WEIGHTS (IN POUNDS) OF FEMALE STATISTICS STUDENTS

120	128	93	130	110	125	118	125
123	129	205	120	112	150	139	117
112	120	130	140	110	116	150	116
117	153	186	118	185	95	125	95
140	132	120	130	105	145	112	
165	111	180	126	112	119	116	
125	117	150	166	132	135	114	

of equal intervals appears to be ridiculously easy. For instance, if a bathroom scale registers 121 pounds after a book is handed to a 120-pound person, then it also should register 176 pounds after the same book is handed to a 175-pound person, and so on, regardless of the original weight of the person on the bathroom scale. Having established that the book, as a standard weight, always registers a one-pound increase, we could also demonstrate that ten identical books always register a ten-pound increase. If you think about it, you'll see that this is just a roundabout way of demonstrating our earlier statement that the difference between 120 and 130 pounds represents the same amount of weight as the difference between 165 and 175 pounds.

To fully appreciate the achievement of a measurement scale with equal intervals, remember that numbers are being used to measure some characteristic. There's no guarantee that the consecutive numbers on a measurement scale really represent equal increments in the characteristic being measured. As will be discussed in the next section, it's particularly difficult to achieve a measurement scale with equal intervals when nonphysical characteristics, such as intellectual aptitude, are being measured.

When measuring weight with a bathroom scale, the achievement of a true zero also is easy. Simply verify that the bathroom scale registers 0 when not in use, that is, when weight is completely absent. In this case, the bathroom scale possesses a true zero, and it's appropriate to express one number as some multiple (or ratio) of another. It can be claimed that 200 pounds is twice the amount of weight of 100 pounds. In other words, when two people, each weighing 100 pounds, occupy the same bathroom scale, it registers 200 pounds.

In the absence of a true zero, numbers—much like the exposed tips of icebergs—fail to reflect the total amount of the characteristic being measured. For example, a zero reading on the Fahrenheit temperature scale does not reflect the complete absence of heat or molecular motion and thus fails to qualify as a true zero. In fact, true zero on this measurement scale equals −459.4°F. It would be inappropriate, therefore, to claim that 80°F is twice as hot as 40°F. An appropriate claim could be salvaged by adding 459.4 to each of these numbers. Thus 80° becomes 539.4° and 40° becomes 499.4°. Clearly, 539.4° is not twice as hot as 499.4°.

When the assignment of numbers to some characteristic, such as weight, *reflects equal intervals and a true zero, the level of measurement is* **interval-ratio.** As will become apparent, this is the most refined level of measurement, and it's associated with quantitative data, particularly when physical characteristics, such as weight, height, age, family size, annual income, are being measured.

1.5 Measurement of Nonphysical Characteristics: A Complication

The assignment of numbers to nonphysical characteristics, such as intellectual aptitude, personality tendency, and academic achievement, usually fails to reflect equal intervals and a true zero and, therefore, falls short of interval-ratio measurement. In the case of IQ scores, there is no external standard (the book in the weight example) to demonstrate that the addition of a fixed amount of intellectual aptitude always produces an equal increase in IQ scores (equal intervals). There also is no instrument (the unoccupied bathroom scale) that registers 0 when intellectual aptitude is completely absent (true zero).

Insofar as numbers, such as IQ scores, indicate an amount or a count, they are treated as quantitative data in statistics. However, when it's suspected that quantitative data reflect less than interval-ratio measurement, as often happens in the behavioral sciences, numerical results should be interpreted cautiously. In the absence of equal intervals, it would be inappropriate to claim that the difference between IQ scores of 120 and 140 represents the same amount of intellectual aptitude as the difference between scores of 100 and 120. By the same token, in the absence of a true zero, it would be inappropriate to claim that 140 represents twice the amount of intellectual aptitude as 70.

Other interpretations are possible. It's often assumed that, although lacking equal intervals, IQ scores contain some interval information—much like crude measures of distance obtained from stretching a rubber yardstick whose length varies from moment to moment. According to this perspective, gross differences between pairs of IQ scores reflect corresponding differences in intellectual aptitude, and therefore, it's probably true that a difference of 20 (for instance, between scores of 120 and 140) represents a greater amount of intellectual aptitude than a difference of 10 (for instance, between scores of 100 and 110).

At this point, you might wish that a person could be injected with 10 points worth of intellectual aptitude (or personality tendency or academic achievement) as the first step toward an IQ scale with equal intervals and a true zero. Lacking this alternative, however, train yourself to look at numbers as products of measurement and to temper your claims accordingly—particularly when, as above, quantitative data appear to fall short of interval-ratio measurement.

1.6 Ranked Data

When observations are described by numbers that indicate relative standings (within the set of observations), the data are **ranked.** The numbers in Table 1.2 represent the responses of one person to the question, What's important in your life? Each number is a rank indicating relative importance, beginning with 1 for that which is most important and ending with 12 for that which is least important. Love (2) is viewed as relatively more important than family (3) and relatively less important than health (1), but the number 2, itself, doesn't represent an amount of importance. The same set of relative standings could be described with an entirely different set of numbers, such as 100, 200, and 300, as long as health is assigned the smallest number, love the next smallest number, and so on. Other examples of ranked data include sets of numbers that indicate the relative standings of the top twenty college football teams, a collection of science projects (ordered from most to least creative), a list of job opportunities (ordered from most to least desirable).

Quantitative data always can be converted to ranked data, and this conversion often occurs in statistics. For instance, to convert the weights in Table 1.1 into ranks, begin with the lightest woman, who weighs 93 pounds, and assign consecutive numerical ranks 1, 2, 3, and so on, up to 53, the rank of the heaviest woman, who weighs 205 pounds. Once in this form, weights are ignored, and the statistical analysis focuses on the numerical ranks.

TABLE 1.2
RANKED DATA

(11)	fame
(1)	health
(7)	wealth
(2)	love
(4)	friendship
(10)	spiritual life
(6)	sex
(9)	adventure
(12)	power
(8)	travel
(3)	family
(5)	job satisfaction

1.7 Ordinal Measurement

Given the numbers in Table 1.2, it's impossible to know whether health (1) is viewed as only slightly more important or considerably more important than love (2). Ranked data supply no information about the size of the differences between consecutive numerical ranks.

Whenever, as with ranked data, *numbers are assigned solely on the basis of a set of observations being ordered from most to least, or vice versa, the level of measurement is* **ordinal.** Ordinal measurement is less refined than interval-ratio measurement, since only the latter possesses equal intervals and a true zero.

1.8 Qualitative Data

When observations are described as words, labels, or numbers that indicate class membership, the data are **qualitative.** Listed in Table 1.3 are the anonymous replies of 83 statistics students to the question, Have you ever smoked marijuana? Each reply, coded as Y and N, indicates membership in either the class of Yes replies or the class of No replies, and the resulting data are qualitative. Other examples of qualitative data include sets of observations based on gender (female, male), ethnic background (Afro-American, Asian-American, Euro-American, Latin-American, Native-American), blood type (O, A, B, AB), and letter grades (A, B, C, D, F).

1.9 Nominal Measurement

Numbers could be assigned to the Yes and No replies in Table 1.3, possibly as a coding device to permit computer processing. For example, 1 might be assigned to Yes and 2 to No, or since the choice of numbers is arbitrary, 234 might be assigned to Yes and 57 to No. In any event, these numbers serve merely as convenient labels to classify observations as either the same (all Yes replies are 1) or different (all No replies are not 1, but 2). It wouldn't be appropriate to claim, because No is 2 and Yes is 1, that No is more of a reply than Yes, or that No is twice as much of a reply as Yes. *Whenever, as with qualitative data, numbers are assigned solely on the basis of whether observations are classified as the same or different, the level of measurement is* **nominal.** Nominal measurement is the least refined level of measurement; it lacks order, as in ordinal measurement, and both equal intervals and a true zero, as in interval-ratio measurement.

1.10 Ordered Qualitative Data

Sometimes an ordinal level of measurement can be claimed for sets of data that otherwise appear to be qualitative. This possibility is present whenever classes of qualitative data can be ordered from most to least—the basis for ordinal measurement. For instance, the classes of ratings for current motion pictures (X, R, PG, G) can be ordered from most restrictive (X) to least restrictive (G), and therefore, these letters reflect ordinal measurement. Other examples of ordered qualitative data (and an ordinal level of measurement) include the ranks of military officers (Brig. Gen., Col., Lt. Col.), and letter grades in a course (A, B, C, D, F).

Although, strictly speaking, ordered qualitative data can be treated as ranked data, this fact is minimized in the present book. When there are many more observations than classes, as is typical for ordered qualitative data, the assignment of numerical ranks (solely on the basis of class membership) produces an excessive number of ties in rank. For instance, among the set of ratings for current motion pictures, all X-rated films would be tied for the same rank, as also would be all R-rated films, and so on. In practice, statistical techniques for ranked data usually are reserved for those situations, as described in Section 1.6, where observations can assume many different values, and therefore, ties in rank are relatively rare.

TABLE 1.3
QUALITATIVE DATA: "HAVE YOU EVER SMOKED MARIJUANA?" YES (Y) OR NO (N) REPLIES OF STATISTICS STUDENTS

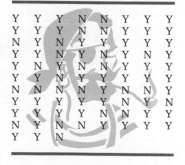

Y	Y	Y	N	N	Y	Y	Y
Y	Y	Y	N	N	Y	Y	Y
N	Y	N	Y	Y	Y	Y	Y
Y	Y	N	Y	N	Y	Y	N
Y	N	Y	N	N	Y	Y	Y
Y	Y	N	Y	Y	Y	Y	Y
N	N	N	N	Y	N	N	Y
Y	Y	Y	Y	Y	Y	Y	N
Y	Y	Y	N	Y	N	Y	Y
N	Y	N	N	Y	Y	Y	Y
Y	Y	N					

In general, then, the same type of statistical analysis is performed for both ordered qualitative data and unordered qualitative data. Nevertheless, the distinction between the two types of data is worth maintaining because, as will be seen, a few extra statistical procedures can be used when qualitative data are ordered.

1.11 An Overview

The identification of data as quantitative, ranked, or qualitative represents an important first step toward a successful statistical analysis. Ordinarily, this shouldn't be too difficult. As a matter of fact, entire sections—and, later, entire chapters—are devoted exclusively to statistical procedures for quantitative data. Although not nearly as extensive, other sections—and, later, the last two chapters of the book—are devoted exclusively to statistical procedures for ranked and qualitative data.

Table 1.4 summarizes the more important properties of each type of data. As has been noted, some quantitative data fail to achieve interval-ratio measurement, and some qualitative data (ordered) achieve more than nominal measurement. Although not mentioned previously, notice that the properties of less refined levels of measurement always appear in the more refined levels as well. In effect, numbers that reflect equal intervals and a true zero also must reflect order and classification, and numbers that reflect order also must reflect classification.

1.12 Summary

Statistics consists of two main subdivisions: descriptive statistics, which is concerned with organizing and summarizing information for sets of observations, and inferential statistics, which is concerned with generalizing information from samples to populations.

It's important to distinguish between quantitative, ranked, and qualitative data.

When observations are described by numbers that indicate an amount or a count, the data are quantitative. Quantitative data tend to reflect equal intervals and a true zero—that is, an interval-ratio level of

"WHAT'S THE NUMBER MEAN?"

TABLE 1.4
IMPORTANT PROPERTIES OF THREE TYPES OF DATA

TYPE OF DATA	NUMBERS INDICATE . . .	TYPICAL LEVEL OF MEASUREMENT (AND PROPERTIES)
Quantitative	Amount or count	*Interval-ratio* True zero Equal intervals Order Classification
Ranked	Relative standing	*Ordinal* Order Classification
Qualitative	Class membership	*Nominal* Classification

measurement—when physical characteristics are being measured. Otherwise, when nonphysical characteristics are being measured, quantitative data tend to fall short of interval-ratio measurement.

When observations are described by numbers that indicate relative standing, the data are ranked. Ranked data reflect an ordering of observations from most to least—that is, an ordinal level of measurement. Quantitative data always can be converted to ranked data, and this conversion often occurs in statistics.

When observations are described as words, labels, or numbers that indicate class membership, the data are qualitative. Qualitative data reflect a classification of observations on the basis of same or different—that is, the level of measurement is nominal.

When the classes of qualitative variables can be ordered, the level of measurement is ordinal rather than nominal. In general, the same type of statistical analysis is performed whether qualitative data are ordered or unordered.

Not only does the type of data dictate the form of the statistical analysis, but the interpretation of the numerical results depends, in part, on the level of measurement of the data. Interpretations are usually straightforward when data achieve one of the three well-defined levels of measurement, but they tend to be tricky when data can't be identified with one of the three levels of measurement. This is particularly true when quantitative data fail to achieve interval-ratio measurement. Under these circumstances, numerical results must be interpreted cautiously.

Important Terms

Descriptive statistics   Qualitative data
Inferential statistics   Interval-ratio measurement
Quantitative data   Ordinal measurement
Ranked data   Nominal measurement

1.13 Exercises

(Answers are listed in Appendix B.)

1. Each of the following terms or phrases describes a set of observations. Decide whether the resulting data are quantitative, ranked, or qualitative.

   (a) eye color
   (b) temperature
   (c) height
   (d) academic rank of college teachers
   (e) blood pressure
   (f) marital status
   (g) twelve heaviest humans
   (h) astrological sign
   (i) ten best-selling books
   (j) test score
   (k) annual income
   (l) reaction time
   (m) political affiliation
   (n) final standings of baseball teams
   (o) place of birth
   (p) religious preference

2. Identify each of the above sets of qualitative data that reflect ordinal measurement (because classes can be ordered from most to least).

3. For each of the following observations, specify whether the level of measurement is nominal, ordinal, or interval-ratio. If, as with IQ scores, measurement falls short of some level, such as interval-ratio, place a question mark next to your answer.

   (a) time spent under water (without breathing)
   (b) grade of eggs
   (c) miles per gallon
   (d) street address
   (e) score indicating degree of anxiety (50-point scale)
   (f) number of alarms for a fire
   (g) zip code
   (h) grade point average (GPA)
   (i) TV show ratings (estimated percent of viewing audience)
   (j) number of editions of a book
   (k) social security number
   (l) student ratings of instructors (5-point scale)
   (m) stock market average
   (n) score on verbal part of Scholastic Aptitude Test (SAT)
   (o) number on soccer jersey
   (p) rank among job applicants
   (q) bar-pressing rate of rat in Skinner box

4. Identify each statement as either true or false.

   (a) When numbers are used as labels to classify observations, qualitative data are transformed into quantitative data.
   (b) If observations can be ordered from least to most, then measurement is at least ordinal.
   (c) Ranked data are associated with interval-ratio measurement.
   (d) The mere fact that a scale contains a zero and consecutive numbers guarantees interval-ratio measurement.
   (e) Quantitative data automatically achieve interval-ratio measurement.
   (f) Quantitative data always can be converted into ranked data.
   (g) If two students score 100 percent on a sociology test, they are equally competent in sociology.

(h)   A grade point average of 4.0 represents twice as much academic achievement as a grade point average of 2.0.

(i)   Sandy ranks higher than Karen on the verbal part of the Scholastic Aptitude Test (SAT), while Karen ranks higher than Sandy on the math part of the same test. Therefore, on the average. Sandy and Karen have about the same SAT scores.

(j)   John and Jim both increased their grade point average by .50 during the previous year. It can be concluded, therefore, that both made equal gains in academic achievement.

(k)   A race horse, wearing number *three,* finishes *second,* with a time of *1.50* minutes. All three levels of measurement—interval-ratio, ordinal, and nominal—are represented in this statement.

(l)   In my group dynamics class, I arrived *last* but spoke *most often* during the lengthy *three*-hour session. All three levels of measurement are represented in this statement.

(m)   When the cashier's window opened, I was *first* in line but couldn't cash my check for *$50* because I forgot my student identification *number*. All three levels of measurement are represented in this statement.

Part I
DESCRIPTIVE STATISTICS

Organizing and Summarizing Data

2   Organizing and Summarizing Data with Tables

3   Describing Data with Graphs

4   Describing Data with Averages

5   Describing Variability

6   Normal Distributions and Standard Scores

7   Measures of Relationship: Correlation

8   Prediction

2

Organizing and Summarizing Data with Tables

FREQUENCY DISTRIBUTIONS FOR QUANTITATIVE DATA

2.1 –Ungrouped Data
2.2 –Stem and Leaf Arrangements
2.3 –Grouped Data
2.4 –Minimizing Irregularities and Ambiguities in Format
2.5 –Selection of an Interval Width
2.6 –Gaps Between Boundaries

OTHER TYPES OF FREQUENCY DISTRIBUTIONS

2.7 –Relative Frequency Distributions
2.8 –Cumulative Frequency Distributions
2.9 –Frequency Distributions for Qualitative Data
2.10–Frequency Distributions for Ranked Data
2.11–Summary
2.12–Exercises

A stem and leaf arrangement refers not to a floral display but to a statistical method that produces a clear and complete description of quantitative data.

You can produce a well-constructed frequency distribution every time by attending to seven simple guidelines.

When comparing two or more frequency distributions based on different numbers of observations, it's often helpful to convert frequencies to proportions or percents.

Given a batch of data, as in Table 1.1, how do you make sense out of it—both for yourself and for others? Hidden among all those observations is there an important message, possibly one that supports (or fails to support) some research hypothesis? At this point, especially if you're facing a fresh set of data in which you have a special interest, statistics can be exciting as well as challenging. Your main responsibility is to describe the data as clearly, completely, and concisely as possible. Statistics supplies some tools—for instance, tables and graphs—and some guidelines. Beyond that, it's just the data and you. As in life, there is not one single right way. Equally valid descriptions of the same data might appear in tables with different formats. Relax. By following just a few guidelines, your reward will be a well-organized and well-summarized set of data.

Table 2.1 shows one way to organize the weights of female statistics students listed in Table 1.1. First, arrange a column of consecutive numbers, beginning with the lightest weight (93) at the bottom and ending with the heaviest weight (205) at the top. (Because of the extreme length of this column, many intermediate numbers have been omitted in Table 2.1—a procedure never actually followed in practice.) Then, place a tally next to a number each time its value appears in the original set of data.

The organization of observations according to their frequencies of occurrence, as in Table 2.1, is referred to as a **frequency distribution** *for ungrouped data. Data are ungrouped whenever observations are organized into classes of single values rather than, as with grouped data, into classes of two or more values.*

FREQUENCY DISTRIBUTIONS FOR QUANTITATIVE DATA

2.1 Ungrouped Data

The frequency distribution shown in Table 2.1 is very unwieldy (and hence only partially displayed) because of the more than 100 possible values between the largest and smallest observations. Frequency distributions for ungrouped data are much more informative when the number of possible values is smaller—say less than about 30. Under these circumstances, they serve as a straightforward method for organizing data.

2.2 Stem and Leaf Arrangements

Table 2.2 shows another way to organize the weights in Table 1.1. **Stem and leaf arrangements** are useful as informal, exploratory devices for looking at data. To construct a stem and leaf arrangement for the weight data, first note that, when counting by tens, the weights range from the 90s to the 200s. Arrange a column of numbers, the stems, beginning with 9 at the bottom (representing the 90s) and ending with 20 at the top (representing the 200s). Let a vertical line separate the stems, which represent units of ten, from the space to be occupied by the leaves, which represent units of one.

Now, each weight in Table 1.1 is split into a stem value and a leaf value, and its leaf value is entered in Table 2.2. The first weight listed in Table 1.1 is 120, which has 12 as a stem and 0 as a leaf. Therefore, enter 0 in the leaf space to the right of 12 in Table 2.2. The next weight is 123, which has 12 as a stem and 3 as a leaf. Enter 3 in the second leaf space to the right of 12 (and after the first leaf, which is 0). The third weight is 112, which has a stem of 11 and a leaf of 2. Enter 2 in the leaf space to the right of 11. Continue this procedure until one number has been entered in the leaf space for each observation. Looking again at Table 2.2, notice that all weights in the 90s are listed together; all those in the 100s are listed together, and so on. Although weights have been grouped together in stems of size ten, the identities of all of the original weights have been preserved. Thus, you could convert to other kinds of grouping, say stems of size five, as in Table 2.3, without having to reorganize the data from scratch—a very nice feature. In Table 2.3 each stem value appears twice. The lower member of each pair is associated with leaves of 0, 1, 2, 3, and 4, while the upper member is associated with 5, 6, 7, 8, and 9.

The stem and leaf arrangement in Table 2.2 (or Table 2.3) supplies lots of information about the weight data. It's obvious that weights peak in the 110s, with a very dense concentration among the 110s, 120s, and 130s. Furthermore, the distribution is spread out, with several gaps, in the direction of heavier weights.

TABLE 2.1
FREQUENCY DISTRIBUTION (UNGROUPED DATA)

WEIGHT	FREQUENCY
205	/
204	0
203	0
202	0
201	0
200	0
199	0
198	0
*	
*	
*	
*	
*	
*	
*	
*	
*	
101	0
100	0
99	0
98	0
97	0
96	0
95	//
94	0
93	/

TABLE 2.2
STEM AND LEAF ARRANGEMENT

STEM	LEAF
20	5
19	
18	6 0 5
17	
16	5 6
15	3 0 0 0
14	0 0 5
13	2 0 0 0 2 5 9
12	0 3 5 8 9 0 0 0 6 5 5 5
11	2 7 1 7 8 0 2 0 2 6 9 8 2 6 4 7 6
10	5
9	3 5 5

Stem values are not limited to units of ten (or five). Depending on the data, you might choose another unit that's some variation on 10 (or 5) such as 1, 100, 1000, or even .1, .01, .001, and so on. For instance, an annual income of 23,784 dollars could have a stem of 23 (units of one thousand) and a leaf of 784. A Graduate Record Exam (GRE) test score of 689 could have a stem of 6 (units of one hundred) and a leaf of 89. A grade point average (GPA) of 3.25 could have a stem of 3 (units of one) and a leaf of 25, or if you wanted more stem values, it could have a stem of 3.2 (units of one-tenth) and a leaf of 5.

Stem and leaf arrangements are statistical "best buys," particularly when, as in the above example, there are many possible values between the largest and smallest observations. Often, just a few minutes' work produces a description of the data that is both clear and complete. As a bonus, these arrangements easily can be converted into grouped data.

Table 2.4 shows still another way to organize the weights listed in Table 1.1. *The organization of observations according to their frequencies of occurrence,* as in Table 2.4, *is referred to as a* **frequency distribution** *for grouped data.* In the present example, data are grouped into classes with ten possible values each.

"THANKS LISA, BUT IT'S A STEM AND LEAF ARRANGEMENT THAT I NEED"

Notice how the distribution in Table 2.4 emerges from the stem and leaf arrangement in Table 2.2. Each stem value converts to a well-defined class interval of width ten. Thus, a stem value of 9 converts to a class interval of 90 to 99. Each leaf converts to a frequency of one, and a count of the leaves for any stem yields the frequency for the corresponding class interval. Thus, there are 7 leaves for the stem of 13, yielding a frequency of 7 for the class interval of 130 to 139.

What has been gained by this conversion? First, weights are expressed in their customary form, rather than as stems and leaves. More important, bunches of numbers (the leaves for any stem) are described concisely by a single number (frequency). Finally, the data

TABLE 2.3
STEM AND LEAF ARRANGEMENT

STEM	LEAF
20	5
20	
19	
19	
18	6 5
18	0
17	
17	
16	5 6
16	
15	
15	3 0 0 0
14	5
14	0 0
13	5 9
13	2 0 0 0 2
12	5 8 9 6 5 5 5
12	0 3 0 0 0
11	7 7 8 6 9 8 6 7 6
11	2 1 0 2 0 2 2 4
10	5
10	
9	5 5
9	3

TABLE 2.4
FREQUENCY DISTRIBUTION (GROUPED DATA)

WEIGHT	FREQUENCY
200–209	1
190–199	0
180–189	3
170–179	0
160–169	2
150–159	4
140–149	3
130–139	7
120–129	12
110–119	17
100–109	1
90–99	3
Total	53

2.3 Grouped Data

are in a form where they can be readily converted to still other kinds of tables and to graphs.

Notice one inevitable by-product of grouping data, as in Table 2.4. The identities of individual observations are sacrificed for a more concise description in terms of class intervals. Occasionally, this can be a liability—if you wish more information than is given in the frequency table but lack access to the original data.

Looking at Table 2.4, you can use phrases, originally inspired by the stem and leaf arrangement, to summarize the frequency distribution for weights. In particular, the weights peak at 110 to 119, with a dense concentration at 110 to 139 and with a spread in the direction of heavier weights.

2.4 Minimizing Irregularities and Ambiguities in Format

TABLE 2.5
FREQUENCY
DISTRIBUTION
(GROUPED DATA)

WEIGHT	FREQUENCY
205–209	1
200–204	0
195–199	0
190–194	0
185–189	2
180–184	1
175–179	0
170–174	0
165–169	2
160–164	0
155–159	0
150–154	4
145–149	1
140–144	2
135–139	2
130–134	5
125–129	7
120–124	5
115–119	9
110–114	8
105–109	1
100–104	0
95–99	1
90–94	2
Total	53

There are seven guidelines for producing a well-constructed frequency distribution. The first three specify some essential ingredients in any well-constructed frequency distribution, and they never should be violated. An asterisk denotes their special status. The last four guidelines specify optional ingredients that can be modified or ignored, as circumstances warrant.

The last guideline requires a few more comments. The use of too many class intervals, as in Table 2.5, where the weight data are grouped into twenty-four intervals of width 5 each, tends to defeat the major purpose of a frequency distribution—namely, to provide a reasonably concise description of data. On the other hand, the use of too few intervals, as in Table 2.6, where the weight data are grouped into six intervals of width 25 each, can mask important data patterns, such as the high density of weights in the 110s and, to a lesser extent, in the 120s and 130s.

There is, however, nothing sacred about fifteen, the recommended number of class intervals. When describing large batches of data, you might aim for more than fifteen class intervals in order to portray some of the more fine-grained data patterns that otherwise could vanish. For example, add two zeros after each frequency in Table 2.5 to create a new weight distribution for 5300 students from the original distribution for 53 students. Given this (admittedly unlikely) weight distribution, it would be important to retain the twenty-four intervals in Table 2.5 in order to portray, for instance, the curious presence of several peak frequencies, including the one at 125 to 129 pounds.

On the other hand, when describing small batches of data, you might choose to aim for less than fifteen class intervals in order to spotlight data regularities that otherwise could be blurred. For example, if the weight distribution had been based on only about twelve students, then it would be desirable to use the six intervals in Table 2.6 in order to spotlight the tendency for frequencies to peak, particularly at 100 to 124 pounds. It's best, therefore, to think of fifteen, the recommended number of class intervals, as a rough rule-of-thumb to be applied with discretion rather than with slavish devotion.

2.5 Selection of an Interval Width

When converting data into a frequency table, the width of the class interval, i, can be determined from the following expression:

$$i = \frac{\text{largest observation} - \text{smallest observation}}{\text{desired number of class intervals}} \qquad (2.1)$$

* 1. All observations should be included in one, and only one, class interval.

EXAMPLE: 90–99, 100–109, 110–119, . . .

COUNTER-EXAMPLE: 90–100, 100–110, 110–120, . . . where, because the boundaries of class intervals overlap, an observation of 100 (or 110) could be assigned to either of two intervals.

* 2. List all class intervals—even those with zero frequencies.

EXAMPLE: listed in Table 2.4 is the interval 170–179 and its frequency of zero.

COUNTER-EXAMPLE: not listing this interval because of its frequency of zero.

* 3. All class intervals (with upper and lower boundaries) should be equal in width.

EXAMPLE: 90–99, 100–109, 110–119, . . .

COUNTER-EXAMPLE: 90–99, 100–119, . . . where the second interval is twice the width of the first interval.

4. All class intervals should have both boundaries.

EXAMPLE: 200–209

COUNTER-EXAMPLE: 200–above, where no maximum value can be assigned to observations in this interval.

5. Select the width of class intervals from convenient numbers, such as 1, 2, 3, . . . 10, particularly 5 and 10 or multiples of 5 and 10.

EXAMPLE: 90–99, 100–109, 110–119, . . . where the interval width is 10, a convenient number.

COUNTER-EXAMPLE: 90–102, 103–115, 116–128, . . . where the interval width is 13, an inconvenient number.

6. The lower boundary of each class interval should be a multiple of the interval width.

EXAMPLE: 90–99, 100–109, 110–119, . . . where the lower boundaries of 90, 100, and 110 are multiples of 10, the interval width.

COUNTER-EXAMPLE: 95–104, 105–114, 115–124, . . . where the lower boundaries of 95, 105, and 115 are not multiples of 10, the interval width.

7. In general, aim for a total of approximately fifteen class intervals but settle for any total of between ten and twenty intervals.

EXAMPLE: the distribution in Table 2.4 uses twelve class intervals.

COUNTER-EXAMPLE: the distribution in Table 2.5 has too many class intervals (twenty-four), while the distribution in Table 2.6 has too few class intervals (six).

Thus, to find the proper interval size for the weight data in Table 1.1, proceed as follows. Substitute 205 and 93 for the largest and smallest observations, and 15 for the desired number of class intervals, to obtain

$$i = \frac{205 - 93}{15} = \frac{112}{15} = 7.47$$

Round off 7.47 to a more convenient number, say 10, and adopt this as the interval width. (If 7.47 had been rounded off to 5, this interval width would have generated too many class intervals, namely twenty-four, as has been noted.) Then, by using multiples of 10 to identify the lower boundaries of class intervals, you're well on the way toward the well-constructed frequency distribution shown in Table 2.4.

2.6 Gaps Between Boundaries

**TABLE 2.6
FREQUENCY
DISTRIBUTION
(GROUPED DATA)**

WEIGHT	FREQUENCY
200–244	1
175–199	3
150–174	6
125–149	17
100–124	23
75–99	3
Total	53

In well-constructed frequency tables, the gaps between boundaries of class intervals, such as between 90 to 99 and 100 to 109 in Table 2.4, show clearly that each observation has been assigned to one, and only one, class interval. The size of the gap should always equal one unit of measurement, that is, it should always equal the smallest possible difference between scores within a particular set of data. In the present case, where weights are reported to the nearest pound, one pound is the unit of measurement, and quite appropriately, the gap between boundaries equals one pound. These gaps would not be appropriate if weights had been reported to the nearest tenth of a pound. In this case, one-tenth of a pound is the unit of measurement, and therefore, the gap should equal one-tenth of a pound. The smallest class interval would be 90.0 to 99.9 (not 90 to 99), and the next class interval would be 100.0 to 109.9 (not 100 to 109), and so on. These new boundaries guarantee that any observation, such as 99.6, will be assigned to one, and only one, class interval.

Gaps between class intervals do not signify any disruption in the essentially continuous nature of the data. It would be erroneous to conclude that because of the gap between 99 and 100 in the frequency distribution for Table 2.4, nobody can weigh between 99 and 100 pounds. As a matter of fact, a woman who reports her weight as 99 pounds actually could weigh anywhere between 98.5 and 99.5 pounds, and a woman who reports her weight as 100 pounds actually could weigh anywhere between 99.5 and 100.5 pounds. Thus, the gaps between class intervals are more apparent than real.

OTHER TYPES OF FREQUENCY DISTRIBUTIONS

2.7 Relative Frequency Distributions

An important variation of the frequency distribution is the relative frequency distribution. Relative frequency distributions are useful when, for instance, you wish to compare the distribution of ages for 632,800 residents of a large city with that for 457 residents of a small town (see Exercise 11). The conversion from absolute to relative frequencies, as described below, effectively eliminates any disparities due to the radically different population sizes and permits a straightforward comparison of the two distributions of ages.

*To convert a frequency distribution into a **relative frequency distribution,** divide the frequency for each interval by the total frequency for the entire distribution.* Table 2.7 illustrates a relative frequency distribution based on the weight distribution of Table 2.4. The conversion to proportions is straightforward. For instance, to obtain the proportion of .06 for the interval 90 to 99, simply divide the frequency of 3 for that interval by the total frequency of 53. Repeat this process until a proportion has been calculated for each interval.

WEIGHT	FREQUENCY	RELATIVE FREQUENCY Proportion	Percent
200–209	1	.02	2
190–199	0	.00	0
180–189	3	.06	6
170–179	0	.00	0
160–169	2	.04	4
150–159	4	.08	8
140–149	3	.06	6
130–139	7	.13	13
120–129	12	.23	23
110–119	17	.32	32
100–109	1	.02	2
90–99	3	.06	6
Total	53	1.02*	102*

TABLE 2.7
RELATIVE FREQUENCY
DISTRIBUTIONS

*The sums do not equal 1.00 and 100% because of rounding off errors.

Some people prefer to deal with percents rather than proportions, possibly because percents usually lack decimal points. By definition, a proportion ranges between 0 and 1, whereas a percent ranges between 0 percent and 100 percent. To convert relative frequencies in Table 2.7 from proportions to percents, multiply each proportion by 100. For example, multiply .06, the proportion for the interval 90 to 99, by 100 to obtain 6 percent.

2.8 Cumulative Frequency Distributions

Cumulative frequencies indicate how many observations fall at or below a particular interval. This type of distribution can be used effectively with sets of observations, such as test scores for intellectual or academic aptitude, when relative standing within the distribution assumes primary importance. Under these circumstances, cumulative frequencies are often converted, in turn, to cumulative proportions or percents.

To convert a frequency distribution into a **cumulative frequency distribution,** *add to the frequency of each interval the sum of the frequencies of all intervals ranked below it.* This gives the cumulative frequency for that interval. The most efficient procedure is to begin with the lowest-ranked interval in the frequency distribution and work upward, finding the cumulative frequencies in ascending order. In Table 2.8, the cumulative frequency for the interval 90 to 99 is 3 since

WEIGHTS	FREQUENCY	CUMULATIVE Frequency	Proportion	Percent
200–209	1	53	1.00	100
190–199	0	52	.98	98
180–189	3	52	.98	98
170–179	0	49	.92	92
160–169	2	49	.92	92
150–159	4	47	.89	89
140–149	3	43	.81	81
130–139	7	40	.75	75
120–129	12	33	.62	62
110–119	17	21	.40	40
100–109	1	4	.08	8
90–99	3	3	.06	6
Total	53			

TABLE 2.8
CUMULATIVE
FREQUENCY
DISTRIBUTIONS

there are no intervals ranked lower. The cumulative frequency for the interval 100 to 109 is 4 since 1 is the frequency for that interval and 3 is the frequency of all lower-ranked intervals. The cumulative frequency for the interval 110 to 119 is 17 since 13 is the frequency for that interval and 4 is the sum of the frequencies of all lower-ranked intervals.

As has been suggested, if relative standing within a distribution is particularly important, then cumulative frequencies are converted to cumulative proportions or percents. A glance at Table 2.8 reveals that .81 or 81 percent of all weights are the same as or lighter than the weights between 140 and 149 pounds. To obtain this cumulative proportion of .81, the cumulative frequency of 43 for the interval 140 to 149 should be divided by the total frequency of 53 for the entire distribution. To determine cumulative percent (81%), multiply the cumulative proportion (.81) by 100.

2.9 Frequency Distributions for Qualitative Data

Frequency distributions for qualitative data are relatively easy to construct. Simply determine the frequency for each class of observations, and report these frequencies as shown in Table 2.9 for the marijuana smoking survey. This frequency distribution reveals that Yes replies are approximately twice as prevalent as No replies.

Notice that it's totally arbitrary whether Yes is listed above or below No in Table 2.9. When qualitative data can be ordered from least to most, however, that order should be preserved in the frequency table, as illustrated in Table 2.10 where academic ranks are listed in descending order from professor to instructor.

Frequency distributions for qualitative variables can always be converted into relative frequency distributions, expressed either as proportions or percents, as illustrated in Table 2.10. Furthermore, if observations can be ordered from least to most, cumulative frequencies (and cumulative proportions and percents) can be used, as illustrated in Table 2.10. If observations can't be ordered, as in Table 2.9, a cumulative frequency distribution is meaningless.

**TABLE 2.9
MARIJUANA SMOKING SURVEY**

RESPONSE	FREQUENCY
Yes	56
No	27
Total	83

2.10 Frequency Distributions for Ranked Data

Ordinarily, frequency distributions are not used to describe ranked data. Except for ties, each numerical rank has a frequency of one, and thus the data pattern is always about the same—a list of the consecutive numerical ranks 1, 2, 3, 4, and so on, each with a frequency of one.

2.11 Summary

Frequency distributions organize observations according to their frequencies of occurrence. Data are ungrouped whenever observations

**TABLE 2.10
ACADEMIC RANKS OF FACULTY AT SAN JOSE STATE UNIVERSITY (1978)**

RANK	FREQUENCY	PROPORTION	CUMULATIVE PERCENT
Professor	370	.48	100
Associate professor	209	.27	52
Assistant professor	122	.16	25
Instructor	66	.09	9
Total	767	1.00	

are organized into classes of single values, while data are grouped whenever observations are organized into classes of two or more values.

Frequency distributions for ungrouped data are most informative when there are less than about 30 possible values between the largest and smallest observations.

Stem and leaf arrangements are useful as informal, exploratory devices for looking at data.

Well-constructed frequency distributions should avoid irregularities and ambiguities. Seven guidelines are listed in Section 2.4.

When comparing two or more distributions based on radically different total numbers of observations, it's helpful to convert frequency distributions into relative frequency distributions involving either proportions or percents. When relative standing within the distribution is important, it's helpful to convert frequency distributions into cumulative frequency distributions, particularly those involving cumulative proportions and percents.

Frequency distributions for qualitative data are easy to construct. They also can be converted to relative frequency distributions involving either proportions or percents and, if the data can be ordered from least to most, into cumulative frequency distributions involving frequencies, proportions, or percents.

Important Terms

Frequency distribution (for ungrouped data)
Stem and leaf arrangement
Frequency distribution (for grouped data)
Relative frequency distribution
Cumulative frequency distribution

1. The following results indicate how many seconds individuals in a group of 35 college students can hold their breath.

2.12 Exercises

91	85	84	79	80
83	96	75	82	79
95	71	105	90	60
123	80	100	88	89
98	62	99	95	90
110	114	94	100	79
112	90	90	98	89

(a) Construct a frequency distribution for ungrouped data.
(b) Interpret any data patterns.
(c) Do a stem and leaf arrangement.
(d) Construct a frequency distribution for grouped data.
(e) Which technique or techniques did you find most helpful?

2. Students in a theatre arts appreciation class rate the movie "Gone With The Wind" on a 10-point scale, as follows:

3	6	2	7	8
0	1	4	10	0
2	5	3	5	8
9	7	6	3	7
8	9	6	0	6

(a) Construct a frequency distribution for ungrouped data.
(b) In a few words, describe any data patterns.

3. Note any poor features of the following frequency distribution.

ESTIMATED WEEKLY TV VIEWING TIME (HRS)
FOR 250 SIXTH GRADERS

VIEWING TIME	FREQUENCY
35–above	2
30–34	5
25–30	29
20–22	60
15–19	60
10–14	34
5–9	31
0–4	29
Total	250

4. Assume that we wish to construct frequency distributions with approximately 15 intervals, given each of the following pairs of observations. For each, specify the interval width **i**. Also for each, specify the boundaries of the two lowest class intervals and of the one highest class interval.

PAIR	LARGEST OBSERVATION	SMALLEST OBSERVATION
a	64	29
b	578	156
c	165	21
d	2,854	650
e	19,200	13,612
f	42	0
g	0.961	0.443
h	1,205	864

5. A public transit district maintains records that indicate the average number of miles between breakdowns for each of its fleet of 50 second-hand buses. The rather dismal record looks like this:

654	429	1630	1470	1800
789	1208	1425	1623	1605
961	1644	752	880	1788
893	1798	847	576	1903
763	423	743	889	567
421	821	1026	593	998
562	756	233	431	741
923	543	378	776	632
1059	1789	1754	679	771
677	777	555	660	638

(a) Do a stem and leaf arrangement.
(b) Construct a frequency distribution for grouped data.
(c) Describe any interesting data patterns.

6. We often wish to compare the results for two or more groups measured under different conditions. Let's pretend that student volunteers are assigned arbitrarily (according to a coin toss) to be trained either to use various relaxation techniques, such as meditation, exercise, and so on, or to simply behave as usual. To determine whether relaxation training influences academic achievement, grade point averages are calculated at the end of the one-year experiment, yielding the following results.

MEDITATORS			NONMEDITATORS		
3.25	2.25	2.75	3.67	3.89	3.00
3.56	3.63	2.25	2.50	2.75	2.10
3.57	2.45	3.80	3.80	2.67	2.90
2.95	3.30	3.68	2.80	2.65	2.58
3.56	3.78	3.75	2.83	3.10	3.37
3.85	3.00	3.71	3.25	2.76	2.86
3.10	2.75	2.90	2.90	2.10	2.66
2.58	2.95	3.56	2.34	3.20	2.67
3.30			3.59		

(a) Do separate stem and leaf arrangements for each condition.
(b) Construct separate frequency distributions using grouped data.
(c) Are there any data patterns?

7. (Class exercise) Collect the weights of all members of your class. Organize and summarize this data using a stem and leaf arrangement and a frequency distribution for grouped data.

8. (Class exercise) Determine the total cash value of the loose change in the pockets, purses, and wallets of your classmates. Organize and summarize these data using a stem and leaf arrangement and a frequency distribution for grouped data.

9. Graduate Record Exam (GRE) scores are distributed as follows for a group of applicants.

GRE	FREQUENCY
720–739	2
700–719	2
680–699	10
660–679	14
640–659	30
620–639	34
600–619	32
580–599	26
560–579	20
540–559	15
520–539	10
500–519	4
480–499	1
	Total 200

(a) Convert to a relative frequency distribution, first using proportions, then percents.

(b) Convert to a cumulative frequency distribution.
(c) Convert to a cumulative relative frequency distribution, first using proportions, then percents.

10. Given the following frequency distribution:

DISTRIBUTION OF FAMILIES BY NUMBER OF CHILDREN (1970 U.S. CENSUS)

NO. OF CHILDREN	FREQUENCY
4 or more	5,123,700
3	5,636,070
2	8,710,290
1	9,222,660
0	22,544,280
	Total 51,237,000

(a) Convert to relative frequencies, expressed as percents.
(b) Convert to a cumulative frequency distribution.
(c) Convert to a cumulative relative frequency distribution, expressed as percents.

11. As has been noted, relative frequency distributions are particularly helpful when you're comparing two frequency distributions based on radically different numbers of observations. After converting each of the following distributions to relative frequencies (proportions), compare them, noting any differences in data patterns. (Save these relative frequency distributions for use in Chapter 3, Exercise 3.)

AGE DISTRIBUTIONS IN TWO DIFFERENT COMMUNITIES

LARGE CITY		SMALL TOWN	
Age	Frequency	Age	Frequency
90–99	1,100	90–99	10
80–89	2,200	80–89	33
70–79	20,500	70–79	34
60–69	59,000	60–69	44
50–59	74,300	50–59	55
40–49	93,400	40–49	60
30–39	105,000	30–39	68
20–29	103,600	20–29	71
10–19	83,000	10–19	40
0–9	90,700	0–9	42
Total	632,800	Total	457

12. Motion picture ratings can be ordered from most to least restrictive, that is, X, R, PG, and G. The ratings of some pictures being shown recently in San Francisco are as follows:

PG PG PG   R   G
G   R   R   PG PG
R   PG R   PG   R
X   X   PG   G   R

(a) Construct a frequency distribution.
(b) Interpret any data patterns.
(c) Convert to relative frequencies, expressed as percents.
(d) If appropriate, construct a cumulative frequency distribution.

13. (a) Construct a frequency distribution for the blood types of a group of prospective donors.

$$
\begin{array}{ccccc}
\text{O} & \text{O} & \text{A} & \text{A} & \text{AB} \\
\text{A} & \text{O} & \text{O} & \text{A} & \text{O} \\
\text{A} & \text{A} & \text{A} & \text{B} & \text{A} \\
\text{A} & \text{O} & \text{A} & \text{O} & \text{O} \\
\text{O} & \text{O} & \text{A} & \text{O} & \text{A} \\
\text{B} & \text{O} & \text{A} & \text{O} & \text{O}
\end{array}
$$

(b) Interpret any data patterns.

(c) Convert to relative frequencies, expressed as proportions and percents.

(d) If appropriate, construct a cumulative frequency distribution.

GRAPHS FOR QUANTITATIVE
DATA

3.1–Histograms
3.2–Frequency Polygons

GRAPHS FOR QUALITATIVE
DATA

3.3–Bar Graphs
3.4–Pie Charts

MISLEADING GRAPHS

3.5–Some Tricks
3.6–An Overview
3.7–Summary
3.8–Exercises

3
Describing Data with Graphs

You don't have to be a magician to transform a histogram into a frequency polygon—just fill in the dotted line.

The next time you see a pie, you'll probably think of a common statistical technique known as a pie chart.

What type of statistical trick could have inspired the 19th-century statesman Disraeli to complain, "There are three kinds of lies—lies, damned lies, and statistics"?

A clear, concise, and complete description of data can be realized with the aid of a well-constructed frequency distribution. An even more vivid description of data often can be realized—particularly when you're attempting to communicate with a general audience—by converting frequency distributions into graphs. This chapter describes some of the most common types of graphs for quantitative and qualitative data. In general, graphs are not used to describe ranked data, since the graphic display would usually be about the same—a monotonous progression of bars or points representing the consecutive numerical ranks 1, 2, 3, and so on, each with a frequency of one, except for ties in rank.

GRAPHS FOR QUANTITATIVE DATA

3.1 Histograms

The weight distribution described in Table 2.4 appears as a **histogram** in Figure 3.1. A casual glance at this histogram confirms previous conclusions: a dense concentration of weights among the 110s, 120s, and 130s with a spread in the direction of heavier weights.

Let's pinpoint some of the more important features of histograms. Equal segments along the horizontal axis (the X axis or abscissa) reflect the various equal-sized class intervals of the frequency distribution. Equal segments along the vertical axis (the Y axis or ordinate) reflect constant increases in frequency. The intersection of the two axes defines the origin, where both weight and frequency equal 0. Numbers always increase from left to right along the horizontal axis and from bottom to top along the vertical axis. It's considered good practice to use wiggly lines, such as the one along the horizontal axis in Figure 3.1, to spotlight breaks in scale between the origin of 0 and the smallest class interval of 90 to 99. The body of the histogram consists of a series of bars whose heights reflect the frequencies for the various class intervals. Notice how adjacent bars in histograms share common boundaries. The

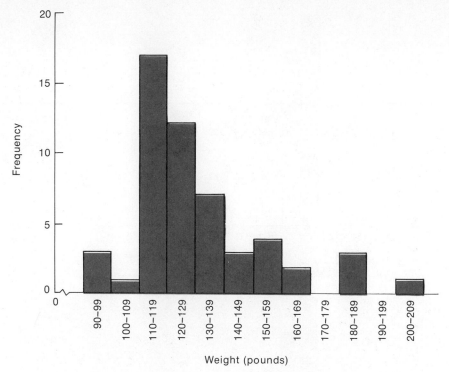

FIGURE 3.1
Histogram.

introduction of gaps between adjacent bars—never done in histograms—would suggest an artificial disruption in the data.

Whenever convenient, the extensive set of numbers along the horizontal scale of Figure 3.1 can be replaced with a few convenient numbers, as in Figure 3.2A. This concession helps to avoid excessive cluttering along the horizontal axis.

An important variation on a histogram is the **frequency polygon** or line graph. Frequency polygons may be constructed directly from a frequency distribution. However, let's follow the step-by-step transformation of a histogram into a frequency polygon as described in panels A, B, C, and D of Figure 3.2.

A. This shows the histogram for the weight distribution.

B. Place dots at the midpoints of each bar top (or in the absence of bar tops, at midpoints on the horizontal axis) and connect with straight lines. Notice that the resulting line tends to smooth out the frequency pattern by cutting protruding corners and filling recessed corners—at the expense of a completely accurate portrait of the original frequency distribution.

C. Anchor the frequency polygon to the horizontal axis. First, extend the upper tail to the midpoint of the first unoccupied interval (210–219) on the upper flank of the histogram. Then, extend the lower tail to the midpoint of the first unoccupied interval (80–89) on the lower flank of the histogram. Notice the smoothing effect here, too. In addition, all area under the frequency polygon is enclosed completely.

D. Finally, erase all of the histogram bars, leaving only the frequency polygon.

Frequency polygons are particularly useful when two or more frequency distributions are included in the same graph. Figure 3.3

3.2 Frequency Polygons

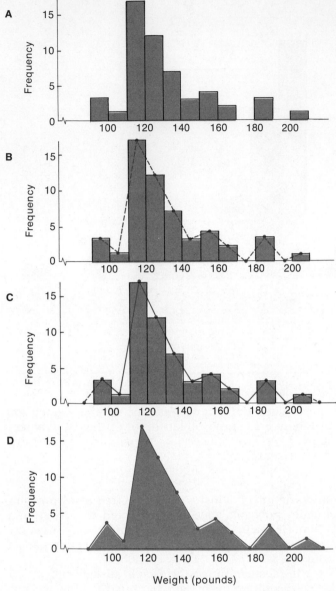

FIGURE 3.2
Transition from histogram to frequency polygon.

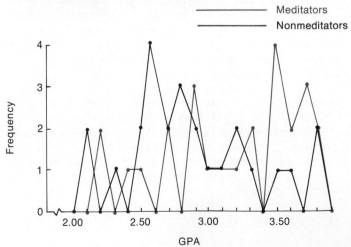

FIGURE 3.3
Two frequency polygons.

depicts two distributions of grade point averages (GPA), one for a group of meditators and the other for a group of nonmeditators, described in Chapter 2, Exercise 6. Considerable variability characterizes both distributions. The most conspicuous difference is the tendency for the GPAs of the meditators to peak about one point higher than the GPAs of nonmeditators. Should we emphasize this difference or should we ignore it? The techniques for inferential statistics (described in Part II of this book) will help us to answer this type of question.

GRAPHS FOR QUALITATIVE DATA

3.3 Bar Graphs

The distribution in Table 2.6, based on replies to the question Have you ever smoked marijuana? appears as a **bar graph** in Figure 3.4. A glance at this graph confirms that Yes replies occur approximately twice as often as No replies.

As with histograms, equal segments along the horizontal axis are allocated to the different words or classes that appear in the frequency distribution for qualitative data. Likewise, equal segments along the vertical axis reflect constant increases in frequency. The body of the bar graph consists of a series of bars whose heights reflect the frequencies for the various words or classes.

Gaps are placed between adjacent bars in bar graphs to emphasize the essentially discontinuous nature of qualitative data. That's why we rarely use frequency polygons to depict qualitative data.

3.4 Pie Charts

A common variation on the bar graph, particularly appropriate for a general audience, is a **pie chart.** Figure 3.5 shows a pie chart of the blood types described in Chapter 2, Exercise 13.

To construct a pie chart, first convert data to proportions. Then use these proportions to slice up the entire pie. Probably the easiest way to accomplish this is to visualize the pie as a 60-minute clock. To determine how many minutes should be awarded to each blood type, multiply the proportion for each category by 60. For example, first multiply the proportion of prospective donors with type O blood, $14/30 = .47$, by 60 to obtain 28 minutes. Then, starting on the hour, award type O a slice of pie that ends at 28 minutes past the hour. Repeat this process for type A. Multiply the proportion of prospective donors with type A, $13/30 = .43$, by 60 to obtain 26 minutes. Award type A a slice of pie that goes from 28 minutes to 54 minutes ($28 + 26 = 54$). Continue the same way for the two remaining blood types.

If you want to construct a fairly accurate pie chart, use a pro-

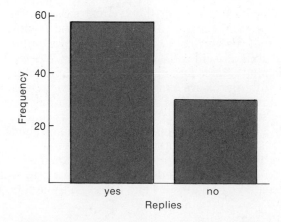

FIGURE 3.4
Bar graph.

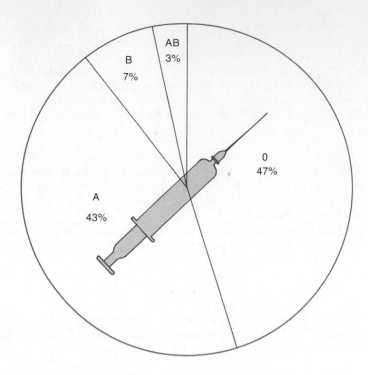

FIGURE 3.5
Pie chart.

tractor. To use the scale on a protractor, multiply the proportion observed for each category by the 360 degrees of a circle rather than, as
suggested above, by the 60 minutes of an hour.

MISLEADING GRAPHS

3.5 Some Tricks

Graphs can be constructed in an unscrupulous manner to support
a particular point of view. Indeed, this type of statistical fraud gives
credibility to a variety of popular quotations, including "Numbers don't
lie, but statisticians do," as well as the one attributed to Disraeli in the
preview for this chapter.

For example, to exaggerate the impression that relatively many
students responded Yes to the marijuana smoking question, an unscrupulous person might resort to various tricks embedded in Figure 3.6.
First, the width of the Yes bar is more than three times that for the No
bar—in violation of the custom that bars be equal in width. Second, the
lower end of the frequency scale is omitted—in violation of the custom
that the entire scale be reproduced, beginning with zero. (Otherwise a
broken scale should be highlighted by a wiggly line, as in Figures 3.1,
3.2, and 3.3.) Third, the height of the vertical axis is several times the
width of the horizontal axis—in violation of the custom, heretofore
unmentioned, that the vertical axis should be *approximately* as tall as
the horizontal axis is wide. (Beware of graphs where, because the
vertical axis is many times larger than the horizontal axis, as in Figure
3.6, frequency differences are exaggerated, or where, because the vertical axis is many times smaller than the horizontal axis, frequency
differences are suppressed.) Finally, the Yes bar is shaded darker than
the No bar—in violation of the custom that the same shade be used for
all bars in any single distribution.

The combined effect of Figure 3.6 is to create the impression that
virtually all students responded Yes. Notice the radically different
impressions created by Figures 3.4 and 3.6, even though both are based

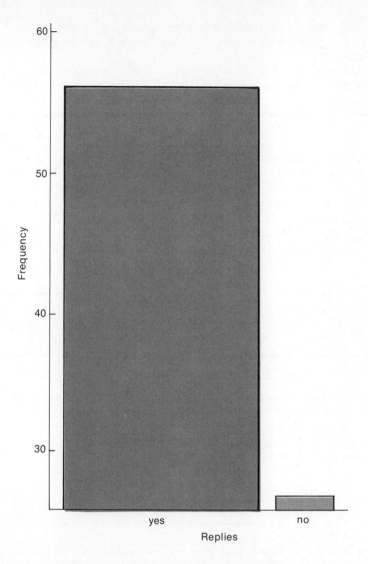

FIGURE 3.6
Distorted bar graph.

on exactly the same data. To heighten your sensitivity to this type of distortion—and to other types of statistical frauds—read the highly entertaining book by Huff.*

3.6 An Overview

When you yourself are constructing a graph, attempt to depict the data as clearly, concisely, and completely as possible. The blatant distortion shown in Figure 3.6 can be avoided easily by complying with the several customs described in the previous section. Otherwise, equally valid graphs of the same data might appear in different formats. It's often a matter of personal preference whether, for instance, a histogram or a frequency polygon should be used with quantitative data. Once again, as with the construction of frequency distributions, relax. With just a little practice—such as that provided by the exercises at the end of this chapter—you should become a competent constructor of graphs.

*Huff, D., *How to Lie With Statistics*. New York: W.W. Norton & Co., 1954.

3.7 Summary

Frequency distributions can be converted to graphs.

If the data are quantitative, histograms or frequency polygons are often used.

Histograms may be transformed into frequency polygons by constructing a line that skims the peaks of the bars, then erasing the bars. Frequency polygons are particularly useful when two or more frequency distributions are to be included in the same graph.

If the data are qualitative, bar graphs and pie charts are often used.

Bar graphs resemble histograms, except that gaps separate adjacent bars in bar graphs. A common variation on the bar graph is the pie chart.

When constructing a graph, it's considered good practice to

(1) allocate equal space to each class interval and

(2) erect a vertical axis that's approximately as tall as the horizontal axis is wide.

Important Terms

Histogram	Bar graph
Frequency polygon	Pie chart

3.8 Exercises

1. The following frequency distribution shows the annual incomes in dollars for a group of college graduates.

   (a) Construct a histogram.
   (b) Construct a frequency polygon.
   (c) Describe any data patterns.

INCOME	FREQUENCY
65,000–69,999	1
60,000–64,999	0
55,000–59,999	1
50,000–54,999	3
45,000–49,000	1
40,000–44,999	5
35,000–39,999	7
30,000–34,999	10
25,000–29,999	14
20,000–24,999	23
15,000–19,999	17
10,000–14,999	10
5,000– 9,999	8
0 – 4,999	3
Total	103

2. Construct a histogram and a frequency polygon for the data in Chapter 2, Exercise 5.

3. Construct frequency polygons for the two relative frequency distributions obtained in Chapter 2, Exercise 11.

4. Construct a histogram for the frequency distribution in Chapter 2, Exercise 10.

5. Construct a histogram and a frequency polygon for the distribution of loose change obtained in Chapter 2, Exercise 8.

6. The most recent census reveals the following distribution of ethnic backgrounds (rounded to the nearest thousand) among residents of San Francisco.

   (a) Construct a bar graph.
   (b) Construct a pie chart.
   (c) Describe any data patterns.

BACKGROUND	FREQUENCY
Afro-American	96,000
Asian-American	97,000
Euro-American	409,000
Latin-American	102,000
Native-American	3,000
Unclassified	8,000
Total	715,000

7. Criticize the following graphs. (Ignore the inadequate labeling of both axes.)

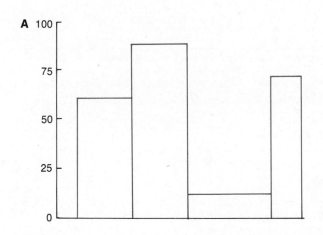

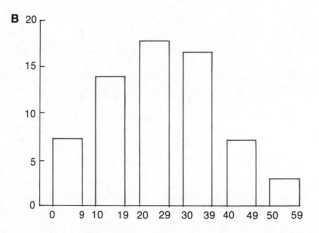

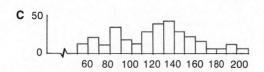

D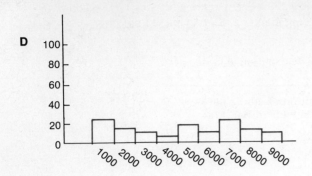

E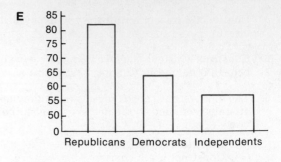

4

Describing Data with Averages

AVERAGES FOR QUANTITATIVE DATA

4.1 –Mode
4.2 –Median
4.3 –Mean
4.4 –Which Average?
4.5 –Positively and Negatively Skewed Distributions
4.6 –A Special Property of the Mean
4.7 –Averages for Grouped Data

AVERAGES FOR QUALITATIVE AND RANKED DATA

4.8 –Averages for Qualitative Data
4.9 –Averages for Ranked Data
4.10–Using the Word "Average"
4.11–Summary
4.12–Exercises

There are three common averages: the most fashionable (mode), the middle (median), and the balance point (mean).

Distributions with two peaks often have a special significance.

You can tell whether a distribution is balanced or lopsided without even looking at it—as long as you know the mean and the median.

The type of average used depends on a number of factors, but mainly whether data are quantitative, qualitative, or ranked.

You might give up a bad habit, such as smoking, because, *on the average,* heavy smokers have a poor life expectancy. You might buy a particular car because, *on the average,* this type of car gives better gas mileage. Averages occur regularly in our everyday life, and they are important tools in statistics. A well-chosen average consists of a single number (or word) about which the data are, in some sense, centered. Actually, even for a given set of data, there can be several different types of averages or, as they're sometimes called, *measures of central tendency.* This chapter describes both how to calculate and how to interpret three commonly employed averages.

"But officer I was averaging 55 miles per hour"

AVERAGES FOR QUANTITATIVE DATA

4.1 Mode

The **mode** *reflects the value of the most frequently occurring observation.* Table 4.1 shows the number of years served by the first 37 United States' Presidents, beginning with Washington (8 years) and ending with Nixon (6 years). Four years is the modal term since the greatest number of presidents, 17, served this term. Note that the mode equals the value of the most frequently occurring observation, 4, *not* the frequency of that observation, 17.

It's very easy to assign a value to the mode. If data are organized, a glance is often enough. If data are not organized, some counting might be required. The mode is readily understood as the value of the most prevalent or "fashionable" observation.

Distributions can have more than one mode. Distributions with two obvious peaks, even though they're not exactly the same height, are referred to as **bimodal.** Distributions with more than two peaks are referred to as **multimodal.** The presence of more than one mode, particularly in large sets of data, might reflect important differences among subsets of data. For instance, the distribution of weights for both male and female statistics students would most likely be bimodal, reflecting the combination of two separate weight distributions—one for males and the other for females. Notice that even the distribution of presidential terms in Table 4.1 is bimodal, with peaks at 4 and 8 years reflecting the normal term of office.

4.2 Median

TABLE 4.1
TERMS IN YEARS OF THE FIRST 37 U.S. PRESIDENTS, LISTED CHRONOLOGICALLY*

8 (Washington)	1
4	4
8	4
8	4
8	4
4	4
8	8
4	4
1	8
4	3
4	6
2	4
3	12
4	8
4	8
4	3
4	6
8	6 (Nixon)
4	

*Source: 1978 Informa-tion Please Almanac.

*The **median** reflects the middle value when observations are ordered from least to most.* The median splits a set of ordered observations into two approximately equal parts—the upper and lower halves.

The procedure for determining the median depends on whether the total number of observations is odd or even. *When the total number of observations is odd, the median equals the value of the middle-ranked observation.* In a set of 37 observations, such as those in Table 4.1, the nineteenth-ranked observation occupies the middle rank, with 18 observations both above and below it. The nineteenth-ranked observation can be found by counting up the 19 *shortest* terms (or by counting down the 19 *longest* terms), once the observations have been arranged from largest to smallest, as in Table 4.2. Four years is the value of the nineteenth-ranked observation, and therefore, this is the value of the median. Note that the median equals the value of the middle observation, 4, *not* the position of that observation, 19, among the ordered observations.

Table 4.3 shows the death rates from car accidents in 20 countries. (Ignore the column of numerical ranks until a later section.) There is no single middle-ranked observation with which to identify the median. *When the total number of observations is even, the median equals a value midway between the two middle-ranked observations.* Inspection of Table 4.3 reveals that observations already have been ordered from least to most and that the two middle-ranked observations, the tenth- and eleventh-ranked countries, France and Finland, have death rates of 61 and 66, respectively. When the two middle-ranked observations have different values, as in the present case, add these values and divide by 2 to find the value of the median. The median death rate equals 63.5 (found by adding 61 and 66, then dividing by 2).

Observations always must be ordered from least to most before the value of the middle-ranked observation(s) can be identified. This task is fairly simple with small sets of data, but it becomes increasingly cumbersome with larger sets of data, particularly when you must start with unorganized data.

4.3 Mean

The mean is the most common average—one that doubtless you have calculated many times. It can be found by adding all observations, then dividing by the number of observations, that is

$$\text{mean} = \frac{\text{sum of all observations}}{\text{number of observations}}$$

To find the mean term for the first 37 presidents, add all 37 terms ($8 + 4 + \ldots + 6 + 6$) to obtain a sum of 191 years, and then divide this sum by 37, the number of presidents, to obtain a mean of 5.16 years.

Note that observations need not be ordered from least to most before calculating the mean. Even when large sets of unorganized data are involved, the calculation of the mean is usually straightforward, particularly with the aid of a calculator.

In the long run, it's usually more efficient to substitute symbols for words in statistical formulas. When symbols are used, $\bar{X}$ designates the mean, and the above formula becomes

$$\bar{X} = \frac{\Sigma X}{n} \tag{4.1}$$

and reads: "X-bar equals the sum of the variable X divided by n."

In formula 4.1, the symbol X represents any unspecified observation within a set of observations, and X can be replaced, in turn, by each of these observations. Thus, X can be replaced, in turn, by each of the 37 presidential terms in Table 4.1, beginning with 8 and ending with 6. The symbol Σ, the capital Greek letter sigma, specifies that all observations represented by the variable X should be added. In the above example ΣX specifies that all presidential terms should be added ($8 + 4 + \ldots + 6 + 6$) to find the sum of 191. Then this sum should be divided by n, the number of observations, that is, 37, in the present example, to obtain the mean presidential term of 5.16 years.

The **mean ($\bar{X}$)** *may be viewed as the "balance point" for a frequency distribution.* Imagine that the histogram in Figure 4.1, showing the terms of the first 37 presidents, has been constructed out of some rigid material, such as wood. Furthermore, imagine that you wish to lift the histogram, using only one finger placed under its base, without disturbing the horizontal balance of the histogram. To accomplish this, your finger should coincide with 5.16, the value of the mean, as shown in Figure 4.1. If your finger were to the right of this point, the entire histogram would seesaw down to the left; if your finger were to the left of this point, the histogram would seesaw down to the right.

All observations enter directly into the calculation of the mean. In contrast, only the middle-ranked observations enter directly into the determination of the median, and only the most frequently occurring observation determines the value of the mode.

When a distribution of observations is not too lopsided, the values of the mode, the median, and the mean are similar, and any of them may be used to describe the central tendency of the distribution. This was the case in Figure 4.1, where the mode and median are 4, and the mean is 5.16. But extreme observations cause a distribution to become lopsided, as in the case for the death rates from car accidents. Here the values of the three averages differ appreciably and, therefore, do not consistently describe the central tendency of the distribution. Recall that the median death rate from car accidents in 20 countries equals 63.5. Further inspection of the data in Table 4.3 reveals that the modal death rate equals 44 (since the largest number of countries (2) had this death rate), while the mean death rate equals 75.85 (from 1517

TABLE 4.2
TERMS OF THE FIRST 37 U.S. PRESIDENTS, ARRANGED BY LENGTH OF TERM

12
8
8
8
8
8
8
8
8
8
6
6
6
4
4
4
4
4 (Median)
4
4
4
4
4
4
4
4
4
4
4
3
3
3
2
2
1

4.4 Which Average?

**TABLE 4.3
DEATH RATES (& RANKS)
FROM CAR ACCIDENTS:
UNITED STATES AND
EUROPE†**

COUNTRY	DEATH RATE*	RANK
Poland	164	20**
Greece	160	19
Hungary	134	18**
Czechoslovakia	117	17**
Austria	104	16
Ireland	97	15
Spain	83	14
West Germany	71	13
Belgium	68	12
Finland	66	11**
France	61	10
Denmark	53	9
Italy	50	8
Switzerland	46	7
Norway	44	5.5
East Germany	44	5.5**
Netherlands	43	4
Sweden	40	3
United Kingdom	38	2
United States	34	1

*For each 100,000 registered cars.
**Warsaw Pact countries.
†Source: *National Highway Traffic Safety Administration.*

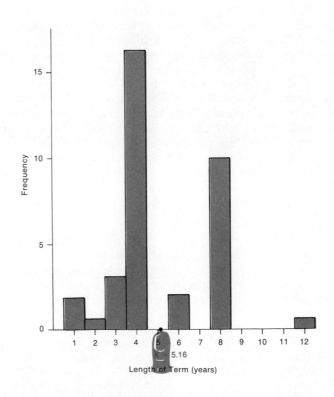

FIGURE 4.1

Mean as balance point for distribution of presidential terms.

divided by 20). A statistician's report on the values of the mode (44), the median (63.4), and the mean (75.85) would state that they "do not consistently measure the central tendency of the distribution of death rates for the 20 countries."

The mode and the median are not very sensitive to extreme observations. Relative immunity to extreme observations characterizes the mode, with its emphasis on the most frequently occurring observation, and the median, with its emphasis on the middle-ranked observation. On the other hand, total sensitivity to extreme observations characterizes the mean. Any extreme observations, such as the death rates of 160 and 164 listed for Greece and Poland in Table 4.3, contribute directly to the calculation of the mean and, with arthmetic inevitability, sway the value of the mean in their direction.

Ideally, when a distribution is lopsided, both the mean and the median should be reported. Appreciable differences between the values of the mean and median signal the presence of a lopsided distribution. If the mean exceeds the median, as it does slightly for presidential terms and more so for the death rates, we can see that the underlying distribution is lopsided because of one or more relatively large observations—specifically, for the above distributions, Franklin Roosevelt's 12-year presidential term and Greece and Poland's death rates of 160 and 164. On the other hand, if the median exceeds the mean, this tells us that the underlying distribution is lopsided because of one or more relatively small observations.

4.5 Positively and Negatively Skewed Distributions

Lopsided distributions are often called skewed distributions. Lopsided distributions caused by a few extremely large observations are called **positively skewed** distributions, since the few extreme observations are in the positive direction (to the right of) the majority of observations. Lopsided distributions caused by a few extremely small observations are called **negatively skewed** distributions, since the few extreme observations are in the negative direction (to the left of) the majority of observations. Some people have difficulty with this terminology because an entire distribution is labeled positively or negatively skewed to reflect the relative location of a minority, rather than a majority, of observations. When labeling lopsided distributions, therefore, force yourself to focus on the few extreme observations and their direction, whether positive or negative, relative to the majority of observations. Figure 4.2 summarizes this important and often confusing concept. Study it well—it will help clear up the confusion.

The distribution of incomes among U.S. families has a pronounced positive skew, with most family incomes under $50,000, and relatively few family incomes spanning a wide range of values above $50,000. On the other hand, the distribution of ages at retirement among U.S. job holders has a pronounced negative skew, with most retirement ages at 60 years or older, and relatively few retirement ages spanning the wide range of ages younger than 60. Hence, median retirement age exceeds mean retirement age.

4.6 A Special Property of the Mean

As has been seen, sometimes the mean fails to typify a set of observations and, therefore, should be used in conjunction with another average, such as the median. In the long run, however, *the mean serves as the single most preferred average for quantitative data because of its smaller sampling variability.*

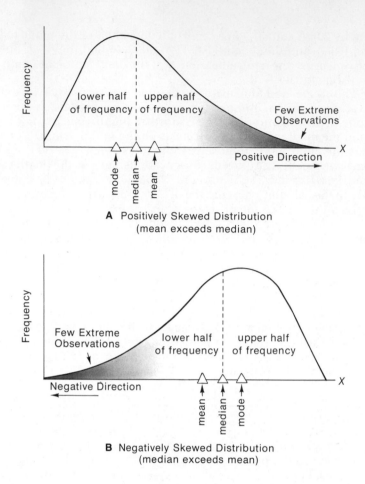

FIGURE 4.2

Mode, mean, and median in two
lopsided distributions.

Cooks often generalize from the taste of one spoonful to an entire
pot of soup, and with the aid of inferential statistics (in Part II of this
book), we'll generalize from a single sample to an entire population.
Never foolproof, these generalizations are particularly risky whenever
a series of samples from the same pot or the same population convey
radically different impressions. To reduce this excessive variability as
much as possible, cooks carefully taste each spoonful from a thoroughly
stirred pot, while statisticians capitalize on a number of factors, including the selection of the one average (from among the three possible
averages) that varies least from sample to sample.

Let's look more closely at the statistician's problem. When several
samples are selected from the same population, some observations
appear in one sample, but not in another. In other words, just by chance,
no two samples contain exactly the same observations. Thus, the value
of the average—whether the mean, median, or mode—shifts from one
sample to another, and these chance differences are referred to as
sampling variability. The sampling variability among the means of
several samples tends to be smaller (and therefore allows more risk-free
generalizations) than the sampling variability among either the medians
or modes of several samples. Because of its smaller sampling variability,
we will use the mean almost exclusively with quantitative data in this
book.

Techniques are available for finding the value of the mode, median, or mean even when you only have access to grouped data in a frequency distribution, instead of original, ungrouped observations. We will not discuss the relatively involved computational techniques for the median and mean because in such cases the original observations are almost always accessible. In those rare situations where only grouped data are accessible, you may simply specify the modal interval (the interval in the frequency distribution that contains the largest frequency) and the median interval (the interval in the frequency distribution that contains the middle-ranked observation). For example, if only grouped data in Table 2.8 were available for the distribution of weights of female statistics students, the modal interval would be 110 to 119 pounds since this interval possesses the largest frequency (17); the median interval would be 120 to 129 pounds since this interval contains the middle-ranked (twenty-seventh) observation for the 53 weights. The latter fact can be verified by cumulating frequencies upward, as shown in Table 2.8, until the cumulative frequency first equals or exceeds a value of 27, as it does for the interval 120 to 129.

4.7 Averages for Grouped Data

So far, we've been talking about quantitative data where, in principle, all three averages can be used. When data are qualitative, however, your choice among averages is restricted. *The mode can be used with qualitative data.* For instance, Yes qualifies as the modal response for the marijuana smoking question since Yes occurred more frequently than No. By the same token, it would be appropriate to report that type O is the modal blood type of prospective donors (Chapter 2, Exercise 13) and that Euro-American is the modal ethnic background of residents of San Francisco (Chapter 3, Exercise 6).

The median can be used whenever it's possible to order qualitative data from least to most. To determine the median for this type of data, simply locate the class that contains the middle-ranked observation (or observations), that is, the observation that splits the frequency distribution into two equal halves. For instance, given that faculty ranks can be ordered from lowly instructor to exalted professor, the median rank of faculty at San Jose State University corresponds to associate professor; as shown in Table 2.10, the class of associate professor contains the middle-ranked 384th observation from among the total of 767 observations.

It would not be appropriate to report a median for unordered qualitative data, such as the blood types of prospective donors or the ethnic backgrounds of San Francisco residents. Nor would it be appropriate to report a mean for *any* qualitative data, such as the faculty ranks at San Jose State University. After all, words can't be added and then divided as required by the formula for the mean.

AVERAGES FOR QUALITATIVE AND RANKED DATA

4.8 Averages for Qualitative Data

A traffic engineer wishes to compare averages of the automobile death rates listed in Table 4.3 for Warsaw Pact countries (allied with the Soviet Union) and for non-Warsaw Pact countries. There's a concern, however, about the caliber of these figures because of the sloppy date-gathering techniques used by some of the twenty countries. Rather than treat death rates as quantitative data, it's decided to transform death rates into ranks and to compare the average rank for Warsaw Pact countries with that for non-Warsaw Pact countries. The switch to ranks

4.9 Averages for Ranked Data

implies that, under the circumstances, the original death rates should be viewed as indicating relative standing rather than amount—in other words, ordinal rather than interval-ratio measurement.

Assign numerical ranks to the twenty countries, beginning with a rank of 1 for the United States and ending with a rank of 20 for Poland, as shown in Table 4.3. When ties appear, as for East Germany and Norway, award these countries the mean (5.5) of the ranks (5 and 6) that would have been assigned if their death rates had been different. Finally, collect the ranks for the five Warsaw Pact countries (5.5, 11, 17, 18, 20) and for the fifteen non-Warsaw Pact countries (1, 2, 3, 4, 5.5, 7, 8, 9, 10, 12, 13, 14, 15, 16, 19), and deal only with these ranks in subsequent analyses.

There is no modal rank for either group of countries. Even if modal ranks had existed, because of at least one tie among ranks within each group, these averages would not necessarily have represented the central tendencies within the two groups. In fact, the mode for ranked data often is either nonexistent or uninformative.

This leaves two other possible comparisons—one based on the mean rank for each group of countries or one based on the median rank for each group. When dealing with ranked data, the median, whose value can be identified by counting up (or down) to the middle rank for each group, is more appropriate than the mean, whose value must be calculated by adding the ranks of each group into a total, then dividing. Strictly speaking, the latter operation implies that ranked data reflects equal intervals (and interval-ratio measurement) when in fact, as discussed in Section 1.7, they reflect merely order (and ordinal measurement). Therefore, the *identification* of a median, rather than the *calculation* of a mean, is more appropriate for ranked data, and *the median is the preferred average for ranked data.*

To find the median rank for the five Warsaw Pact countries (5.5, 11, 17, 18, 20), simply identify the value of the middle (or third) rank, namely 17. Likewise, to find the median rank for the fifteen non-Warsaw Pact countries (1, 2, 3, 4, 5.5, 7, 8, 9, 10, 12, 13, 14, 15, 16, 19), identify the value of the middlemost (or eighth) rank, which, coincidentally, happens to be 8. Thus, an analysis based on ranked data reveals that, on the average, Warsaw Pact countries rank higher (have higher death rates) than non-Warsaw Pact countries. In particular, when the entire twenty countries are assigned numerical ranks, median ranks of 17 and 8 are obtained for the Warsaw Pact and non-Warsaw Pact countries, respectively. Although this conclusion lacks the impact of one based directly on the original death rates, it could, under the circumstances, represent the results of the only appropriate analysis.

4.10 Using the Word "Average"

Strictly speaking, an "average" could refer to the mode, median, or mean—or even to some more exotic average, such as the geometric mean or the harmonic mean. Conventional usage prescribes that "average" usually signifies "mean," and this connotation is often reinforced by the context. For instance, "grade point average" is virtually synonymous with "mean grade point." To my knowledge, not even the most enterprising, grade point impoverished student has attempted to satisfy graduation requirements by exchanging a more favorable modal or median grade point for the customary mean grade point. Unless context and usage make it clear, however, it is a good policy to specify the particular average with which you are dealing, even at the cost of a slight explanation. When dealing with controversial topics, it is always wise to insist that the exact nature of the average be specified.

4.11   Summary

The mode equals the value of the most frequently occurring observation.

The median equals the value of the middle-ranked observation (or observations), and it splits frequencies into upper and lower halves.

The mean is the balance point of a frequency distribution, and it can be calculated from Formula 4.1.

When frequency distributions are not lopsided, the values of all three averages tend to be similar and equally representative of central tendencies within the distributions. When frequency distributions are lopsided, however, the values of the three averages differ appreciably. Ideally, in this case, both the mean and the median should be reported.

Whenever possible, use original ungrouped observations to determine the value of any average. In those rare situations where only grouped data are available, simply specify the modal interval or the median interval.

The mean is the preferred average for quantitative data because of its smaller sampling variability. The mode can be used with qualitative data. If it's possible to order qualitative data from least to most, the median also can be used. The median is the preferred average for ranked data.

Conventional usage prescribes that "average" usually signifies "mean," but when dealing with controversial topics, it's wise to insist that the exact nature of the average be specified.

Important Terms & Symbols ──────────────────

Mode	Bimodal
Median	Positively skewed
Mean ($\overline{X}$)	Negatively skewed

4.12   Exercises

1. During their first swim through a water maze, fifteen laboratory rats make the following number of errors (blind alleyway entrances): 2, 17, 5, 3, 28, 7, 5, 8, 5, 6, 2, 12, 10, 4, 3.

    (a) Find the mode, median, and mean for these data.
    (b) Without constructing a frequency distribution or graph, would you characterize the shape of this distribution as balanced, positively skewed, or negatively skewed?

2. The owner of a new car conducts a series of six gas mileage tests and obtains the following results, expressed in miles per gallon: 20.3, 22.7, 21.4, 20.6, 21.4, 20.9. Find the mode, median, and mean for these data.

3. Calculate the mode, median, and mean for the weight data originally listed in Table 1.1. (Hint: a stem and leaf arrangement, as in Table 2.2, is helpful in determining the mode and median.)

4. Determine the mode, median, and mean for the following distribution of retirement ages: 60, 63, 45, 63, 65, 70, 55, 63, 60, 65, 63.

5. College students are surveyed to determine where, if money were not a consideration, they would most like to spend Spring vacation: Hawaii (H), Miami Beach (M), Mexico (Me), Spain (S), and Other (O). Results are as follows:

H	H	Me	S	H
Me	M	S	H	O
O	M	Me	H	S
H	Me	H	O	H

Find the mode and, if possible, the median.

6. Find the mode and, if possible, the median for the distribution in Chapter 3, Exercise 6.

7. Find the mode and, if possible, the median for the movie ratings listed in Chapter 2, Exercise 12.

8. (Identify all correct alternatives.) The mean exceeds the median—and, therefore, the shape is positively skewed—in which of the following distributions?

   (a) Test scores on an easy test, with most students scoring high.
   (b) Ages of college students, with most students in their late teens or early twenties.
   (c) Loose change in pockets and purses of classmates, with most carrying less than $1.00.
   (d) Attendance at a popular movie theater, with most audiences at or near capacity.

9. (Sentence completion)

   (a) The mean is the preferred average for quantitative data because
   _____

   (b) The median is the preferred average for ranked data because
   _____

10. At a salary bargaining session, the school board claims that salaries of district teachers average $25,000 per year, while the teachers' association claims that district teachers average only $22,000 per year. What's probably the cause of this discrepancy?

11. Determine the modal and median intervals for the following relative frequency distribution.

DISTRIBUTION OF FAMILIES BY NUMBER OF CHILDREN*

NO. OF CHILDREN		PERCENT (%)
4 or more		10 %
3		11 %
2		17 %
1		18 %
0		44 %
	Total	100 %

*Based on 51,237,000 families, 1970 U.S. Census.

12. Determine the modal and median intervals for the frequency distribution of weights given in Table 2.4.

13. The two hundred and seventy-three respondents to a poll of news editors and directors, conducted by the Associated Press, ranked the biggest events in U.S. history in descending order of importance.

*The American Revolution
 Drafting the U.S. Constitution
*World War II ✔
 The U.S. moon landings ✔
*Development of the atomic bomb ✔
*The 1929 crash and the ensuing Great Depression ✔
*Watergate and the resignation of Richard M. Nixon ✔
*World War I ✔
 Henry Ford, the Model T, and the rise of the automobile ✔
*Lincoln's assassination
 The development of television ✔
*Kennedy's assassination ✔
 Thomas Edison and the electrification of the nation
*Vietnam ✔
 Franklin Roosevelt's New Deal ✔
 The changing role of women
 The Wright brothers and the growth of aviation ✔
 The Louisiana Purchase
 The 1954 Supreme Court decision outlawing segregation in schools ✔

(a) Assign numerical ranks and determine the median rank for events judged to be more negative than positive (and marked with an asterisk) and the median rank for the remaining events.

(b) Find the median rank for events of the twentieth century (marked with a check) and the median rank for the remaining events.

MEASURES OF VARIABILITY
FOR QUANTITATIVE DATA

5.1–Range
5.2–Standard Deviation
5.3–Deviation Formula
5.4–A Measure of Distance
5.5–Raw Score Formula
5.6–Variance

MEASURES OF VARIABILITY
FOR QUALITATIVE AND
RANKED DATA

5.7–General Comments
5.8–Summary
5.9–Exercises

5
Describing Variability

The simplest measure of variability, the range, suffers from some serious deficiencies.

The statistician's yardstick is a measure of variability known as the standard deviation.

A majority of observations are usually within one standard deviation of the mean—at least for most distributions in the behavioral sciences. By the same token, relatively few observations are more than three standard deviations from the mean.

MEASURES OF VARIABILITY FOR QUANTITATIVE DATA

Although the mean annual temperature tends to be the same—57°—for both San Francisco and Wichita, it would be most erroneous to conclude that these two cities have comparable weather. As a matter of fact, monthly temperatures for San Francisco register close to 57° throughout most of the year, to the delight of the tourist industry, while those in Wichita fluctuate radically, from a warm 81°F to a chilly 32°F, depending on the season. Clearly, prospective residents of these cities should assess their wardrobe needs in terms of the "dispersion," "scatter," or "variability" of monthly temperatures about the mean annual temperature.

Ordinarily, when summarizing data in descriptive statistics, we specify not only a measure of central tendency, such as the mean, but also a measure of variability, that is, a measure of the amount by which observations deviate or vary about the measure of central tendency. This chapter focuses on two measures of variability—the range and particularly the standard deviation.

You probably already possess an intuitive feel for gross differences in variability. In Figure 5.1, each of the three frequency distributions consists of seven observations with the same mean (10), but different variabilities. Before reading on, rank the three distributions from least to most variable. Your intuition is correct if you concluded that distribution (a) has the smallest variability, distribution (b) has intermediate variability, and distribution (c) has the greatest variability.

5.1 Range

More exact measures of variability not only enhance communication but are essential for subsequent quantitative work in statistics. One such exact measure is the range. The **range** is the distance from the largest observation to the smallest observation in a distribution. In Figure 5.1, the least variable distribution (a) has a range of 0 (from 10 to

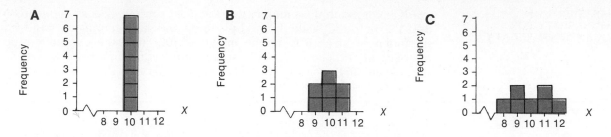

FIGURE 5.1

Three distributions with the same mean (10) but different amounts of variability.

10), (b) has a range of 2 (from 11 to 9), and the most variable distribution (c) has a range of 4 (from 12 to 8), in agreement with the intuitive judgments about differences in variability. The range is a handy measure of variability; it can be readily calculated and understood.

The range is not the statistician's preferred measure of variability because of several shortcomings. First, its value depends on only two observations—the largest and smallest—and thus it fails to use the information provided by the remaining observations. Furthermore, the value of the range tends to increase with increases in the total number of observations. For instance, the range of adult heights might be 6 or 8 inches for a distribution of half a dozen people, while it might be 14 or 16 inches for a distribution of six dozen people. Larger sets of observations are more likely to include very short and very tall people who, of course, inflate the value of the range. Instead of being a relatively "pure" measure of variability, the size of the range is determined, in part, by how many observations are included in the distribution.

5.2 Standard Deviation

The statistician's preferred measure of variability is the **standard deviation(S).** The standard deviation avoids the shortcomings of the range. It reflects the contributions of all (not just two) observations, and it doesn't increase in value with larger sets of observations.

As its name implies, the standard deviation supplies us with a standard, much like a yardstick, that reflects the deviations of observations about the mean. The size of the standard deviation depends on the variability in each distribution. In distributions with little variability, observations tend to hover in the vicinity of the mean, and their deviations about the mean are correspondingly small, as also is the value of the standard deviation. In distributions with much variability, observations tend to scatter from the mean, and their deviations about the mean are correspondingly large, as also is the value of the standard deviation. For instance, even though two classes of fourth graders have the same mean IQ of 100, one class might be easy to teach because of the similar intellectual aptitudes of students, as reflected in a standard deviation of 5 IQ points, while the other class might be very difficult to teach because of the dissimilar intellectual aptitudes of students, as reflected in a standard deviation of 15 IQ points.

For most frequency distributions in the behavioral sciences, a majority (usually 50% to 85%) of observations are within a range of from one standard deviation above the mean to one standard deviation below the mean. An overwhelming majority (usually 90% to virtually 100%) are within a range of from three standard deviations above the mean to three standard deviations below the mean. In the difficult class of fourth graders, for example, a majority of the class probably score between 85 and 115 on the IQ scale (found by adding and subtracting one standard deviation of size 15 from the mean of 100), and the overwhelming majority of the class probably score—hold your breath—between 55 and 145 on the IQ scale (found by adding and subtracting three standard deviations of size 15 from the mean of 100).

"Doctor, I can't continue living with a man who is more than three standard deviations from the mean."

5.3 Deviation Formula

There are several formulas for the standard deviation, and two of them will be described in this chapter. Known as the **deviation formula**, the first formula requires that each observation be expressed as a deviation, that is, a positive or negative distance, from the mean. After each deviation has been squared, that is, multiplied by itself, all squared deviations are added. The resulting total is divided by the total number of observations to obtain the mean for all squared deviations. Finally, a square root is extracted from this mean.*

The deviation formula for the standard deviation, S, takes the form:

$$S = \sqrt{\frac{\Sigma(X - \bar{X})^2}{n}} \qquad (5.1)$$

and may be read as "The standard deviation equals the square root of the sum of the squared deviations of observations from the mean, divided by the number of observations." On the right side of Formula 5.1, there's only one new symbol, $\sqrt{}$, which specifies the extraction of a square root. All other terms should be familiar.

At this point, you might be mystified about the precise origin of the formula for the standard deviation. Don't conclude, therefore, that you've missed something. As a matter of fact, the formula for the standard deviation emerges from mathematical reasoning beyond the scope of this book. For our purposes, it's sufficient if you can both calculate and interpret the standard deviation—even though, in a pinch, you might still be on friendlier terms with a more transparent (but less preferred) measure of variability, such as the range.

Table 5.1 illustrates how to calculate the standard deviation for the seven observations (one 8, two 9s, one 10, two 11s, and one 12) of

---

*The square root of a number is that number which, when multiplied by itself, yields the original number. For example, the square root of 16 is 4, since 4 times 4 equals 16. To extract the square root of any number, consult a calculator with a square root key.

TABLE 5.1
**CALCULATION OF THE
STANDARD DEVIATION
(DEVIATION FORMULA)**

A. COMPUTATIONAL SEQUENCE

Assign a value to n ① representing the number of X scores.

Add all X scores ② , and obtain the mean of these scores ③ .

Express each X score, one at a time, as a deviation from the mean, that is, subtract the mean from each X score ④ .

Square each deviation score ⑤ , one at a time, and then add all squared deviation scores ⑥ .

Substitute numbers into the formula ⑦ , and solve for S.

B. DATA AND COMPUTATIONS

	④	⑤
X	$X - \bar{X}$	$(X - \bar{X})^2$
12	2	4
10	0	0
11	1	1
8	−2	4
9	−1	1
11	1	1
9	−1	1

① $n = 7$ ② $\Sigma X = 70$ ⑥ $\Sigma(X - \bar{X})^2 = 12$

③ $\bar{X} = \dfrac{70}{7} = 10$

⑦ $S = \sqrt{\dfrac{\Sigma(X - \bar{X})^2}{n}} = \sqrt{\dfrac{12}{7}} = \sqrt{1.71} = 1.31$

distribution (c) in Figure 5.1. In Table 5.1, the numbers in red circles cross-reference descriptions of the computational sequence in the top panel with the actual computations in the bottom panel. (This coding device appears in all of the remaining computational tables.) In the bottom panel, notice that observations needn't be ranked in any special order when calculating the standard deviation—just as when calculating the mean.

Following the computational procedures outlined in Table 5.1, you could calculate a standard deviation for the other two distributions in Figure 5.1. These calculations would reveal that the relative size of the standard deviation tends to mirror the intuitive judgments about differences in variability. More specifically, the value of the standard deviation equals 0.00 for distribution (a), judged to be the least variable; 0.75 for distribution (b), judged to be moderately variable; and 1.31 for distribution (c), judged to be the most variable. As has been suggested, the greater the variability of the data, the larger the value of the standard deviation.

5.4 A Measure of Distance

A key to understanding the standard deviation rests with the following distinction: *The mean is a measure of position, but the standard deviation is a measure of distance.* Figure 5.2 describes the weight distribution for female statistics students, originally shown in Figure 3.1. Note that the mean of 130 pounds has a particular position or location along the horizontal axis—it's located at the point, and only at the point, corresponding to 130 pounds. On the other hand, the standard deviation of 23 pounds for the same distribution has no particular location along the horizontal axis. As a measure of distance, the standard deviation functions much like a yardstick. Using a yardstick as our

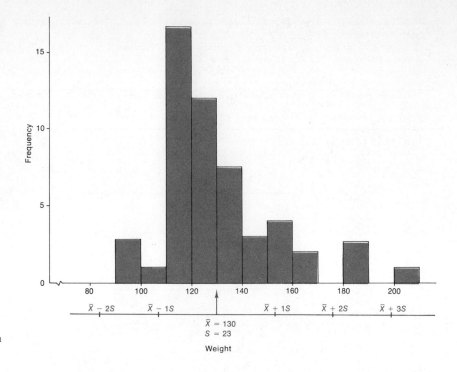

FIGURE 5.2

Weight distribution with mean
and standard deviation.

unit of measurement, we might find the width of a classroom to be 13 yards; similarly, using the standard deviation as our unit of measurement, we might find some person's weight to be two standard deviations above the mean, $\bar{X} + 2S$, another person's weight to be two thirds of one standard deviation below the mean, $\bar{X} - 2/3S$, and so on.

Our yardstick analogy is useful so far as it goes, but it's important to remember that yardsticks and standard deviations differ in several ways. Yardstick distances originate from any point, while standard deviation distances always originate from the mean and are expressed as positive or negative deviations above or below the mean. Practically speaking, yardsticks are constant, while standard deviations, of course, vary from one distribution to another, reflecting different amounts of variability.

5.5 Raw Score Formula

The deviation formula is cumbersome in cases where the mean equals some complex number, such as 129.45, or where the number of observations is large. In these cases, which often occur in practice, use the more efficient **raw score formula,** defined as follows:

$$S = \sqrt{\frac{n\Sigma X^2 - (\Sigma X)^2}{n^2}} \tag{5.2}$$

Note that no mean appears in the raw score formula. Table 5.2 illustrates the use of this formula to calculate the standard deviation for the same seven observations as in Table 5.1. It's wise to adopt the computational format shown in the bottom panel.

Not unexpectedly, because of their mathematical equivalence, the deviation and raw score formulas yield the same standard deviation of 1.31 for distribution (c). The tremendous efficiency of the raw score formula becomes more apparent when dealing with large sets of observations, as in Exercise 7 at the end of this chapter.

TABLE 5.2
CALCULATION OF THE
STANDARD DEVIATION
(RAW SCORE FORMULA)

A. COMPUTATIONAL SEQUENCE

Assign a value to n ①, representing the number of X scores.
Sum all X scores ②.
Square each X score ③, one at a time, and then add all squared X scores ④.
Substitute numbers into the formula ⑤, and solve for S.

B. DATA AND COMPUTATIONS

X	③ X^2
12	144
10	100
11	121
8	64
9	81
11	121
9	81

① $n = 7$ ② $\Sigma X = 70$ ④ $\Sigma X^2 = 712$

⑤ $S = \sqrt{\dfrac{(n)(\Sigma X^2) - (\Sigma X)^2}{(n)^2}} = \sqrt{\dfrac{(7)(712) - (70)^2}{(7)^2}}$

$= \sqrt{\dfrac{4984 - 4900}{49}} = \sqrt{\dfrac{84}{49}} = \sqrt{1.71} = 1.31$

A most common confusion involves $(\Sigma X)^2$, obtained by first adding all scores, then squaring the total, and ΣX^2, obtained by first squaring each score and then adding all squared scores. It's a helpful computational check to remember that neither the square of the summed scores, $(\Sigma X)^2$, nor the sum of the squared scores, ΣX^2, can be negative numbers. Nor can the expression $n\Sigma X^2 - (\Sigma X)^2$ be a negative number. As an almost foolproof method for avoiding computational errors, calculate everything twice, and proceed only when computational results agree.

5.6 Variance

The square of the standard deviation, S^2, is known as the **variance.** This measure of variability often assumes a very special role, even to the exclusion of the standard deviation, in more advanced statistical work. The formula for the variance can be obtained from any formula for the standard deviation by simply eliminating the square root sign. For instance, the deviation formula for the variance reads:

$$S^2 = \frac{\Sigma(X - \bar{X})^2}{n} \tag{5.3}$$

In this book, the variance assumes an important role in Chapters 17 and 18. Until then, however, we'll have very little use for it, except as a symbol that occasionally appears in some formulas.

MEASURES OF VARIABILITY FOR QUALITATIVE AND RANKED DATA

5.7 General Comments

Measures of variability are virtually nonexistent for these two types of data. In the case of qualitative data, it's probably adequate to note merely whether observations are evenly divided among the various classes (maximum variability), concentrated mostly in one class (minimum variability), or unevenly divided among the various classes (intermediate variability). For example, since Yes replies to the marijuana smoking question are twice as prevalent as No replies, variability can be described as intermediate. If Yes and No replies had been almost equally frequent, variability could have been described as maximum. If No replies had been rare, variability could have been described as minimum.

If qualitative data can be ordered, then it would be appropriate to describe variability by identifying extreme observations. For instance, the active membership of an officers' club might include no one with a rank below first lieutenant or above brigadier general.

No measures of variability exist for ranked data, other than a simple specification of how many ranks have been assigned. This, of course, supplies no information about the variability of the original observations, prior to the assignment of numerical ranks.

5.8 Summary

The simplest measure of variability, the range, suffers from several shortcomings.

The standard deviation is the preferred measure of variability. It measures the amount by which observations deviate about the mean. In most frequency distributions in the social and behavioral sciences, a majority of observations are within one standard deviation of the mean, and an overwhelming majority are within three standard deviations of the mean.

This book describes two formulas for the standard deviation. The deviation formula (5.1) presents the standard deviation in its most comprehensible form, while the raw score formula (5.2) presents the standard deviation in its most convenient computational form.

In its role as a measure of distance (rather than position), the standard deviation functions as the statistician's yardstick.

Measures of variability are virtually nonexistent for qualitative and ranked data.

Important Terms & Symbols

Range	Raw score formula
Standard deviation(S)	Variance(S^2)
Deviation formula	

5.9 Exercises

1. Identify the alternatives that reflect shortcomings of the range.
   (a) difficult to calculate
   (b) based on only two observations
   (c) influenced by sample size
   (d) difficult to understand

2. (Class exercise) Find the range of heights for people sitting in the front row of the classroom and the range of heights for all of the remaining members of the class. Just by chance, the range for the larger group should tend to exceed that for the smaller group. Verify this.

3. Using the deviation formula, calculate the standard deviation for this distribution of four numbers: 1, 3, 4, 4.

4. Add 10 to each of the observations above to produce a new distribution (11, 13, 14, 14). Would you expect the value of S to be the same for both the original and present distributions? Explain your answer, and then calculate S for the present distribution.

5. Add 10 to only the smallest observation in Exercise 3 to produce still another distribution (11, 3, 4, 4). Would you expect the value of S to be the same for both this distribution and the original distribution in Exercise 3? Explain your answer, and then calculate S for the present distribution.

6. Verify that the value of S is the same for the set of scores (12, 6, 3, 4, 5), regardless of whether the deviation or raw score formula is used.

7. Using the raw score formula, calculate the standard deviation for the distribution of weights in Table 1.1.

8. Assume that the distribution of IQ scores for all college students has a mean of 120 with a standard deviation of 15. These two bits of information imply that:

   (a) All students have IQs of either 105 or 135 because everybody in the distribution is either one standard deviation above or below the mean. True or False.
   (b) All students score between 105 and 135 because everybody is *within* one standard deviation of the mean. True or False.
   (c) Some students deviate more than one standard deviation above or below the mean. True or False.
   (d) All students deviate more than one standard deviation above or below the mean. True or False.
   (e) A student who has an IQ of 150 deviates two standard deviations above the mean. True or False.

9. Employees of Corporation A earn annual salaries described by a mean of $30,000 and a standard deviation of $2000, while employees of Corporation B earn annual salaries described by a mean of $30,000 and a standard deviation of $5000. Which corporation would you prefer to join? Explain.

10. Can the value of the standard deviation ever be negative? Explain.

11. Referring to Chapter 3, Exercise 6, would you describe the distribution as having maximum, intermediate, or minimum variability?

NORMAL DISTRIBUTIONS

6.1 –The Normal Curve
6.2 –*z* Scores
6.3 –Standard Normal Curve
6.4 –Standard Normal Table
6.5 –Finding Areas
6.6 –Finding Scores

STANDARD SCORES

6.7 –More About *z* Scores
6.8 –Other Standard Scores
6.9 –Percentile Ranks
6.10–Summary
6.11–Exercises

6 Normal Distributions and Standard Scores

Amaze your friends by interpreting standardized test scores with the aid of one of the oldest and most useful tables in statistics—the standard normal table.

What links Bill Walton, the professional basketball player who stands 6 feet 11 inches tall, and a midget who stands only 4 feet 5 inches tall? See Exercise 5, this chapter.

Shades of alchemy! You'll learn to convert a distribution of original scores into a distribution of standard scores having any mean and standard deviation that you desire (without, however, destroying the essential information contained in the original distribution).

NORMAL DISTRIBUTIONS

As portrayed in the movie "The President's Analyst," the director of the Federal Bureau of Investigation, rather short himself, encourages the recruitment of similarly short FBI agents. If, in fact, FBI agents are to be selected only from applicants who are no taller than exactly 65 inches, what proportion of all of the original applicants will be eligible? This question is difficult to answer without more information.

One source of additional information is the observed distribution of heights for 3091 men shown in Figure 6.1. If these data can be viewed as representative of the men who apply for FBI jobs, they can be used to estimate the proportion of applicants who will be eligible. To obtain the estimated proportion of .165 from Figure 6.1, merely add the values associated with the shaded bars. (Only half of the bar at 65 inches is shaded to adjust for the fact that any height between 64.5 and 65.5 inches is reported as 65 inches, whereas prospective applicants must be no taller than *exactly* 65 inches, that is, 65.0 inches.)

6.1 The Normal Curve

Another source of information is supplied by the idealized red curve, known as the normal curve, which is superimposed on the observed distribution of heights in Figure 6.2. Obtained from a mathematical equation, *the* **normal curve** *is a theoretical curve noted for its symmetrical, bell-shaped form.* Being symmetrical, the lower half of the

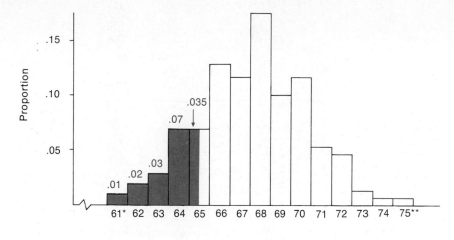

*61 inches or shorter.

**75 inches or taller.

FIGURE 6.1

Relative frequency distribution for heights of 3091 men. (Source: National Center for Health Statistics, Series 11, No. 14.)

normal curve is simply the mirror image of the upper half of the curve. Being bell-shaped, the normal curve peaks above a point midway along the horizontal spread and then tapers off gradually in either direction from the peak (without actually touching the horizontal axis, since, in theory, the tails of a normal curve extend out infinitely far).

Many observed frequency distributions in the real world approximate a normal curve. Other examples include scores on IQ tests, slight measurement errors made by a succession of people who attempt to measure the same distance, and even the heights of corn stalks and the useful lives of 100-watt electric light bulbs. Having established that a given set of quantitative data approximates a normal curve, you can use the normal curve as a basis for calculating the answers to a wide variety of questions, including the one about eligible FBI applicants. In the long run this proves to be a much more efficient procedure than dealing directly with each observed frequency distribution.

The normal curve can be used much as you would a road map—as an abstract guide that, when coordinated with a few bits of outside information, leads to your destination. To continue the road map analogy, you must, of course, be traveling through the area de-

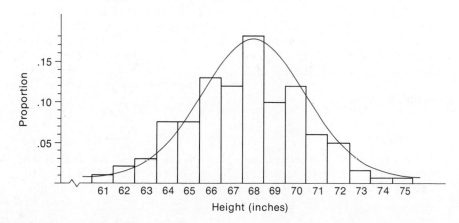

FIGURE 6.2

Normal curve superimposed on distribution of heights.

55

"Now ladies and gentlemen, a normal American family."

scribed on the map. By the same token, you must be dealing with a set of data that adequately approximates a normal curve. (In this chapter, we'll always assume that the approximation is adequate. More advanced statistics books supply techniques for testing this assumption.) Given a relevant map, you must coordinate it with one crucial bit of outside information—your present position as determined from some landmark, such as a town, highway intersection, or freeway sign. In the case of the normal curve, two bits of outside information are indispensible—values from the original distribution for the mean (about which the normal curve always is centered) and for the standard deviation. Thus, before the normal curve can be used to answer the question about eligible FBI applicants, it must be established that, for the original distribution of 3091 men, the mean height equals approximately 68 inches and the standard deviation equals approximately 3 inches.

6.2 z Scores

To interpret distances on a road map, you must use the specified scale, for instance, 25 miles to the inch. To interpret the tables for the normal curve, you also must use a special scale involving z scores. *To obtain a **z score,** express any original score,* such as the maximum height of 65 inches for FBI applicants, *as a distance from the mean (by subtracting the mean), and split this distance into standard deviation units (by dividing with the standard deviation).* In effect, the resulting z score always indicates the position of the original score relative to its mean and standard deviation. Regardless of whether the original units of measurement were inches, miles, pounds, or dollars, a z score of 2.00 always signifies that the original score exceeds its mean by exactly two standard deviations. Similarly, a z score of −1.27 signifies that the original score deviates below its mean by exactly one and twenty-seven hundredths standard deviations. A z score of 0 signifies that the original score coincides with the mean.

To convert any original observation into a z score, use the following formula:

$$z = \frac{X - \bar{X}}{S} \tag{6.1}$$

where X is the original score, and $\bar{X}$ and S are the mean and standard deviation for the distribution of original scores. Thus, to answer the question about eligible FBI applicants, replace X with 65, the maximum permissible height, $\bar{X}$ with 68, the mean height, and S with 3, the standard deviation of heights, and solve for z as follows,

$$z = \frac{65 - 68}{3} = \frac{-3}{3} = -1.00$$

This informs us that the cutoff height is exactly one standard deviation below the mean. Armed with a value for z, we can refer to the standard normal table. First, however, a few comments about the standard normal curve.

6.3 Standard Normal Curve

If the original distribution approximates a normal curve, then the shift to z scores always produces a new distribution that approximates the standard normal curve. This is the one normal curve for which tables are actually available. *The **standard normal curve** always has a mean of 0 and a standard deviation of 1.* Mathematical proof that a distribution of z scores has a mean of 0 and a standard deviation of 1 is not given in this book, but Table 6.1 demonstrates that the z score formula equals 0 when an original score coincides with the mean and that the z score formula equals 1 when an original score is located exactly one standard deviation above the mean.

6.4 Standard Normal Table

Essentially, the standard normal table consists of columns of z scores coordinated with columns of proportions. In a typical problem, access to the table is gained through a z score, such as -1.00, and the answer is read as a proportion (or relative frequency), such as the proportion of eligible FBI applicants.

Let's look at the table of the standard normal curve as presented briefly in Table 6.2 and more completely in Table A of Appendix C. Columns are arranged in sets of three, designated as A, B, and C in the legend at the top of each page. *When using the top legend, all entries refer to the upper half of the standard normal curve.* The entries in column A are z scores, beginning with 0.00 and ending with ∞ (infinity). Given a z score of zero or more, columns B and C indicate how the z score splits the area in the upper half of the normal curve. As suggested by the shading in the top legend of the table, column B indicates the proportion of area between the mean and the z score, while column C

TABLE 6.1

DEMONSTRATION WITH z SCORE FORMULA

(i) *Given:* $z = \dfrac{X - \bar{X}}{S}$

(ii) *Verify:* that $z = 0$ when the original score X is located at the mean $\bar{X}$.
Substitute $\bar{X}$ for X in the z score formula:

$$z = \frac{\bar{X} - \bar{X}}{S} = \frac{0}{S} = 0$$

(iii) *Verify:* that $z = 1$ when the original score X is located at one standard deviation unit above the mean, $\bar{X} + S$.
Substitute $\bar{X} + S$ for X in the z score formula:

$$z = \frac{(\bar{X} + S) - \bar{X}}{S} = \frac{S}{S} = 1$$

TABLE 6.2
AREAS UNDER
STANDARD NORMAL
CURVE FOR VALUES OF z
(FROM TABLE A OF
APPENDIX C)

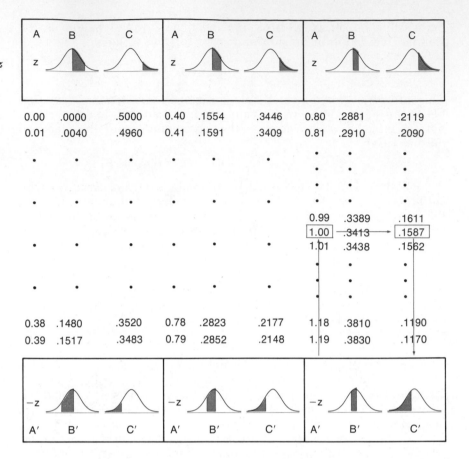

A	B	C	A	B	C	A	B	C
z			z			z		
0.00	.0000	.5000	0.40	.1554	.3446	0.80	.2881	.2119
0.01	.0040	.4960	0.41	.1591	.3409	0.81	.2910	.2090
·	·	·	·	·	·	·	·	·
·	·	·	·	·	·	·	·	·
						·	·	·
						0.99	.3389	.1611
						1.00	.3413	.1587
·	·	·	·	·	·	1.01	.3438	.1562
						·	·	·
·	·	·	·	·	·	·	·	·
0.38	.1480	.3520	0.78	.2823	.2177	1.18	.3810	.1190
0.39	.1517	.3483	0.79	.2852	.2148	1.19	.3830	.1170
A′	B′	C′	A′	B′	C′	A′	B′	C′
−z			−z			−z		

indicates the proportion of area beyond the z score—in the upper tail of the standard normal curve.

Because of the symmetry of the normal curve, the entries in Table 6.2 and Table A also can refer to the lower half of the normal curve. Now, columns are designated as A′, B′, C′ in the legend at the bottom of each page. *When using the bottom legend, all entries refer to the lower half of the standard normal curve.*

Imagine that the nonzero entries in column A′ are negative z scores, beginning with −0.01 and ending with −∞ (negative infinity). Given a negative z score, columns B′ and C′ indicate how that z score splits the lower half of the normal curve. As suggested by the shading in the bottom legend of the table, column B′ indicates the proportion of area between the mean and the negative z score, while column C′ indicates the proportion of area beyond the negative z score—in the lower tail of the standard normal curve.

When using these tables, it's important to remember that, for any z score, the corresponding proportions in columns B and C (or columns B′ and C′) always sum to .5000. By the same token, the total area under the normal curve always equals 1.0000, the sum of the proportions in the lower and upper halves, that is, .5000 + .5000. Finally, although a z score can be either positive or negative, proportions of area under the curve (or relative frequencies) are always positive or zero but never negative.

A wide variety of problems can be solved with the aid of normal curve tables. The next two sections contain representative examples.

Let's use the normal curve to answer the question at the beginning of this chapter: What proportion of all original applicants will stand no taller than exactly 65 inches, given that the distribution of heights approximates a normal curve with a mean of 68 inches and a standard deviation of 3 inches?

When first solving normal curve problems, it's wise to sketch a normal curve and shade in the target proportion, as in the top panel of Figure 6.3. In this case, the target proportion identifies the proportion of men who are no taller than 65 inches.

Before consulting Table 6.2 (or Table A in the Appendix), 65 must be expressed as a z score, and you might recall it equals -1.00. Given a negative z score, use the bottom legend of the standard normal table to interpret all entries. The red lines in Table 6.2 indicate how to use the normal curve table in the present example. Look up column A' to a z score of 1.00 (representing -1.00), and note the corresponding proportion of .1587 or .16 in column C'. As suggested in the bottom panel of Figure 6.3, the entry from column C' reveals the proportion of area below a z score of -1.00. Thus, it can be concluded that only .16 of all of the original applicants will, in fact, be eligible.

What's the relative standing of a student with a score of 640 on the Graduate Record Exam (GRE), given that these scores approximate a normal curve with a mean of 500 and a standard deviation of 100?

Once again, sketch a normal curve and shade in the target proportion, as in the top panel of Figure 6.4. In this case, the target proportion is identified with the proportion of scores smaller than 640.

Before consulting Table A, express 640 as a z score:

$$z = \frac{X - \bar{X}}{S} = \frac{640 - 500}{100} = \frac{140}{100} = 1.40$$

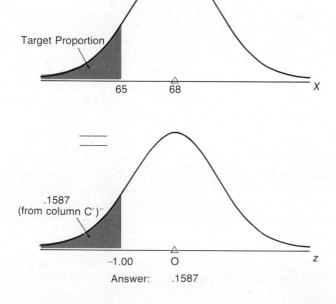

Less than 65

Target Proportion

65   68   X

.1587
(from column C')

−1.00   O   z

Answer: .1587

FIGURE 6.3

Finding area.

59

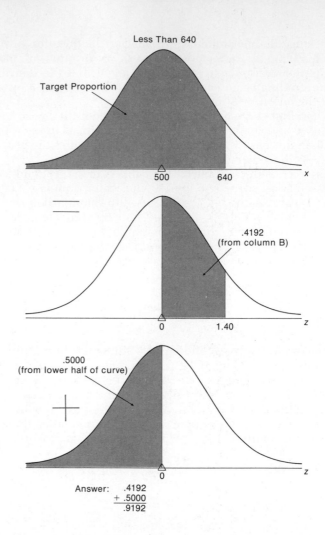

FIGURE 6.4
Finding area.

Given a positive z score of 1.40, use the top legend of Table A to interpret all entries. Look down column A to a z score of 1.40, and note the corresponding proportion of .4192 in column B. As suggested in the middle panel of Figure 6.4, the entry from column B reveals the proportion of area between the mean of 0 and a z score of 1.40. To this proportion must be added, as suggested in the bottom panel of Figure 6.4, the proportion of .5000, representing the lower half of the normal curve. The proportion of scores smaller than 640 equals .9192, or .92. In other words, a student with a GRE score of 640 does better than 92 percent of those students in the distribution.

IMPORTANT REMINDER

When read from left to right, the numerical scale along the base of the normal curve always increases in value. Accordingly, the proportion of area to the left of a given score represents the relative frequency of smaller scores, while the proportion of area to the right of a given score represents the relative frequency of larger scores.

EXAMPLE

Assume that new FBI agents must stand between 63 and 65 inches. Now, what proportion of all applicants will be eligible, assuming, as before, a normal distribution of heights with a mean of 68 inches and a standard deviation of 3 inches?

Figure 6.5 illustrates the solution of this problem. The top panel identifies the target proportion—the proportion of applicants who are between 63 and 65 inches in height.

Before consulting Table A, convert each height into a z score:

$$z = \frac{65 - 68}{3} = \frac{-3}{3} = -1.00$$

$$z = \frac{63 - 68}{3} = \frac{-5}{3} = -1.67$$

Given negative z scores of -1.00 and -1.67, use the bottom legend of Table A to interpret all entries. Look up column A′ to a negative z score of -1.00 (remember, you must imagine the negative sign) and note the corresponding proportion of .1587 in column C′. As suggested in the middle panel of Figure 6.5, this proportion represents all applicants shorter than a negative z score of -1.00.

Next, look up column A′ to a negative z score of -1.67 and note the corresponding proportion of .0475 in column C′. As suggested in the bottom panel of Figure 6.5, this proportion represents all applicants

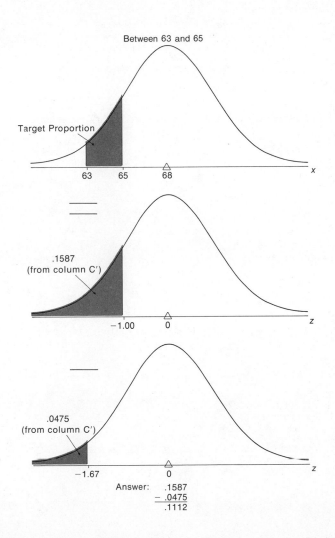

FIGURE 6.5

Finding area.

shorter than a negative z score of −1.67. If the smaller proportion is subtracted from the larger proportion, the remaining proportion, .1112, or .11, represents the answer. Thus, the proportion of all applicants who satisfy these stringent height requirements equals only .11 (or 11%).

Incidentally, the above problem could have been answered using column B′ rather than column C′. To check your understanding of Table A, do the problem again using information from column B′, along with rough graphs similar to those in Figure 6.5.

EXAMPLE

School district officials believe that their students' intellectual aptitudes approximate a normal distribution with a mean of 100 and a standard deviation of 15. Assuming that their belief is correct, what proportion of student IQs should be more than 30 points from the mean?

Figure 6.6 illustrates the solution to this problem. The top panel identifies the two target proportions—the proportions of students with IQs more than 30 points from the mean.

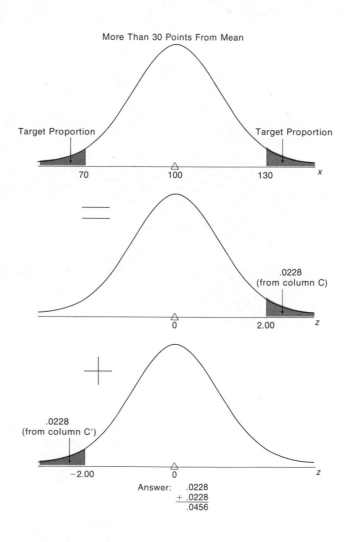

FIGURE 6.6
Finding area.

Expressing the two relevant IQs of 130 and 70 as z scores, we have

$$z = \frac{130 - 100}{15} = \frac{30}{15} = 2.00$$

$$z = \frac{70 - 100}{15} = \frac{-30}{15} = -2.00$$

Given a positive z score of 2.00, use the top legend of Table A. Locate a z score of 2.00 in column A, and note the corresponding proportion of .0228 in column C. As suggested in the middle panel of Figure 5.6, this proportion represents all students with IQs greater than 130.

Given a negative z score of −2.00, use the bottom legend in Table A. Locate a z score of −2.00 in column A′, and note the corresponding proportion of .0228 in column C′. As suggested in the bottom panel of Figure 6.6, this proportion represents all students with IQs smaller than 70.

To find the proportion of IQs more than 30 points from the mean, simply add the two above proportions, yielding a value of .0456 or .05.

This type of normal curve problem becomes particularly important in inferential statistics. Because of the symmetry of the normal curve, this problem can be dealt with most efficiently by merely doubling the proportion in one tail—without laboring through all of the steps shown in Figure 6.6. Use this more efficient procedure to answer any future questions about equal deviations in either direction from the mean.

Heretofore, we've concentrated on normal curve problems where Table A must be consulted to find the unknown area associated with some score or pair of scores. For instance, given a GRE score of 640, we found that the proportion of scores smaller than 640 equals .92. Now, we'll concentrate on the opposite type of normal curve problem where Table A must be consulted to find the unknown score or scores associated with some area.

6.6 Finding Scores

Exam scores for a large biology class approximate a normal curve with a mean of 230 and a standard deviation of 25. Furthermore, students are graded "on a curve," with the upper 20 percent being awarded grades of A. What's the lowest score on the exam to earn a grade of A?

Figure 6.7 illustrates the solution to this problem. The top panel identifies the target score—the score that separates the upper 20 percent (or .2000, when expressed as a proportion) from the remainder of the distribution. Since the unknown score is located in the upper tail, use the top legend of Table A to interpret all entries. Reading in column C, locate the desired proportion .2000. If the desired proportion fails to appear in column C, as typically will be the case, locate the entry closest to the desired proportion. (If adjacent entries are equally close to the desired proportion, use either entry—it's your choice!) As shown in the second panel in Figure 6.7, the closest entry equals .2005, and the corresponding z score in column A equals 0.84. (Verify by checking Table A.) This z score approximately splits the area under the normal curve into two proportions: an upper .2000 and a lower .8000.

EXAMPLE

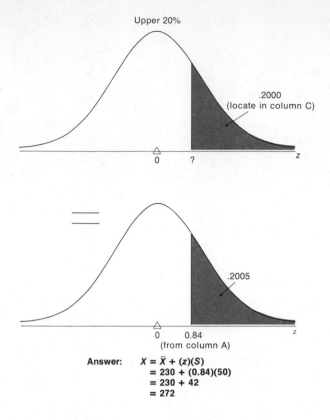

.2000
(locate in column C)

.2005

0   0.84
(from column A)

Answer: $X = \bar{X} + (z)(S)$
          $= 230 + (0.84)(50)$
          $= 230 + 42$
          $= 272$

FIGURE 6.7

Finding scores.

Finally, the z score must be translated into a score on the biology exam. Use the following equation (derived from the z score equation listed in the previous section):

$$X = \bar{X} + (z)(S) \qquad (6.2)$$

where X is the target score, expressed in original units of measurement; and $\bar{X}$ and S are the mean and standard deviation for the normal curve; and z is the standard score read from column A (or A′) of Table A. When appropriate numerical substitutions are made, as shown in the bottom of Figure 6.7, 272 represents the answer—the smallest score on the exam to qualify for a grade of A.

EXAMPLE

Assume that annual rainfall in the San Francisco area approximates a normal curve with a mean of 22 inches and a standard deviation of 4 inches. What are the maximum and minimum rainfalls for the more "typical" years, defined as those that exclude the wettest $2\frac{1}{2}$ percent of all years and the driest $2\frac{1}{2}$ percent of all years?

Figure 6.8 illustrates the solution of this problem. The top panel identifies the pair of target scores—the scores that separate the upper and lower $2\frac{1}{2}$ percent (or .0250, when expressed as a proportion) from the remainder of the distribution. To find the target score in the upper tail, locate the desired proportion of .0250 in column C, and note the corresponding z score of 1.96 in column A, as shown in the second panel of Figure 6.8. Although you could repeat this procedure, using column C′ to identify a z score of −1.96 in column A′, you can capitalize on the symmetry of the normal curve by noting that both target scores involve the same numbers with different signs.

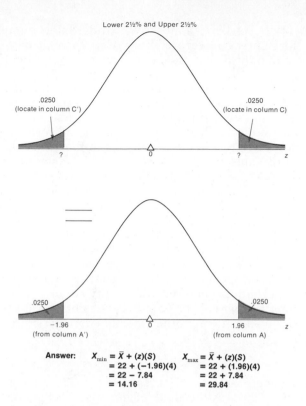

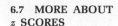

FIGURE 6.8

Finding scores.

Finally, both z scores must be translated into inches of annual rainfall. When appropriate substitutions are made, as shown in the bottom of Figure 6.8, the maximum and minimum rainfalls for "typical" years are found to range from 29.84 to 14.16 inches.

STANDARD SCORES

6.7 More About z Scores

Whenever scores are expressed relative to a known mean and standard deviation, they are referred to as **standard scores.** Thus, z scores qualify as standard scores because they are expressed relative to a mean of 0 and a standard deviation of 1.

Heretofore, z scores have been used only in the context of the normal curve, and throughout the remainder of this book, the reappearance of z scores—in one form or another—will usually prompt an interpretation based on the standard normal table. It would be erroneous to conclude, however, that z scores are limited to distributions that approximate a normal curve. As a matter of fact, non-normal distributions also can be transformed into sets of z scores. In this case, the standard normal table can't be consulted, since the shape of the distribution of z scores is the same as that for the original non-normal distribution. For instance, if the original distribution is positively skewed, the distribution of z scores also is positively skewed. *Regardless of the shape of the distribution,* however, *the shift to z scores always produces a distribution with a mean of 0 and a standard deviation of 1.*

Under most any circumstances, z scores provide efficient descriptions of relative performance on one or more tests. Without additional information, it's meaningless to know that Betty earned a raw score of 159 on a math test, but it's very informative to know that she earned a z score of 1.80. The latter score suggests that she did relatively well on the math test, being almost two standard deviation units above the mean. (More precise interpretations of this score could be made, of course, given that the distribution of test scores approximates a normal curve.)

The use of z scores can help you identify a person's relative strengths and weaknesses on several different tests. For instance, Table 6.3 shows the scores earned by Betty on college achievement tests in three different subjects. Evaluation of her test performance is greatly facilitated by converting her raw scores into the z scores listed in the final column of Table 6.3. Now, a glance at the z scores suggests that, while she did relatively well on the math test, her performance on the English test was only slightly above average, as indicated by a z score of 0.50, and her performance on the psychology test was slightly below average, as indicated by a z score of −0.67.

Remember that z scores reflect performance relative to some group, rather than relative to an absolute standard. A meaningful interpretation of z scores requires, therefore, that the nature of the reference group be specified. In the present example, for instance, it would be important to know whether Betty's scores were relative to those of students at her college or to those of students at a wide variety of colleges, as well as to any other special characteristics of the reference group.

6.8 Other Standard Scores

When reporting test results, z scores are often converted to other types of standard scores that lack negative signs and decimal points. These conversions change neither the shape of the original distribution nor the relative standing of any test score within the distribution. Thus, a test score located one standard deviation below the mean might be reported not as a z score of −1.00 but as a T score of 40 (in a distribution of T scores with a mean of 50 and a standard deviation of 10). The important point to realize is that, although reported as a score of 40, this T score accurately reflects the relative location of the score—a T score of 40 is located at a distance of one standard deviation (of size 10) below the mean (of size 50). Figure 6.9 shows the values of some of the more common types of standard scores relative to the various portions of area under the normal curve. (Also illustrated in Figure 6.9 are percentile ranks, to be discussed in the next section.)

To convert any z score (from a distribution with a mean of 0 and a standard deviation of 1) into a new z' score (having a distribution with any desired mean and standard deviation), use the following formula:

$$z' = \overline{X}' + (z)(S') \qquad (6.3)$$

where z' is the new standard score, $\overline{X}'$ is the desired mean, S' is the desired standard deviation, and z is the original standard score.

For instance, if you wish to convert a z score of −1.50 into a new distribution of z' scores for which the mean $\overline{X}'$ equals 500 and the

TABLE 6.3
BETTY'S ACHIEVEMENT TEST SCORES

SUBJECT	RAW SCORE	MEAN	STANDARD DEVIATION	z SCORE
Math	159	141	10	1.80
English	83	75	16	0.50
Psych	23	27	6	−0.67

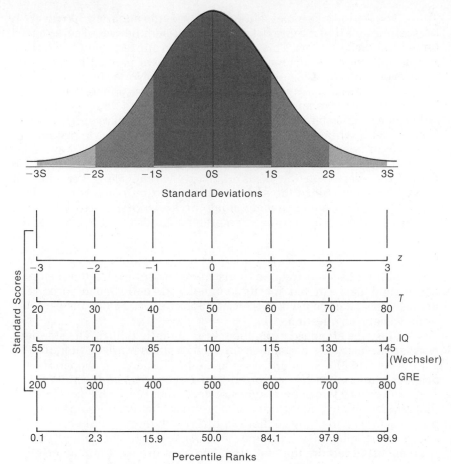

FIGURE 6.9

Common standard scores (and percentile ranks) associated with normal curves.

standard deviation S' equals 100, substitute these numbers into the above formula to obtain

$$z' = 500 + (-1.50)(100)$$
$$= 500 - 150$$
$$= 350$$

Again notice that the new standard score accurately reflects the location of the original standard score—it's located at a distance of one and one-half standard deviation units (each of size 100) below the mean (of size 500). The change from a z score of -1.50 to a z' score of 350 eliminates negative signs and decimal points without distorting the relative location of the original score, expressed as a distance from the mean in standard deviation units.

6.9 Percentile Ranks

Sometimes original scores are interpreted as percentile ranks. *The* **percentile rank** *of a score indicates the percentage of smaller scores in the distribution.* For example, a score has a percentile rank of 89 if smaller scores constitute 89 percent of the entire distribution. The use of percentile ranks are ideally suited for exchanges, such as between a parent and teacher (or a patient and doctor), that often require the interpretation of test scores (or lab findings) with a minimum of numerical explanation.

The limitations of percentile ranks become apparent, however, in those situations that require detailed comparisons between scores based on normal distributions. This problem can be appreciated by referring to the bottom scale of Figure 6.9, which shows percentile ranks for various portions of area under the normal curve. Notice that the scale of percentile ranks lacks the orderly increase in values apparent in each of the other scales for standard scores. Instead, in the vicinity of the mean, a distance of one standard deviation unit separates percentile ranks of 50.0 and 84.1, while in the upper tail of the normal curve, a distance of one standard deviation unit separates percentile ranks of 97.7 and 99.9. Clearly, differences between percentile ranks must be interpreted cautiously—partly on the basis of their location along the scale of percentile ranks. Since the same problem doesn't occur for standard scores, with their equal spacing along the horizontal scale, those who regularly deal with test scores often prefer standard scores to percentile ranks.

6.10 Summary

Many observed frequency distributions approximate a theoretical normal curve noted for its symmetrical, bell-shaped form. The normal curve can be used as the basis for calculating the answers to a wide variety of questions.

Only the standard normal curve is actually tabled. The standard normal curve always has a mean of 0 and a standard deviation of 1. To gain access to the standard normal table (Table A in Appendix C), original scores must be converted into z scores, using Formula 6.1.

There are two general types of normal curve problems: (1) those that require you to find the unknown area associated with some score or pair of scores, and (2) those that require you to find the unknown score or scores associated with some area. Answers to the latter type of problem often require that you translate a z score back into an original score, using Formula 6.2.

Even when distributions fail to approximate normal curves, z scores can provide efficient descriptions of relative performance on one or more tests.

When reporting test results, z scores are often converted to other types of standard scores that lack negative signs and decimal points. These conversions change neither the shape of the original distribution nor the relative standing of any test score within the original distribution.

The percentile rank of a score indicates the percentage of smaller scores in the distribution.

Important Terms

Normal curve	Standard score
z score	Percentile rank
Standard normal curve	

6.11 Exercises

1. Suppose the burning times of electric light bulbs approximate a normal curve with a mean of 2200 hours and a standard deviation of 600 hours. What proportion of lights . . .
    (a) fail before 1400 hours?
    (b) survive 4000 hours?
    (c) fail within 200 hours of the mean?

If new lights are installed at the same time and burn continuously, at what time have . . .

(d) 10 percent failed?

(e) 50 percent failed?

(f) 99 percent failed?

(g) If a new inspection procedure eliminates the weakest 8 percent of all lights before being marketed, the manufacturer could safely offer a money-back guarantee on all lights that fail before _____ hours of burning time?

2. Scores on the Wechsler Intelligence Scale approximate a normal curve with a mean of 100 and a standard deviation of 15.

What proportion of IQ scores . . .

(a) exceed 125?

(b) are exceeded by 82?

(c) are within 9 points of the mean?

(d) are more than 40 points from the mean?

What IQ score is identified with . . .

(e) the upper 2 percent?

(f) the lower 10 percent?

(g) the middle 95 percent?

(h) the middle 99 percent?

3. An investigator surveys common cold sufferers, asking them to estimate the number of hours of physical discomfort. Their estimates approximate a normal curve with a mean of 83 hours and a standard deviation of 20 hours.

What proportion of estimates . . .

(a) exceed two days?

(b) exceed one week?

(c) are between two and four days?

(d) are between one and three days?

What are the estimated number of hours for . . .

(e) the shortest 5 percent?

(f) the longest 20 percent?

(g) the middle 20 percent?

(h) the extreme 1 percent in either direction from the mean?

4. Admission to a state university depends partially on the applicant's high school grade point average (GPA). Assume that the GPAs of applicants approximate a normal curve with a mean of 3.20 and a standard deviation of 0.30.

(a) If applicants with GPAs of 3.50 or above are automatically admitted, what proportion of applicants are in this category?

(b) If applicants with GPAs of 2.50 or below are automatically denied admission, what proportion of applicants are in this category?

(c) A special honors program is open to all applicants with GPAs of 3.75 or better. What proportion of applicants are eligible?

(d) If the special honors program were limited to students whose GPAs rank in the upper 10 percent, what GPA would be required for admission to this program?

5. Given that the heights of adult males approximate a normal curve with a mean of 68 inches and a standard deviation of 3 inches, do you see any conection between the height of Bill Walton, who stands 6 feet, 11 inches tall, and a midget, who stand 4 feet, 5 inches tall?

6. It has been claimed that the distribution of z scores always has a mean of 0 and a standard deviation of 1. To convince yourself of this—short of a mathematical proof—transform each of the following raw scores into z scores, and then verify that the mean of the z scores does, in fact, equal 0

and that the standard deviation of the z scores does, in fact, equal 1. The raw scores are: 5, 7, 8, 9, 11.

> *Hint:* First find the mean and standard deviation of the raw scores. Then convert each raw score to a z score. Finally, calculate the mean and standard deviation of the z scores.

If, after doing this exercise, you suspect that these five numbers might be special, use any set of arbitrary numbers to verify that a set of z scores always has a mean of 0 and a standard deviation of 1. (Your answers might deviate slightly from 0 or 1 because of rounding errors.)

7. If you earned a raw score of 64 on some test, which one of the following distributions would permit the most favorable interpretation of this score?

(a) $\bar{X} = 50, S = 2$ (c) $\bar{X} = 30, S = 30$
(b) $\bar{X} = 42, S = 10$ (d) $\bar{X} = 20, S = 20$

8. Assume that each of the raw scores listed below originate from a distribution that approximates a normal curve. Convert each raw score, in turn, to a series of standard scores with means and standard deviations of 0 and 1; 50 and 10; 100 and 15; and 500 and 100, respectively. Also convert each raw score to a percentile rank (using the standard normal table).

	RAW SCORE	MEAN	STANDARD DEVIATION
(a)	24	20	5
(b)	37	42	3
(c)	346	310	20
(d)	1263	1400	74

9. When describing test results, someone objects to the conversion of raw scores into standard scores, claiming that this constitutes an arbitrary change in the value of the test score. How might you respond to this objection?

10. Assume that all of the standard scores listed below originate from distributions that approximate normal curves. (Refer to Figure 6.9 for details about the means and standard deviations of these standard scores.)
(a) What's the percentile rank of a GRE score of 640?
(b) Given that a score on the Wechsler IQ scale has a percentile rank of 93, what's the value of the IQ score?
(c) What's the percentile rank of a T score of 35?
(d) Given a percentile rank of 60 for some GRE score, what's the value of that score?

7

Measures of Relationship: Correlation

7.1 –An Intuitive Approach
7.2 –Scatterplots
7.3 –A Correlation Coefficient for Quantitative Data: r
7.4 –z Score Formula for r
7.5 –Raw Score Formula for r
7.6 –Interpretation of r
7.7 –Correlation Not Necessarily Cause-effect
7.8 –A Correlation Coefficient for Ranked Date: r_s
7.9 –Summary
7.10–Exercises

Some people contend that everything is related to everything else. This chapter supplies you with some tools with which to check out this claim.

With the aid of scatterplots, you'll often be able to determine at a glance whether or not two variables are related, and if so, some of the more important properties of the relationship.

An extensive correlation study has identified a number of factors associated with long life. What's wrong with a newspaper account that begins: "Men can add 11 years to their lives if they follow seven 'golden' rules of behavior—including moderate drinking, no smoking, regular meal times, and eight hours of sleep each night."? See Exercise 8.

The familiar saying "You get what you give" has many ramifications, including, for instance, the exchange of holiday greeting cards. A researcher suspects that there is a relationship between the number of greeting cards given and the number of greeting cards received by individuals. Prior to a full-fledged survey, he obtains estimates for the most recent holiday season from five friends, as shown in Table 7.1.

7.1 An Intuitive Approach

You don't need elaborate statistical tests to determine whether two variables are related when, as in the present example, the analysis involves only a few pairs of observations. If, in fact, the suspected relationship exists between cards given and received, then an inspection of the data might reveal, as one possibility, that "big givers" tend also to be "big receivers." In other words, the pair of observations for any single person ranks about the same among both sets of observations.

Trends among pairs of observations can be detected most easily by first ranking observations along one variable and then assigning consecutive numerical ranks 1, 2, 3, 4, 5 to each set of observations. In Table 7.2A, the five friends are ranked according to the number of cards given, and the red numbers in parentheses designate each person's rank—beginning with a rank of 1 for the smallest observation within each set—for cards given or for cards received. For example, Doris' 13 ranks highest (5) among cards given, while her 14 ranks next-to-highest (4) among cards received. Notice the considerable regularity among

TABLE 7.1
GREETING CARDS GIVEN AND RECEIVED BY FIVE FRIENDS

Friend	NUMBER OF CARDS Given	Received
Andrea	5	10
Mike	7	12
Doris	13	14
Steve	9	18
John	1	6

TABLE 7.2
THREE TYPES OF
RELATIONSHIPS

A. POSITIVE RELATIONSHIP

| FRIEND | GIVEN | | RECEIVED | |
	Number	Rank	Number	Rank
Doris	13	(5)	14	(4)
Steve	9	(4)	18	(5)
Mike	7	(3)	12	(3)
Andrea	5	(2)	10	(2)
John	1	(1)	6	(1)

B. NEGATIVE RELATIONSHIP

| FRIEND | GIVEN | | RECEIVED | |
	Number	Rank	Number	Rank
Doris	13	(5)	6	(1)
Steve	9	(4)	10	(2)
Mike	7	(3)	14	(4)
Andrea	5	(2)	12	(3)
John	1	(1)	18	(5)

C. LITTLE OR NO RELATIONSHIP

| FRIEND | GIVEN | | RECEIVED | |
	Number	Rank	Number	Rank
Doris	13	(5)	10	(2)
Steve	9	(4)	18	(5)
Mike	7	(3)	12	(3)
Andrea	5	(2)	6	(1)
John	1	(1)	14	(4)

pairs of ranks; each of the five friends tends to rank about the same among both sets of observations. We conclude, therefore, that the two variables are related. This relationship implies that "You get what you give." *When high values are paired with high values, and low values are paired with low values, the relationship is* **positive.**

"BUREAU OF POSITIVE RELATIONSHIPS"

In Tables 7.2B and 7.2C, the original pairs of observations for the five friends are rearranged to illustrate two other possibilities—a negative relationship and no relationship. (In practice, of course, data cannot be rearranged.) Notice the regularity among pairs of ranks in Table 7.2B.

Now reverse ranks tend to be paired; high ranks tend to be paired with low ranks, and vice versa. For instance, although John ranks lowest (1) among card givers, he ranks highest (5) among card receivers. Given this regularity, we conclude that the two variables are related. This relationship implies that "You get the opposite of what you give." *When high values are paired with low values, and low values are paired with high values, the relationship is* **negative.**

No regularity is apparent among pairs of ranks in Table 7.2C. Now ranks are paired in seemingly haphazard fashion. For instance, although Mike ranks in the middle (3) among both card givers and receivers, Doris ranks highest (5) among card givers, but next-to-lowest (2) among card receivers. Given this lack of regularity, we now conclude that little, if any, relationship exists between the two variables and that "What you get has no bearing on what you give."

This method of searching for regularity among pairs of ranks is cumbersome and inexact when the analysis involves more than a few pairs of observations. Hence, although this technique has much intuitive appeal, it must be abandoned in favor of several other more efficient and exact statistical techniques, namely a special graph known as a scatterplot and some new measures known as correlation coefficients.

7.2 Scatterplots

Scatterplots *depict relationships as clusters of dots.* With a little training, you can use any dot cluster as a preview of the fully measured relationship.

To construct a scatterplot, as in Figure 7.1, scale each of the two variables along the horizontal (X) and vertical (Y) axes and use each pair of observations to locate a dot within the scatterplot. For example, the pair of numbers for Mike, 7 and 12, define points along the X and Y

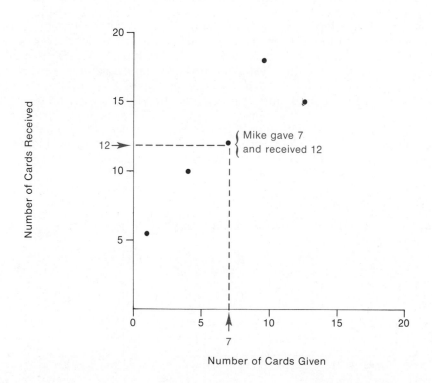

FIGURE 7.1
Scatterplot for greeting card exchange.

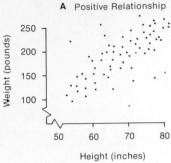

A Positive Relationship

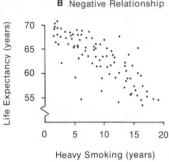

B Negative Relationship

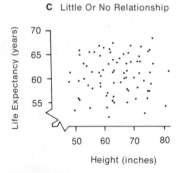

C Little Or No Relationship

FIGURE 7.2

Three types of relationships.

axes, respectively. Using these points to anchor lines perpendicular (at right angles) to each axis, locate Mike's dot where the two lines intersect. Repeat this process, with imaginary lines, for each of the four remaining pairs of numbers to create the scatterplot of Figure 7.1.

Our simple example about greeting cards has shown the basic idea of correlation and the construction of a scatterplot. We will now examine more complex sets of data in order to learn how to interpret scatterplots.

The first step is to note the orientation or tilt, if any, of a dot cluster. *A dot cluster that tilts from the upper right to the lower left,* as in Figure 7.2A, *reflects a positive relationship.* Large values of one variable are paired with large values of the other variable, and small values are paired with small values. In Figure 7.2A, tall people tend to be heavy, while short people tend to be light.

On the other hand, *a dot cluster that tilts from the upper left to the lower right,* as in Figure 7.2B, *reflects a negative relationship.* Large values of one variable tend to be paired with small values of the other variable, and vice versa. In Figure 7.2B, people with many years of heavy smoking tend to live shorter lives, while people with few or no years of heavy smoking tend to live longer.

Finally, *a dot cluster that lacks any apparent tilt,* as in Figure 7.2C, *reflects little or no relationship.* Large values of one variable are just as likely to be paired with small, medium, or large values of the other variable. In Figure 7.2C, notice how the dots are strewn about in an irregular shotgun fashion, suggesting that there is little or no relationship between the height of young adults and their life expectancies.

Having established that a relationship is either positive or negative, next note how closely the dot cluster approximates a straight line. *The more closely the dot cluster approximates a straight line, the stronger (the more regular) the relationship.* Figure 7.3 shows a series of scatterplots, each representing a different positive relationship between IQ scores for pairs of people whose backgrounds reflect different degrees of overlap, ranging from minimum overlap, as with fosterparents and fosterchildren, to maximum overlap, as with identical twins. Ignore the parenthetical expressions involving r, to be discussed later. Notice that the dot cluster more closely approximates a straight line for people with greater degrees of genetic overlap—for parents and children in Figure 7.3B, and even more so, for identical twins in Figure 7.3C.

A dot cluster that equals (rather than merely approximates) a straight line reflects a perfect relationship between two variables. In practice, perfect relationships are most unlikely, and none appears in the scatterplots of this book.

FIGURE 7.3

Three positive relationships*

*Scatterplots simulated from a 50-year literature survey. (Source: Erlenmeyer-Kimling, L., and Jarvik, L.F.: Genetics and intelligence: a review. *Science,* 1963, **142,** 1477–1479.

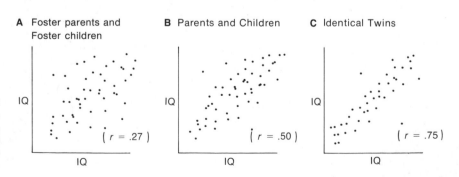

A Foster parents and Foster children (r = .27)

B Parents and Children (r = .50)

C Identical Twins (r = .75)

The previous discussion assumes that, if a dot cluster has a discernible tilt, it approximates a *straight* line and, therefore, reflects a **linear relationship.** This is not always the case. Sometimes a dot cluster approximates a *bent* or *curved* line, as in Figure 7.4, and, therefore, reflects a **curvilinear relationship.** Descriptions of these relationships are more complex than those for linear relationships. For instance, we see in Figure 7.4 that physical strength, as measured by the force of a person's hand grip, is less for youngsters, more for adults, then less again for older people. Otherwise, the scatterplot can be interpreted as before. Specifically, the more closely the dot cluster approximates a curved line, the stronger the curvilinear relationship.

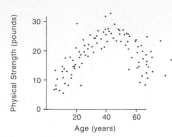

FIGURE 7.4
Curvilinear relationship.

Take another look at the scatterplot in Figure 7.1 for the greeting card data. Although any interpretation is hindered by the small number of dots in Figure 7.1, the dot cluster appears to approximate a straight line, stretching from lower left to upper right. This suggests a positive relationship between greeting cards given and received in agreement with the earlier intuitive analysis for these data.

7.3 Correlation Coefficient for Quantitative Data: *r*

A **correlation coefficient** *is a number that measures the relationship between variables.* The next few sections concentrate on the type of correlation coefficient, designated as *r*, that measures linear relationships between pairs of variables for quantitative data. Section 7.8 describes a correlation coefficient for ranked data. Many other types of correlation coefficients have been introduced to handle specific types of data, including qualitative data, but these are not described in this book.

Named in honor of British scientist Karl Pearson, *the* **Pearson correlation coefficient,** *r,* *can equal any value between* $+1.00$ *and* -1.00. A number with a plus sign indicates a positive relationship, while a number with a minus sign indicates a negative relationship. For example, an *r* with a plus sign would describe the positive relationship between height and weight shown in Figure 7.2A, while an *r* with a minus sign would describe the negative relationship between heavy smoking and life expectancy shown in Figure 7.2B.

A value of *r* in the direction of either $+1.00$ or -1.00 indicates a strong (highly regular) relationship, while a value of *r* in the neighborhood of 0 indicates a weak (relatively irregular) relationship. For example, an *r* of $-.90$ indicates a stronger relationship than an *r* of $-.70$, while an *r* of $-.70$ indicates a stronger relationship than an *r* of .50. (If no sign appears, it's understood to be plus.) In Figure 7.3, notice how the values of *r* shift from .75 to .27 as the analysis for pairs of IQ scores shifts from a strong relationship for identical twins to a relatively weak relationship for foster parents and foster children.

Although unlikely in practice, as mentioned previously, an *r* of either 1.00 or -1.00 indicates a perfect positive or perfect negative relationship. An *r* of 0 indicates the absence of any relationship.

7.4 *z* Score Formula for *r*

The simplest formula for *r* reads:

$$r = \frac{\Sigma z_X z_Y}{n} \qquad (7.1)$$

where z_X and z_Y are the z score equivalents for each pair of original observations, X and Y, and *n* refers to the number of paired observations. The term $\Sigma z_X z_Y$ is found by first multiplying each pair of z_X and z_Y scores, then adding the products for all pairs.

Table 7.3 illustrates the calculation of r for the original greeting card data in Table 7.2A. As in previous computational tables, the numbers in red circles are used to cross-reference the top and bottom panels in Table 7.3. An inspection of this table reveals that r equals .80, a strong positive correlation, as you might have anticipated from our previous discussion of these data.

Let's look more closely at how three different types of correlations—a strong positive correlation, a strong negative correlation, and a zero correlation—are processed by Formula 7.1. If there is a strong positive correlation, such as that illustrated in Table 7.3, pairs of X and Y scores tend to have similar relative locations in their respective distributions, and therefore, pairs of z scores tend to have the same sign. As can be seen in Table 7.3, positive z scores are paired with positive z scores, while negative z scores are paired with negative z scores. Consequently, and this is a crucial point, the products of paired z scores, $z_X z_Y$, are positive because multiplication involves pairs of numbers with like signs, either both plus or both minus. As a result, the numerator term in Formula 7.1, $\Sigma z_X z_Y$, becomes a relatively large positive number, which, when divided by n, the number of pairs, yields a positive r—in the present case, an r of .80.

If pairs of z scores had the same magnitude, as well as the same signs (for instance, one pair might be 2.34 and 2.34, while another might be -1.25 and -1.25) r assumes its maximum value of 1.00, indicating a perfect positive correlation. Under these circumstances, the relationship displays total regularity, with pairs of z scores occupying exactly the

TABLE 7.3
CALCULATION OF r:
z **SCORE FORMULA**

A. COMPUTATIONAL SEQUENCE

Assign a value to n ①, representing the number of paired scores.
Obtain a mean ② and standard deviation ③ for the X scores.
Obtain a mean ④ and standard deviation ⑤ for the Y scores.
Determine z_X scores ⑥, one at a time, by subtracting $\bar{X}$ from each X and dividing the difference by S_X.
Determine z_Y scores ⑦, one at a time, by subtracting $\bar{Y}$ from each Y and dividing the difference by S_Y.
Find the product of each pair of z scores ⑧, one at a time, then add all of these products ⑨.
Substitute numbers into the formula ⑩ and solve for r.

B. DATA AND COMPUTATIONS

Friend	Cards Given, X	Cards Received, Y	⑥ z_X	⑦ z_Y	⑧ $z_X z_Y$
Doris	13	14	1.50	0.50	0.75
Steve	9	18	0.50	1.50	0.75
Mike	7	12	0.00	0.00	0.00
Andrea	5	10	-0.50	-0.50	0.25
John	1	6	-1.50	-1.50	2.25

① $n = 5$ $\quad$ $\Sigma X = 35$ $\quad$ $\Sigma Y = 60$ $\qquad$ ⑨ $\Sigma z_X z_Y = 4.00$

$$② \; \bar{X} = \frac{35}{5} = 7 \qquad ④ \; \bar{Y} = \frac{60}{5} = 12$$

$$③ \; S_X = 4^* \qquad ⑤ \; S_Y = 4^*$$

$$⑩ \; r = \frac{\Sigma z_X z_Y}{n} = \frac{4.00}{5} = .80$$

*Computations not shown. Verify, if you wish, using Formula 5.1.

same relative locations in their respective distributions. Returning to Table 7.3, you might verify that r would have been equal to 1.00 if the top two entries in the z_X column had been reversed, causing each pair of z scores to have the same magnitude, as well as the same sign.

If there is a strong negative correlation, such as that between cards given and received in Table 7.2B, pairs of X and Y scores tend to have relative locations that are reversed in their respective distributions, and therefore, pairs of z scores tend to have unlike signs. Positive z scores are paired with negative scores, and vice versa. Consequently, the products of paired z scores, $z_X z_Y$, are negative because multiplication involves pairs of numbers with unlike signs, one plus and the other minus. As a result, the numerator term in Formula 7.1, $\Sigma z_X z_Y$, is a relatively large negative term, which, when divided by n, yields a negative r.

If pairs of z scores had the same magnitude, but unlike signs (for instance, one pair might be 1.75 and -1.75, while another pair might be .53 and $-.53$) r assumes its minimum value of -1.00, indicating a perfect negative correlation. Under these circumstances the relationship also displays perfect regularity, with pairs of z scores occupying relative locations that are exactly reversed in their respective distributions.

If there is a zero or nearly zero correlation, such as that between cards given and received in Table 7.2C, no consistent pattern describes the relative locations of pairs of X and Y scores in their respective distributions. Therefore, a positive z score is equally likely to be paired with either a positive or negative z score, and vice versa. Consequently, about half of all products of paired z scores, $z_X z_Y$, are positive because multiplication involves numbers with like signs, and about half of all products of paired z scores, $z_X z_Y$, are negative because multiplication involves numbers with unlike signs. Since positive and negative products tend to cancel each other, the numerator term in Formula 7.1, $\Sigma z_X z_Y$, tends toward a small positive or negative number that, when divided by n, yields a value of r in the vicinity of 0.

To summarize, an understanding of correlation, as measured by r, can be gained from the z score formula (7.1). The pattern among pairs of z scores telegraphs the value of r. If pairs of z scores are similar both in magnitude and sign, the value of r tends toward 1.00, indicating a strong positive correlation. If pairs of z scores are similar in magnitude but opposite in sign, the value of r tends toward -1.00, indicating a strong negative correlation. As the pattern among pairs of z scores becomes less apparent, the value of r tends toward 0, indicating a weak or nonexistent correlation.

The z score formula also pinpoints another important property of r—its immunity to the original units of measurement. As a matter of fact, the same value of r would describe the correlation between height and weight for a group of adults, regardless of whether height was measured in inches or centimeters, or whether weight was measured in pounds or grams. In effect, the value of r depends only on the pattern among pairs of z scores, which in turn show no traces of the units of measurement for the original X and Y observations.

7.5 Raw Score Formula for r

Although the z score formula for r supplies the most intuitive picture of correlation, it doesn't win any awards for computational efficiency. It would be most laborious first to convert each X and Y score into their z score equivalents, using means and standard deviations that typically aren't whole numbers, and then to calculate r using the z score formula. It's usually more efficient to calculate a value for r using the raw score formula, which reads:

$$r = \frac{n\Sigma XY - (\Sigma X)(\Sigma Y)}{\sqrt{[n\Sigma X^2 - (\Sigma X)^2][n\Sigma Y^2 - (\Sigma Y)^2]}} \quad (7.2)$$

All terms in this expression have been encountered previously, except for ΣXY, which is found first by multiplying each pair of X and Y scores and then by adding the products for all pairs.

Table 7.4 illustrates the computation of r for the original greeting card data, using the raw score formula. As expected, this formula yields the same value of .80 for r as did the z score formula.

7.6 Interpretation of r

Avoid interpretations of r as a proportion or percent. An r of .30 doesn't signify that the strength of an existing relationship is .30 or 30 percent of the maximum possible strength of a perfect relationship. There are other more sophisticated interpretations of r, but we'll limit our interpretation to statements such as "an r of .30 reflects a stronger relationship than one of .25 but a weaker relationship than one of .45."

Even this interpretation of r can be misleading in the absence of information about sampling variability. (Sampling variability was discussed briefly in Section 4.6. When several samples are taken from the same population, no two samples will contain exactly the same set of observations. Thus, the value of any measure, including r, shifts from sample to sample, and these chance differences are referred to as

**TABLE 7.4
CALCULATION OF r:
RAW SCORE FORMULA**

A. COMPUTATIONAL SEQUENCE

Assign a value to n (1), representing the number of paired scores.
Sum all scores for X (2) and for Y (3).
Find the product of each pair of X and Y scores (4), one at a time, then add all of these products (5).
Square each X score (6), one at a time, then add all squared X scores (7).
Square each Y score (8), one at a time, then add all squared Y scores (9).
Substitute numbers into the formula (10) and solve for r.

B. DATA AND COMPUTATIONS

Friend	Cards Given, X	Cards Received, Y	(4) XY	(6) X^2	(8) Y^2
Doris	13	14	182	169	196
Steve	9	18	162	81	324
Mike	7	12	84	49	144
Andrea	5	10	50	25	100
John	1	6	6	1	36

(1) $n = 5$ (2) $\Sigma X = 35$ (3) $\Sigma Y = 60$ (5) $\Sigma XY = 484$ (7) $\Sigma X^2 = 325$ (9) $\Sigma Y^2 = 800$

$$(10) \quad r = \frac{(n)(\Sigma XY) - (\Sigma X)(\Sigma Y)}{\sqrt{[(n)(\Sigma X^2) - (\Sigma X)^2][(n)(\Sigma Y^2) - (\Sigma Y)^2]}}$$

$$= \frac{(5)(484) - (35)(60)}{\sqrt{[(5)(325) - (35)^2][(5)(800) - (60)^2]}}$$

$$= \frac{2420 - 2100}{\sqrt{[1625 - 1225][4000 - 3600]}} = \frac{320}{\sqrt{[400][400]}}$$

$$= \frac{320}{\sqrt{160000}} = \frac{320}{400} = .80$$

sampling variability.) If sampling variability is large, as tends to be the case when values of r are based on relatively small numbers of paired observations, then the difference between an r of .45 and an r of .25 might be attributable to chance and, therefore, best ignored. As will be seen in Chapter 13, when values of r are based on very small samples, then even an r of .80 might originate from a population where the true correlation is close to zero. Accordingly, the r of .80 for the greeting card data should not be taken seriously. In practice, much larger samples should be employed.

Given that the value of r is based on a relatively large number of paired observations, an r value in excess of approximately .50, in either the positive or the negative direction, would qualify as a strong correlation in most areas of behavioral and educational research. There are exceptions to this statement. For example, when correlation coefficients measure "test reliability," as determined from the scores of people who take the same test twice, a value of less than .80 for r would be considered rather weak.

Don't conclude that since a correlation exists between the prevalence of poverty and crime in U.S. cities, poverty causes crime. Other explanations are equally plausible. For instance, you might speculate that poverty and crime are produced by a common cause, such as racial oppression, inadequate education, population density, or some combination of these factors. Regardless of its size, *a correlation coefficient never provides information about whether an observed relationship reflects a simple cause-effect relationship or some more complex state of affairs.*

Within recent years the interpretation of the correlation between cigarette smoking and lung cancer was vigorously disputed. American Cancer Society spokesmen interpreted the correlation as a causal relationship; smoking causes lung cancer. On the other hand, tobacco industry spokesmen interpreted the correlation as, at most, an indication that both cigarette smoking and lung cancer are caused by some more basic but as yet unidentified factor or factors, such as the body metabolism or personality of some people. According to this line of reasoning, people with high body metabolism might be both more prone to smoke and also, quite independent of their smoking, more vulnerable to lung cancer.

Sometimes experimentation can resolve this kind of controversy. In the present case, laboratory animals were trained to inhale different amounts of tobacco tars and then were destroyed. Autopsies revealed that the observed incidence of lung cancer varied directly with the amount of inhaled tobacco tars—even though possible "contaminating" factors, such as different body metabolisms or personalities, had been neutralized either through experimental control or random assignment of subjects to different test conditions. Experimental confirmation of a correlation provides strong evidence in favor of a cause-effect interpretation of the observed relationship; indeed, in the smoking-cancer controversy, cumulative experimental findings have provided overwhelming support for the conclusion that smoking causes lung cancer.

7.7 Correlation Not Necessarily Cause-Effect

At the beginning of this chapter, ranks were introduced as a means to clarify the nature of a relationship. If only ranks are available, or if for some reason, numbers are converted to ranks, it's still possible

7.8 A Correlation Coefficient for Ranked Data: r_s

to calculate a correlation coefficient for these ranked data. Named in honor of British psychologist Charles Spearman, the **Spearman correlation coefficient** for ranked data, r_s, behaves much like r. In particular, r_s *varies between 1.00 and −1.00, reflecting the degree of regularity among pairs of ranks* rather than, as in the case of r, pairs of numbers. When based on the same set of original observations, first in the form of quantitative data, then converted to ranked data, the values of r and r_s tend to be quite similar. The formula for r_s reads:

$$r_s = 1 - \frac{6\Sigma D^2}{n(n^2 - 1)} \tag{7.3}$$

where D refers to the difference between each pair of ranks, and 6 is a constant that always appears in the formula. As usual, n refers to the number of pairs of ranks.

Table 7.5 illustrates the computation of r_s for the original greeting card data, expressed as ranks in Table 7.2A. The resulting value of .90 for r_s suggests, as expected, a strong positive correlation between pairs of ranks. Notice the similarity between this value for r_s and the value of .80 for r, obtained previously when the data in Table 7.2A were analyzed as numbers rather than ranks. The difference between these two correlation coefficients can be attributed to the fact that, in contrast to r, r_s responds only to the rank, rather than also to the amount, of each original observation in Table 7.2A. In effect, r is more responsive than r_s to the detailed information contained in the original quantitative data.

The derivation of Formula 7.3 assumes that no ties appear among ranks. In the event of ties, assign to each tied observation the average of the ranks that would have been assigned if the tie had not occurred. In Table 7.5, for instance, if Steve and Mike both had given the same number of cards, each of them would have been assigned a rank of 3.5,

TABLE 7.5
CALCULATION OF r_s

A. COMPUTATIONAL SEQUENCE

Assign a value to n ①, representing the number of paired ranks.
Determine the difference between each pair of ranks ②.
Square each difference between pairs of ranks ③, one at a time, and then add all squared differences ④.
Substitute numbers into the formula ⑤ and solve for r_s.

B. DATA AND COMPUTATIONS

	Ranks		②	③
Friend	Given	Received	**D**	**D²**
Doris	5	4	1	1
Steve	4	5	−1	1
Mike	3	3	0	0
Andrea	2	2	0	0
John	1	1	0	0

① $n = 5$ ④ $\Sigma D^2 = 2$

$$⑤ \; r_s = 1 - \frac{(6)(\Sigma D^2)}{(n)[(n)^2 - 1]} = 1 - \frac{(6)(2)}{(5)[(5)^2 - 1]}$$

$$= 1 - \frac{12}{(5)[25 - 1]} = 1 - \frac{12}{(5)[24]} = 1 - \frac{12}{120}$$

$$= 1 - .10 = .90$$

the average of ranks 3 and 4. Once average ranks have been assigned, treat all ranks as though they were numbers and calculate the correlation using Formula 7.2 instead of Formula 7.3. This procedure yields a correlation coefficient for ranked data that is corrected for ties, and the result should be reported as a value of r_s.

7.9 Summary

The presence of regularity among pairs of X and Y scores indicates that the two variables are related, while the absence of any regularity suggests that the two variables are, at most, only slightly related. When the regularity consists of relatively high X scores being paired with relatively high Y scores, and relatively low X scores being paired with relatively low Y scores, the relationship is positive. When it consists of relatively high X scores being paired with relatively low Y scores, and vice versa, the relationship is negative.

Scatterplots depict relationships as clusters of dots. A dot cluster that tilts from the upper right to the lower left reflects a positive relationship, and a dot cluster that tilts from the upper left to the lower right reflects a negative relationship. A dot cluster that lacks any apparent tilt reflects little or no relationship.

Given a positive or negative relationship, the more closely the dot cluster approximates a straight line, the stronger the relationship.

When the dot cluster approximates a straight line, the relationship is linear; when it approximates a bent line, the relationship is curvilinear.

The Pearson correlation coefficient, r, measures linear relationships between pairs of variables for quantitative data. An understanding of correlation, as measured by r, can be gained from the z score formula (Formula 7.1). In practice, it's usually more efficient to calculate r by using the raw score formula (Formula 7.2).

Values of r vary between +1.00 and −1.00. Plus and minus signs indicate positive and negative relationships, respectively. Values of r in the direction of either +1.00 or −1.00 indicate a strong relationship, while values of r in the neighborhood of 0 indicate a weak relationship.

Avoid straightforward interpretations of r as a proportion or percent. Also avoid interpretations of r that ignore its sampling variability.

Regardless of its size, a correlation coefficient never provides information about whether an observed relationship reflects a simple cause-effect relationship or some more complex state of affairs.

The Spearman correlation coefficient, r_s, measures the degree of regularity among pairs of ranks. It varies in value from +1.00 to −1.00 and can be determined from Formula 7.3.

The value of r_s tends to be similar to that for r when both measures are determined for the same data.

Important Terms & Symbols

Positive relationship	Linear relationship
Negative relationship	Curvilinear relationship
Scatterplot	Pearson correlation coefficient (r)
Spearman correlation coefficient (r_s)	

1. Couples who attend a clinic for first pregnancies are asked to estimate (independently of each other) the ideal number of children. Results are as follows:

COUPLE	X	Y
A	1	2
B	3	4
C	2	3
D	3	2
E	1	1
F	2	3

(a) Using the intuitive approach, estimate whether these pairs of observations reflect a positive relationship, a negative relationship or no relationship.
(b) Construct a scatterplot. Verify that the scatterplot doesn't describe a pronounced curvilinear trend.
(c) Calculate a value for r. (Use Formula 7.2.)

2. (a) Using the intuitive approach of Section 7.1, determine whether the following pairs of observations reflect a positive relationship, a negative relationship, or no relationship.

X	Y
64	66
40	79
30	98
71	65
55	76
31	83
61	68
42	80
57	72

(b) Construct a scatterplot. Verify that the scatterplot doesn't describe a pronounced curvilinear trend.
(c) Calculate r, using the original numbers given above in Formula 7.2.

3. Verify that a strong negative relationship is depicted in Table 7.2B by using (a) the numerical observations to calculate a value of r (Formula 7.2), and by using (b) the ranks to calculate a value of r_s.

4. Verify that little or no relationship exists between the observations in Table 7.2C by using (a) the numerical observations to calculate a value of r (Formula 7.2), and by using (b) the ranks to calculate a value of r_s.

5. Pretend that there's a perfect positive (+1.00) relationship between height and weight for adults. (Actually, it's in the general vicinity of .75 for college students.) In this case, if someone stands 2 standard deviations above the mean height, this person's weight would be __(a)__ standard deviation units __(b)__ the mean.
If someone stands $1\frac{1}{2}$ standard deviation units below the mean height, this

person's weight would be __(c)__ standard deviation units __(d)__ the mean.

If someone is ⅓ of a standard deviation above the mean weight, this person's height would be __(e)__ of a standard deviation __(f)__ the mean.

6. Repeat Exercise 5, assuming a perfect negative (−1.00) relationship between height and weight.

7. Assume that an r of .80 describes the relationship between daily food intake, measured in ounces, and body weight, measured in pounds, for a group of adults. Would a shift in the units of measurement from ounces to grams and from pounds to kilograms produce a change in the value of r? Justify your answer.

8. As indicated in Figure 7.3, the correlation between IQ scores of identical twins is .75, while that between IQ scores of parents and children is .50. Does this signify, therefore, that the relationship between identical twins is 50 percent stronger than the relationship between parents and children?

9. An extensive correlation study indicates that a longer life is experienced by people who follow seven "golden rules" of behavior—including moderate drinking, no smoking, regular meals, some exercise, and eight hours of sleep each night. Can we conclude, therefore, that this type of behavior causes a longer life?

10. Return to the data in Exercise 2. Change each of the bottom three Y scores (68, 80, 72) to 73, creating three observations with identical values. Assign ranks to the X and Y scores. Because of the ties in ranks, follow the procedure recommended in Section 7.8, that is, treat all ranks as though they were numbers and calculate r (Formula 7.2) rather than r_s.

11. (Class Exercise) Collect estimates of height and weight for all members of your class. Construct a scatterplot and calculate r (Formula 7.2). Save your results for use in Chapter 8, Exercise 4.

8.1 –A "Rough" Prediction
8.2 –A Prediction Line
8.3 –Least Squares Prediction
    Line
8.4 –Least Squares Equation
8.5 –Prediction: An Overview
8.6 –Standard Error of
    Prediction, $S_{y \cdot x}$
8.7 –Other Sources of Error
8.8 –Regression
8.9 –Assumptions
8.10–Summary
8.11–Exercises

8
Prediction

Statisticians have crystal balls—known as prediction equations.

Not only are predictions made, but the degree of predictive error also can be specified.

There is a fairly strong—but by no means perfect—positive correlation between the height and weight of adult males. As has been noted previously, Bill Walton, the pro basketball player, stands about 4 standard deviations above the average height. Would you predict, therefore, that he also weighs about 4 standard deviations more than the average weight? See Exercise 5.

A correlation analysis of the exchange of greeting cards, as reported by five friends for the most recent holiday season, suggests a strong positive relationship between cards given and received. When informed of these results, another friend, Emma, with a weakness for receiving greeting cards, asks you to predict how many cards she will receive during the next holiday season, given that she plans to send 11 cards.

8.1 A "Rough" Prediction

You can supply Emma with a rough prediction by referring to the scatterplot for the original five friends shown in Figure 8.1. Notice that Emma's plan to send 11 cards locates her along the X axis between Steve's 10 and Doris' 13. Using the data points for Steve and Doris as boundaries, construct a rough prediction for Emma, as shown in Figure 8.1. Focusing on the interval along the Y axis intercepted by the arrows, you could predict that Emma's return should be between Doris' 15 and Steve's 18.

Our rough prediction might satisfy Emma, but it wouldn't win any statistical awards. Although each of the five dots in Figure 8.1 supplies valuable information about the exchange of greeting cards, our prediction for Emma is based only on the two dots for Steve and Doris.

8.2 A Prediction Line

This limitation doesn't exist for the more precise prediction, illustrated in Figure 8.2, that Emma will receive 15.20 cards. Placement of the straight line, designated as the "prediction line" in Figure 8.2, reflects the contributions of all five dots. If all five dots had defined a single straight line, placement of the prediction line would have been simple; one merely would let it pass through all dots. When the dots fail to define a single straight line, as in the scatterplot for the five friends, the prediction line represents a compromise. It passes through the main cluster, possibly touching some dots but missing others.

Look at the discrepancies between solid and open dots in Figure 8.3. Solid dots reflect the number of cards actually received by the five

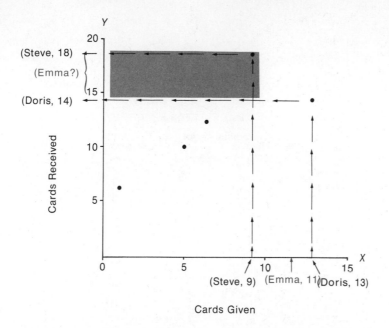

FIGURE 8.1

A "rough" prediction for Emma.

friends, while the open dots reflect the corresponding predictions for the five friends. Discrepancies between pairs of dots signify errors that would have occurred if the prediction line had been used to predict the number of cards received by each of the five friends. The smaller the sum of all of these discrepancies in Figure 8.3, the smaller the total predictive error and the more favorable the prognosis for the predictive effort. Clearly, it is desirable for the prediction line to be placed in a position that *minimizes* these discrepancies and, therefore, the total predictive error.

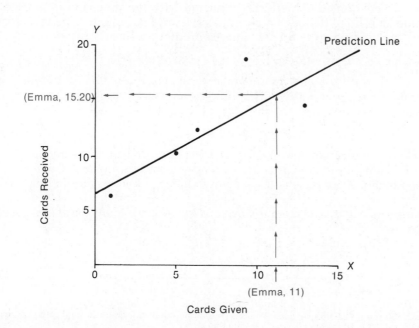

FIGURE 8.2

Prediction of 15.20 for Emma (using prediction line).

85

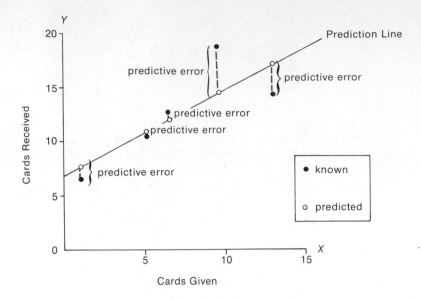

FIGURE 8.3

Predictive errors.

8.3 Least Squares Prediction Line

For mathematical reasons not discussed in this book, the placement of the prediction line minimizes not the total predictive error, but the total *squared* predictive error, that is, the sum of all squared predictive errors. When located in this fashion, the prediction line is often referred to as the **least squares prediction line.** Although more difficult to visualize, this approach is consistent with the original aim—to minimize the total predictive error or some version of the total predictive error, thereby providing a more favorable prognosis for the predictive effort.

Without the aid of mathematics, the search for a least squares prediction line would be a frustrating job. Scatterplots would be proving grounds cluttered with provisional prediction lines, discarded because their sums of squared discrepancies are excessive. Even the most time-consuming effort, conducted with the precision of a draftsman and the patience of a zen master, would culminate only in a close approximation to the least squares prediction line.

8.4 Least Squares Equation

Happily, we have recourse to an equation that pinpoints the exact least squares prediction line for any scatterplot. Most generally, this equation reads

$$Y' = bX + a \tag{8.1}$$

where Y' (pronounced Y prime) represents the predicted value (the predicted number of cards that will be received by any new friend, such as Emma), X represents the known value (the known number of cards given by any new friend, including Emma), and *b* and *a* represent numbers calculated from the original correlation analysis, as described below.*

To obtain a working prediction equation, solve each of the

---

*You might recognize that the least squares prediction equation defines a straight line with a slope of *b* and a Y-intercept of *a*. Since, as will become apparent, you'll never actually have to graph the least squares prediction equation, no additional reference will be made to the mathematics of straight lines.

following expressions, first for b and then for a, using data from the original correlation analysis. The expression for b reads

$$b = \frac{S_Y}{S_X}(r) \qquad (8.2)$$

where S_Y represents the standard deviation for all observations along Y (the cards received by the five friends); S_X represents the standard deviation for all observations along X (cards given by the five friends); and r represents the correlation between X and Y (letters given and received by the five friends).

The expression for a reads

$$a = \bar{Y} - b\bar{X} \qquad (8.3)$$

where $\bar{Y}$ and $\bar{X}$ refer to the means for all observations along Y and X, respectively, while b is defined by the preceding expression.

Values of all terms in expressions for b and a can be obtained from the original correlation analysis either directly, as with the value of r, or indirectly, as with the values of the remaining terms: S_X, S_Y, $\bar{X}$, and $\bar{Y}$. Table 8.1 illustrates the computational sequence that culminates in a least squares prediction equation for the greeting card example, namely

$$Y' = .80(X) + 6.40$$

where .80 and 6.40 represent the values computed for b and a, respectively.

In its present form, the prediction equation can be used to predict the number of cards that Emma will receive, assuming that she plans to give 11 cards. Simply substitute 11 for X and solve for the value of Y' as follows:

$$
\begin{aligned}
Y' &= .80(11) + 6.40 \\
&= 8.80 + 6.40 \\
&= 15.20
\end{aligned}
$$

TABLE 8.1
DETERMINING THE LEAST SQUARES PREDICTION EQUATION

A. COMPUTATIONAL SEQUENCE

Determine values of S_X, S_Y, and r ① by referring to the original correlation analysis in Table 7.4.

Substitute numbers into the formula ② and solve for b.

Assign values to $\bar{X}$ and $\bar{Y}$ ③ by referring to the original correlation analysis in Table 7.4.

Substitute numbers into the formula ④ and solve for a.

Substitute numbers for b and a in the least squares prediction equation ⑤.

B. COMPUTATIONS

① $\quad S_X = 4^*$
$\quad\ \ S_Y = 4^*$
$\quad\ \ \ \ r = .80$

② $\quad b = \left(\dfrac{S_Y}{S_X}\right)(r) = \left(\dfrac{4}{4}\right)(.80) = .80$

③ $\quad \bar{X} = 7^{**}$
$\quad\ \ \bar{Y} = 12^{**}$

④ $\quad a = \bar{Y} - (b)(\bar{X}) = 12 - (.80)(7) = 12 - 5.60 = 6.40$

⑤ $\quad Y' = (b)(X) + a$
$\qquad\ \ = (.80)(X) + 6.40$

---

*Computations not shown. Verify, if you wish, using Formula 5.1.
**Computations not shown. Verify, if you wish, using Formula 4.1.

TABLE 8.2
**PREDICTED CARD
RETURNS (Y') FOR
DIFFERENT CARD
INVESTMENTS (X)**

X	Y'
0	6.40
4	9.60
8	12.80
10	14.40
12	16.00
20	22.40
30	30.40

Notice that the predicted card return for Emma, 15.20, qualifies as a genuine prediction, that is, a forecast of an unknown event based on information about some other event. This prediction appeared earlier in Figure 8.2.

Our working prediction equation provides an inexhaustible supply of predictions for the card exchange situation. Each prediction emerges simply by substituting some value for X and solving the equation for Y', as described above. Table 8.2 lists the predicted card returns for a number of different card investments.

Conceivably, Emma might survey these predicted card returns before committing herself to a particular card investment. This strategy could backfire because of several complications. First, there's no guarantee—on the basis of existing data, where 13 is the maximum number of cards given—that the relationship between cards given and received can be extended to cases where as many as 20, 30, or more cards are given. Second, there's no evidence of a cause-effect relationship between cards given and cards received. The desired effect might be completely missing if, for instance, Emma expands her usual card distribution to include casual acquaintances and even strangers, as well as her friends and relatives.

8.5 Prediction: An Overview

The least squares prediction equation serves as our best prediction method whenever the underlying relationship is linear. Its emergence as a working equation, with the assignment of numbers to b and a as described above, automatically minimizes one estimate of predictive error: the sum of the squared predictive errors for known Y scores in the original correlation analysis.

Encouraged by Figures 8.2 and 8.3, you might be tempted to generate predictions from graphs rather than equations. Unless constructed very skillfully, graphs yield less accurate predictions than equations. In the long run, it's more accurate and easier to generate predictions from equations.

8.6 Standard Error of Prediction, $S_{Y \cdot X}$

Having predicted that Emma's investment of 11 cards will yield a return of 15 cards (when 15.20 is rounded off to the nearest whole number), we would, nevertheless, be surprised if she actually received 15 cards. It's more likely that, because of the imperfect relationship between cards given and cards received, Emma's return will be some number other than the predicted 15. Although designed to minimize predictive error, the least squares equation does not eliminate it. So our next task is to estimate the amount of error associated with our predictions. The smaller the estimated error, the better the prognosis for the predictive effort.

The estimate of error for new predictions is based on known data—the failures to predict the number of cards received by the original five friends, as depicted by the discrepancies between solid and open dots in Figure 8.3. We don't have to deal directly with these discrepancies. We can estimate predictive error from the following expression:

$$S_{Y \cdot X} = S_Y \sqrt{1 - (r)^2} \qquad (8.4)$$

where $S_{Y \cdot X}$ denotes the estimated predictive error, S_Y represents the standard deviation for all observations along Y (cards received by the five friends), and r represents the correlation between X and Y (cards given and received).

TABLE 8.3
CALCULATION OF THE
STANDARD ERROR OF
PREDICTION $S_{Y.X}$

A. COMPUTATIONAL SEQUENCE

Assign values to S_Y and r ① by referring to previous work with the least squares prediction equation in Table 8.1.

Substitute numbers into the formula ② and solve for $S_{Y.X}$.

B. COMPUTATIONS

① $S_Y = 4$
     $r = .80$

② $S_{Y.X} = (S_Y)[\sqrt{1 - (r)^2}] = (4)[\sqrt{1 - (.80)^2}]$
        $= (4)[\sqrt{1 - .64}] = (4)[\sqrt{.36}] = (4)[.60]$
        $= 2.40$

Although not obvious from Formula 8.4, the estimated predictive error or **standard error of prediction, $S_{Y.X}$,** *represents a special kind of standard deviation that reflects the magnitude of predictive error.* The value of 2.40 for $S_{Y.X}$, as calculated in Table 8.3, represents the standard deviation for the discrepancies between known and predicted card returns shown in Figure 8.3. In its role as an estimate of predictive error, the value of $S_{Y.X}$ could be attached to any new prediction. Thus, a concise prediction statement could read: "the predicted card return for Emma equals 15.20 ± 2.40," where the latter term reminds us of the magnitude of predictive error.

Once calculated, the value of $S_{Y.X}$ should estimate the error for all predictions from the same least squares equation. A value of 2.40 should describe the magnitude of predictive errors for other friends, as well as for Emma. Ordinarily, a value of 1 $S_{Y.X}$ should include the predictive errors for a majority of friends, and a value of 2 $S_{Y.X}$ should include the predictive errors for almost all friends.

If it can be assumed that actual card returns (Y) are approximately normally distributed about their respective predicted values (Y'), then even more meaning can be assigned to the value of 2.40 for $S_{Y.X}$. Recalling some of the well-known properties of normal curves, you could claim that approximately 68 percent of all card returns should be within 2.40 cards (1 $S_{Y.X}$) of their predicted card returns. Or after multiplying 2.40 by 1.96 to obtain 4.70, you could claim that approximately 95 percent of all card returns should be within 4.70 cards (1.96 $S_{Y.X}$) of their predicted card returns.

8.7 Other Sources of Error

The standard error of prediction doesn't summarize all errors associated with a predictive effort. Not only is the value of r subject to sampling variability, as noted in the previous chapter, but also the values of b and a, as well as $S_{Y.X}$. Accordingly, even the least squares prediction equation is subject to sampling variability and, therefore, contributes still another source of error to the predictive effort. In practice, error can be minimized by working only with large numbers of paired observations, preferably in the hundreds and certainly many more than the 5 pairs of observations used to illustrate the prediction of greeting card returns.

8.8 Regression

Many statistics books refer to prediction equations as regression equations. Whenever a correlation is less than perfect, predictions from the least squares equation (Formula 8.1) show **"regression toward the mean."** For instance, even though a given X score exceeds its mean by 2 standard deviations, it's predicted that the corresponding Y score will exceed its mean by less than 2 standard deviations. By the same token, even though a given X score falls 1 standard deviation below its mean, it's predicted that the corresponding Y score will fall less than 1 standard deviation below the mean. In other words, whether the original X score is above or below the mean, the predicted values of Y are "regressed toward the mean."

Regression can best be demonstrated by first converting the least squares prediction equation in Formula 8.1 into z scores. The z score version of this equation reads

$$z_Y' = r\, z_X \qquad\qquad (8.5)$$

where z_Y' represents the predicted Y score expressed as a standard score, r is the correlation coefficient, and z_X represents the given X score expressed as a standard score.

Figure 8.4 illustrates the degree of regression for three values of r. Regression is nonexistent when r equals 1.00, that is, when there is a perfect relationship between X and Y; regression is considerable when r equals .50; and it's total when r equals 0, that is, when there is no relationship between X and Y. Thus, given a value of 2 for z_X, Formula 8.5 yields a value of either 2, 1, or 0 for z_Y', depending on whether r equals either 1.00, .50, or 0. Regression increasingly characterizes and ultimately dominates least squares predictions as r approaches 0.

The regression effect spotlights the importance of the correlation coefficient in any predictive effort. If the relationship between X and Y is weak, as indicated by a small value of r, then the predictions from the least squares equation represent only a slight improvement over routinely predicting the mean of the Y scores, that is, $z_Y = 0$, regardless of the value of X. In this case, predictions show considerable regression. If the relationship between X and Y is strong, as indicated by a large value of r, then the predictions from the least squares equation represent a great improvement over routinely predicting the mean of the Y scores, regardless of the value of X. In this case, predictions show little regression.

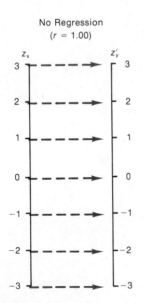

No Regression
(r = 1.00)

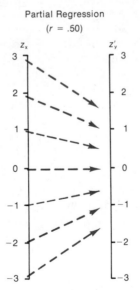

Partial Regression
(r = .50)

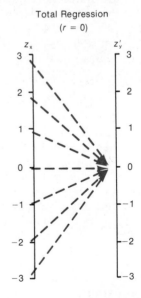

Total Regression
(r = 0)

FIGURE 8.4
Regression toward the mean (three values of r).

Use of the least squares prediction equation requires that the underlying relationship be linear. You need to worry about a violation of this assumption only when the scatterplot for the original correlation analysis reveals an obviously bent or curvilinear dot cluster, such as illustrated in Figure 7.4. In the event that a dot cluster describes a pronounced curvilinear trend, consult more advanced statistics books for appropriate procedures.

Use of the standard error of prediction, $S_{Y \cdot X}$, presumes that, except for chance, dots in the original scatterplot tend to be dispersed equally about all segments of the prediction line or about some straight line that roughly denotes the main trend of the dot cluster. You need to worry about violations of this assumption, officially known by its tongue-twisting designation as the assumption of "homoscedasticity" (pronounced ho-mo-ske-das-ti-ci-ty), only when the scatterplot reveals a dramatically different type of dot cluster, such as that shown in Figure 8.5. At the very least, the standard error of prediction for the data in Figure 8.5 should be used cautiously, since its value overestimates the variability of dots about the lower half of the prediction line, while it underestimates the variability of dots about the upper half of the prediction line.

8.9 Assumptions

If a linear relationship exists between two variables, then one variable can be predicted from the other by using the least squares prediction equation, as described in Formulas 8.1, 8.2, and 8.3.

The least squares equation minimizes a variation on the total predictive error, that is, the total of all squared predictive errors that would have occurred if the equation had been used to predict known Y scores from the original correlation analysis.

An estimate of predictive error can be obtained from Formula 8.4. Known as the standard error of prediction, this estimate is a special

8.10 Summary

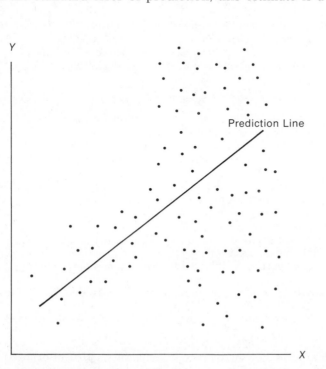

FIGURE 8.5

Violation of homoscedasticity assumption. (Dots lack equal variability about all line segments.)

kind of standard deviation that reflects the magnitude of predictive error.

Regression toward the mean increasingly characterizes, and ultimately dominates, predictions from the least squares equation as r approaches zero. Regression can best be demonstrated by referring to the z score version of the least squares equation shown in Formula 8.5.

Several assumptions underlie any predictive effort. The prediction equation assumes a linear relationship between variables, while the standard error of prediction assumes homoscedasticity—approximately equal dispersion of data points about all segments of the prediction line.

Important Terms & Symbols

Least squares prediction equation
Standard error of prediction ($S_{Y \cdot X}$)
Regression toward the mean

8.11 Exercises

1. Assume that an r of .70 describes the relationship between educational level (highest grade completed) and estimated number of hours spent reading each week. More specifically,

    EDUCATIONAL LEVEL (X)   WEEKLY READING TIME (Y)

    $$\bar{X} = 13 \qquad\qquad \bar{Y} = 8$$
    $$S_X = 2 \qquad\qquad S_Y = 4$$
    $$r = .70$$

    (a) Determine the least squares equation for predicting weekly reading time from educational level.
    (b) Mary's educational level is 15. What is her predicted reading time?
    (c) Jan's educational level is 11. What is her predicted reading time?
    (d) Determine the standard error of prediction.

2. Assume that an r of $-.80$ describes the strong negative relationship between years of heavy smoking (X) and life expectancy (Y). Assume, furthermore, that the distributions of heavy smoking and life expectancy each have the following means and standard deviations:

    $$\bar{X} = 5 \qquad \bar{Y} = 60$$
    $$S_X = 3 \qquad S_Y = 6$$

    (a) Determine the least squares equation for predicting life expectancy from years of heavy smoking.
    (b) Determine the standard error of prediction, $S_{Y \cdot X}$.
    (c) Predict the life expectancy for Sara, given that she has smoked heavily for 8 years.
    (d) Assume that life expectancies approximate a normal distribution about their respective predicted values. Complete the following statements for a group of people who, like Sara, have smoked heavily for 8 years:
         68 percent of this group will have life expectancies between _____ and _____ years.
         95 percent will have life expectancies between _____ and _____ years.
    (e) Predict the life expectancy for Bill, given that he has never smoked heavily.
    (f) Answer 2d, but for a group of people who, like Bill, have never smoked heavily.

3. Each of the following pairs represents the number of licensed drivers (X) and the number of cars (Y) for houses in my neighborhood.

DRIVERS (X)	CARS (Y)
5	4
5	3
2	2
2	2
3	2
1	1
2	2

(a) Construct a scatterplot to verify a lack of pronounced curvilinearity.
(b) Determine the least squares equation for these data. (Remember, you'll first have to calculate r, $\bar{X}$, $\bar{Y}$, S_X, and S_Y.)
(c) Determine the standard error of prediction, $S_{Y \cdot X}$.
(d) Predict the number of cars for each of two new families with 2 and 5 drivers.

4. Using the results (if available) from Chapter 7, Exercise 10, do the following:

   (a) Determine the least squares equation for predicting weight from height.
   (b) Calculate the standard error of prediction.
   (c) Predict the weight for some friend (not in your statistics class) whose height and weight can be obtained.
   (d) Compare the predicted and true weights for this friend. Are they equal?
   (e) If true weights are normally distributed about their respective predicted weights, then about 68 percent of all true weights should be within one standard error of prediction of their predicted weights. First, determine whether or not your friend's true weight is within one standard error of prediction of his or her predicted weight, and then combine your finding with those of your classmates, to check whether approximately 68 percent of all true weights are within $1\,S_{Y \cdot X}$ of the various predicted (Y') values.
   (f) Repeat the previous exercise to determine whether approximately 95 percent of all true weights are within $1.96\,S_{Y \cdot X}$ of Y'—again on the assumption that true weights are normally distributed about their respective predicted weights.

5. Assume that an r of .75 describes the relationship between high school and college grade point averages.

   (a) What is Sam's predicted college GPA, expressed in z score form, given that his high school GPA is 2 standard deviations above the mean?
   (b) What is Tim's predicted college GPA, expressed in z score form, given that his high school GPA is 1 standard deviation below the mean?
   (c) Repeat the above predictions for Sam and Tim, assuming that $r = .25$.
   (d) Repeat the above predictions for Sam and Tim, assuming that $r = -.50$.
   (e) Which of the above pairs of predictions shows the most regression toward the mean?
   (f) Which of the above pairs of predictions shows the least regression toward the mean?

6. In Exercise 2, a least squares equation was used to predict the number of cars (Y) from the number of licensed drivers (X). Would a different equation be required to predict the number of licensed drivers from the number of cars? First, speculate about your answer, then actually calculate the least squares equation for predicting the number of licensed drivers (Y) from the number of cars (X).

Part II

INFERENTIAL STATISTICS

Generalizing from Samples to Populations

9 Populations, Samples, and Probabilities

10 Sampling Distributions

11 Introduction to Hypothesis Testing: the z Test

12 Two Types of Error and Sample Size Selection

13 t Test for One Sample

14 t Test for Two Independent Samples

15 t Test for Two Dependent Samples

16 Estimation

17 Analysis of Variance (One Factor)

18 Analysis of Variance (Two Factors)

19 Chi Square (χ^2) Test for Qualitative Data

20 Tests for Ranked Data

9

Populations, Samples, and Probabilities

9.1–Why Samples?
9.2–Populations
9.3–Samples
9.4–Random Samples
9.5–Tables of Random Numbers
9.6–Some Complications
9.7–Random Assignment of Subjects
9.8–Probability
9.9–Summary
9.10–Exercises

Even though only very small portions of populations are actually sampled, generalizations from samples to populations are often amazingly accurate. Note the accuracy of recent election polls by Gallup, as described in Table 9.1.

In statistics, the word "random" has a special meaning. It doesn't merely signify casual or haphazard. You'll learn the special definition of "random" and ways to guarantee that a sample truly is random.

You'll become familiar with two rules for combining probabilities to determine, for instance, the likelihood that a man will stand at least six feet tall or that all three children in the same family will be girls.

In everyday life, we regularly generalize from a limited set of observations (samples) to broader sets of observations (populations). One sip indicates that the batch of soup is too salty; dipping a toe in the swimming pool reassures us before the first plunge; a test drive triggers suspicions that the used car is less than advertised; and a casual encounter with a stranger stimulates fantasies about a deeper relationship. In inferential statistics we are supplied with more sophisticated tools for generalizing from samples to populations.

Generalizations can, of course, backfire if the sample misrepresents the population. The sip contains unstirred salt; the toe dips too near the warm water outlet; the test drive merely reflects an untuned engine; and the casual encounter was really very contrived. In inferential statistics, we can specify and control the chances that generalizations will backfire.

9.1 Why Samples?

Faced with the possibility of erroneous generalizations, you might prefer to bypass the uncertainties of inferential statistics by surveying an entire population. This is often done if the size of the population is fairly small. For instance, you calculate your college grade point average (GPA) from *all* of your course grades, not just a sample. If the size of the population is large, however, complete surveys are often prohibitively expensive and sometimes impossible. Under these circumstances, you might have to use samples and risk the possibility of erroneous generalizations. For instance, you might have to use a sample to estimate the mean GPA for all students at a large university.

9.2 Populations

Any set of potential observations may be characterized as a **population.** Complete descriptions of populations specify the nature of the observations to be taken. For example, a population might be described as "attitudes toward religion of currently enrolled students at Penn State" or as "SAT scores of currently enrolled students at Penn State."

Populations may be real or hypothetical. A **real** population is one in which all observations are accessible at the time of sampling. Examples of real populations include the two described in the previous paragraph, as well as the ages of all visitors to Disneyland on a given day, the ethnic backgrounds of all current employees of the U.S. Postal Department, gas mileage of all BMWs sold during the previous year, presidential preferences of all currently registered voters in the United States. Incidentally, federal law requires that a complete survey be taken every 10 years of the real population of all United States households—at a considerable expense, involving many thousands of census takers—as a means of revising election districts for the House of Representatives.

Pollsters, such as the Gallup organization, deal with real populations. Research workers tend to deal with hypothetical populations. A **hypothetical** population is one in which all potential observations are not accessible at the time of sampling. In most experiments, subjects are selected from very small real populations—the lab rats housed in the local animal colony or student volunteers from general psycholgy classes. Experimental subjects are viewed, nevertheless, as a sample from a much larger hypothetical population, loosely described as "the scores of all similar animal subjects (or student volunteers) who, either now or *in the future,* could conceivably undergo the present experiment." Experimental findings are generalized to this much larger population that contains many potential observations which, because of their hypothetical status in future experiments, couldn't possibly be included in the present experiment.

According to the rules of inferential statistics, generalizations should be made only to real populations that have, in fact, been sampled. Generalizations to hypothetical populations should be viewed, therefore, as provisional conclusions based on the wisdom of the researcher rather than on any logical or statistical necessity. In effect, it's an open question—sometimes resolved only through additional experimentation—whether or not a given experimental finding merits the generality assigned to it by the researcher.

9.3 Samples

Any subset of observations from a population may be characterized as a **sample.** In typical applications of inferential statistics, sample size is small relative to population size. Less than 1 percent of all United States households are included in the monthly survey of the Bureau of Labor Statistics to estimate the current rate of unemployment. Although only about 3500 voters have been sampled in recent presidential election polls by Gallup, predictions have been amazingly accurate, as can be seen in Table 9.1.

There is no simple rule of thumb for determining the best or optimal sample size for any particular situation. Optimal sample size depends on answers to a number of questions, including "What is the estimated variability among observations?" and "What is an acceptable amount of probable error?". Once these questions have been answered—with the aid of guidelines discussed in Chapter 12—specific

| YEAR | SAMPLE SIZE | WINNER | PERCENT OF MAJOR PARTY VOTE | | ERROR |
			Predicted	Actual	
1952	5385	Eisenhower	51%	55.4%	−4.4%
1956	8144	Eisenhower	59.5%	57.8%	1.7%
1960	8015	Kennedy	51%	50.1%	0.9%
1964	6625	Johnson	64%	61.3%	2.7%
1968	4414	Nixon	43%	43.5%	−0.5%
1972	3689	Nixon	62%	61.8%	0.2%
1976	3439	Carter	49.5%	51.1%	−1.6%

TABLE 9.1
PRESIDENTIAL ELECTION POLLS BY GALLUP SINCE 1952*

* Prior to 1952, Gallup used less refined survey techniques, yielding higher predictive errors—including the erroneous prediction that Dewey would defeat Truman in the 1948 election.

Source: The Gallup Poll (American Institute of Public Opinion).

procedures can be followed to determine the optimal sample size for any situation.

9.4 Random Samples

In order to use techniques from inferential statistics, the analysis should be based on random samples. *A sample is* **random** *if, at each stage of sampling, the selection process guarantees that all remaining observations in the population have equal chances of being included in the sample.* It's important to note that randomness describes the selection process, that is, the conditions under which the sample is taken, and not the particular pattern of observations in the sample. Having established that a sample is random, you still can't predict anything about the particular pattern of observations in that sample. The observations in the sample should tend to be representative of those in the population, but there is no guarantee that this will be true.

A casual or haphazard sample doesn't qualify as a random sample. Not every student at Penn State has an equal chance of being sampled if, for instance, a pollster casually selects students who enter the main library. Obviously excluded from this sample are all those students who never enter the main library. Furthermore, even the final selection of students from among those who do enter the main library might reflect various biases of the pollster, such as an unconscious preference for well-dressed, attractive students who are walking alone.

How, then, do you select a random sample? Several techniques have been used with varying degrees of success. Probably the best known technique is the "fish-bowl" method. This method requires that all potential observations be represented in some fashion on slips of paper. For instance, the names of all students at some college might be written on separate slips of paper. All slips of paper are then deposited in a bowl and stirred. Some person draws a slip from the bowl, and the name on that slip designates the first student to be included in the sample. Drawings continue until the desired sample size has been reached.

The adequacy of the fish-bowl method depends on the thoroughness of stirring. A truly thorough stirring of slips of paper is more difficult to achieve than might be suspected, as evidenced by the selective service lottery in 1969. Essentially, in this lottery the population consisted of the 366 days of the year, representing the birthdays of all draft eligible young men. All of these days (rather than a sample) were to be selected from the fish bowl, with the order of selection from the fish bowl establishing the order of induction into military service.

If the lottery was random, as required by law, then all 366 days should have had an equal chance of being drawn first; all remaining 365 days should have had an equal chance of being drawn second, and so forth. Although anything can happen by chance, the results of this lottery looked suspiciously nonrandom. There was a noticeable tendency for the days of some months—notably those months during the latter part of the year—to be drawn early and, therefore, to be assigned a high draft priority number. One possible explanation of this tendency is that the original slips of paper, each stuffed in a separate plastic capsule, were unintentionally deposited in the bowl in chronological order, and then stirred inadequately. To correct this situation, the subsequent selective service lotteries involved elaborate efforts to guarantee randomization, particularly in terms of the original placement of capsules in the bowl.

9.5 Tables of Random Numbers

The most adequate method for generating a truly random sample involves tables of random numbers—a method used to randomize the initial placement of capsules in post-1969 selective service lotteries. These tables are generated by a computer designed to equalize, as much as possible, the occurrence of any one of the ten digits: 0, 1, 2, . . . , 8, 9. For convenience, many random number tables are spaced in columns of five-digit numbers. Table G in Appendix C shows a specimen page of random numbers from a book devoted almost entirely to random digits.

The size of the real population determines whether you deal with numbers having one, two, three or more digits. The only requirement is that you have at least as many different numbers as you have potential observations within the population. For example, if you are attempting to take a random sample from a real population consisting of 679 students at some college, you could use the 1000 three-digit numbers ranging from 000 to 999. In this case, you could identify each of the potential observations, as represented by a particular student's name, with a single number. For instance, if a student directory were available, the first person, Alice Aakins, might be assigned the three-digit number 001, while the last person in the directory, Zachary Ziegler, might be assigned 679.

Enter the random number table at some arbitrarily determined place. Ordinarily, this should be determined haphazardly. Open a book of random numbers to any page and begin with the number closest to a blind pencil stab. For illustrative purposes, however, let's use the upper left hand corner of the specimen page as our entry point. (Ignore the column of numbers that identify the various rows.) Read in a consistent direction, for instance, down. Then as each column is used up, shift to the right and repeat the entire process, beginning at the top of each new column. As a given number between 001 and 679 is encountered, the person identified with that number is included in the random sample.

Since the first number on the specimen page in Table G is 100 (disregard the fourth and fifth digits in each five-digit number), the person identified with that number is included in the sample. The next three-digit number, 375, identifies the second person, and 084 identifies the third person. Ignore the next number, 990, since none of the numbers between 680 and 999 are identified with any names in the student directory. Also ignore repeat appearances of any numbes between 001 and 679. Continue this process until the specified sample size has been realized.

"THE RANDOM NUMBER IS EVEN --
DO YOU WISH TO KICK OR RECEIVE?"

Even with real populations, it's often difficult to identify all potential observations to be paired with random numbers. Lacking the convenience of an existing population directory, investigators resort to various embellishments of the previous procedure. For instance, the Gallup organization does a separate presidential survey in each of the four geographical areas of the United States: Northeast, South, Midwest, and West. Within each of these areas, a series of random selections culminate in the identification of particular election precincts—small geographical districts with a single polling place. Once household directories have been obtained for each of these precincts, households are randomly selected, and specific members of these households are interviewed.

As has been noted, the researcher, unlike the pollster, tends to deal with hypothetical populations. Unfortunately, it's impossible to take random samples from hypothetical populations. All potential observations can't have an equal chance of being included in the sample if, in fact, some observations aren't accessible at the time of sampling. It's common practice, nonetheless, for researchers to treat samples from hypothetical populations *as if* they were random samples, and to analyze sample results with techniques from inferential statistics.

9.6 Some Complications

Typically, experiments involve at least an experimental condition and a control condition. Even though subjects in experiments can't be selected randomly from a hypothetical population, they should be assigned randomly, that is, with equal likelihood, to these conditions. In effect, this procedure tends to equalize the several samples of subjects at the outset of the experiment. For instance, to determine whether vitamin C improves academic performance, volunteer subjects should be assigned randomly either to the experimental group, whose members receive a daily dose of vitamin C, or to the control group, whose members receive only fake vitamin C. *The purpose of random assignment of subjects is to insure that, except for random differences, groups of subjects are similar with respect to any uncontrolled variables, such*

9.7 Random Assignment of Subjects

as academic preparation, motivation, IQ, and so forth. At the conclusion of the experiment, therefore, any observed differences in academic performance between these two groups—not attributable to random differences—can be attributed to vitamin C.

Random assignment of subjects can be accomplished in a number of ways. For instance, as each new subject arrives to participate in the experiment, a flip of a coin can decide whether that subject should be assigned to the experimental condition (if "heads" turn up) or the control condition (if "tails" turn up). An even better procedure—because it eliminates any biases of a live coin tosser—relies on tables of random numbers. Once the tables have been entered at some arbitrary point, they can be consulted, much like a string of coin tosses, to determine whether each new subject should be assigned to the experimental condition (if, for instance, the random number is odd) or to the control condition (if the random number is even).

For a variety of reasons, it's highly desirable that equal numbers of subjects be assigned to the experimental and control conditions. Given this restriction, random assignment should involve pairs of subjects. Thus, if the table of random numbers assigns the first volunteer to the experimental condition, the second volunteer is *automatically* assigned to the control condition. If the random numbers assign the third volunteer to the control condition, the fourth volunteer is *automatically* assigned to the experimental condition, and so forth. This procedure guarantees that, at any stage of random assignment, equal numbers of subjects will be assigned to the two conditions.

Incidentally, the page of random numbers in Table G serves only as a specimen. For serious applications, refer to a more extensive collection of random numbers, such as that in the book by the Rand Corporation, described at the bottom of Table G.

9.8 Probability

In later chapters, we'll be able to specify the probability that generalizations from samples to populations will be erroneous. **Probability** *refers to the proportion or fraction of times that some outcome will occur.*

The probability of an outcome can be determined in several ways. We could *speculate* that, if a coin is truly fair, heads and tails should be equally likely to occur whenever the coin is tossed, and therefore, the probability of heads should equal .50 or $\frac{1}{2}$. Or we could actually *observe* the outcomes of a long string of coin tosses and conclude, on the basis of these observations, that the probability of heads equals approximately .50 or $\frac{1}{2}$.

Table 9.2 shows the relative frequency distribution of heights for a random sample of 3091 men. These relative frequencies indicate the proportion of men in the total sample who are associated with a particular height. They also indicate the probability that any one randomly selected man from the total sample will have a particular height. For example, the probability is .18 that a randomly selected man will be 68 inches tall. From this perspective, all of the relative frequencies in Table 9.2 are probabilities. Each of these probabilities could vary in value between zero (impossible) and one (certain). Furthermore, probabilities are never less than zero nor greater than one, and an entire set of probabilities sum to one.

There are two simple rules for combining probabilities of various types of outcomes: the addition rule and the multiplication rule. Each will be illustrated, using probabilities from Table 9.2.

TABLE 9.2
RELATIVE FREQUENCY DISTRIBUTION FOR HEIGHTS OF 3091 MEN

HEIGHT (INCHES)	RELATIVE FREQUENCY
75 or taller	.01
74	.01
73	.02
72	.05
71	.06
70	.12
69	.10
68	.18
67	.12
66	.13
65	.07
64	.07
63	.03
62	.02
61 or shorter	.01
Total	1.00

Source: See Figure 6.1.

102

Example   **9.8 PROBABILITY**

What's the probability that a randomly selected man will be at least 72 inches tall? That's the same as asking, "What's the probability that a man will stand 72 inches tall or taller?" To answer this type of question, which involves a cluster of outcomes connected by the word "or," simply add the respective probabilities. The probability that a man will stand at least 72 inches tall equals the sum of the probabilities that a man will stand 72 or more inches tall, that is,

$$.05 + .02 + .01 + .01 = .09$$

Whenever you must find the probability for two or more sets of outcomes connected by the word "or," use the **addition rule.** It tells you to add the probabilities for each of the separate outcomes. Stated generally, for outcomes A and B, the addition rule reads:

$$\text{Pr (A or B)} = \text{Pr (A)} + \text{Pr (B)} \qquad (9.1)$$

where Pr (   ) refers to the probability of the outcome in the parentheses.

Example

Given the probability is .09 that a randomly selected man will be at least 72 inches tall, what's the probability that two randomly selected men will be at least 72 inches tall? That's the same as asking, "What's the probability that the first man stands at least 72 inches tall *and* the second man stands at least 72 inches tall?"

To answer this type of question, which involves clusters of outcomes connected by the word "and," simply multiply the respective probabilities. Thus, the probability that both men stand at least 72 inches tall equals the product of the probabilities that the first man stands at least 72 inches tall *and* the second man stands at least 72 inches tall, that is

$$(.09)(.09) = .0081$$

Whenever you must find the probability for two or more sets of outcomes connected by the word "and," use the **multiplication rule.** It tells you to multiply the probabilities for each of the separate outcomes. Stated generally, the multiplication rule reads

$$\text{Pr (A and B)} = [\text{Pr(A)}][\text{Pr(B)}] \qquad (9.2)$$

where Pr (   ) is defined as above, and A and B are two outcomes.*

Finally, it should be noted that *proportions of area under curves may be interpreted as probabilities.* For instance, in the standard normal table (Table A in Appendix C), column B indicates that .4332 is the proportion of total area between the mean and a z score of 1.50. By the same token, .4332 represents the probability that the value of any randomly selected z score falls between the mean and 1.50. This is a fairly common outcome since it should happen about 43 times in 100. Similarly, column C indicates that .0668 is the proportion of total area above a z score of 1.50. Therefore, .0668 represents the probability that the value of any randomly selected z score exceeds 1.50. This is not a common outcome since it should happen only about 7 times in 100. In

---

*Although not encountered in this book, there are more complex types of outcomes, referred to as nonmutually exclusive and nonindependent outcomes, that require modification of the addition and multiplication rules, respectively.

the remainder of this book, proportions of area under various curves, including the standard normal curve, will be routinely interpreted as probabilities. As will be seen, probability considerations form the basis of inferential statistics, and statistical conclusions will inevitably be couched in terms of probability statements.

9.9 Summary

Any set of potential observations may be characterized as a population. Any subset of observations constitutes a sample.

Populations are either real or hypothetical, depending on whether or not all observations are accessible at the time of sampling.

The valid application of techniques from inferential statistics requires that samples be random. A sample is random if, at each stage of sampling, the selection process guarantees that all remaining observations in the population have equal chances of being included in the sample.

Tables of random numbers provide the most effective method both for taking random samples in surveys and also for randomly assigning subjects to various conditions in experiments. Some type of randomization always should occur during the early stages of any investigation.

The probability of some outcome specifies the proportion of times that this outcome will occur.

Whenever you must find the probability of sets of outcomes connected with the word "or," use the addition rule—add the probabilities of each of the separate outcomes. Whenever you must find the probability of sets of outcomes connected with the word "and," use the multiplication rule—multiply the probabilities of each of the separate outcomes.

Proportions of area under curves may be interpreted as probabilities.

Important Terms

Population	Probability (of some outcome)
Sample	Addition rule
Random sample	Multiplication rule

9.10 Exercises

1. For each of the following pairs, indicate with a Yes or No whether the relationship between the first and second expressions could describe that between a sample and its population, respectively.

   (a) spoonful of soup; cupful of soup
   (b) students in the last row; students in class
   (c) students in class; students in college
   (d) students in the last row; a student in the last row
   (e) citizens of Wyoming; citizens of New York
   (f) every tenth California income tax return; every California income tax return.
   (g) 20 lab rats in some experiment; all lab rats, similar to those used, that could undergo the same experiment.
   (h) all U.S. presidents; all registered Republicans
   (i) two tosses of a coin; all possible tosses of a coin

2. List all expressions in Exercise 1 that involve a hypothetical population.

3. Indicate whether each of the following statements is true or false. A random selection of 10 playing cards from a deck of 52 cards implies that

   (a) the random sample of 10 cards accurately represents the important features of the whole deck.
   (b) each card in the deck has an equal chance of being selected.
   (c) it is impossible to get ten cards from the same suit (for example, ten hearts).
   (d) any outcome, however unlikely, is possible.

4. Describe how you would use the tables of random numbers to take

   (a) a random sample of size 5 from your statistics class.
   (b) a random sample of size 40 from the student body at your school.

5. As subjects arrive to participate in an experiment, tables of random numbers are used to make random assignments to either an experimental or a control group. Indicate with a Yes or No whether each of the following rules would work:

   (a) Assign the subject to the experimental group if the random number is even and to the control group if the random number is odd.
   (b) Assign the subject to the experimental group if the first digit of the random number is between 0 and 4, and to the control group if the first digit is between 5 and 9.
   (c) Assign the subject to the experimental group if the first two digits of the random number are between 00 and 40, and to the control group if the first two digits are between 41 and 99.
   (d) Assign the subject to the experimental group if the first three digits of the random number are between 000 and 499, and to the control group if the first three digits are between 500 and 999.

6. (a) Sometimes experiments involve more than two conditions. Describe a rule, similar to those listed in Exercise 5, that could be used to assign subjects randomly to *four* experimental conditions.
   (b) It was suggested earlier that random assignment should involve pairs of subjects if equal numbers of subjects are to be assigned to the experimental and control conditions. How might the above rule be modified if equal numbers of subjects are to be assigned to all four experimental conditions?

7. The probability of a boy being born equals .50 or $\frac{1}{2}$, as also does the probability of a girl being born. For a randomly selected family with two children, what's the probability of

   (a) two boys?
   (b) two girls?
   (c) either two boys or two girls?

8. Assume the same probabilities as in Exercise 7. For a randomly selected family with three children, what's the probability of

   (a) three boys?
   (b) three girls?
   (c) either three boys or three girls?
   (d) neither three boys nor three girls?
   (Hint: 8d can be answered indirectly by remembering that probabilities always sum to 1.)

9. Referring to Table 9.2, find the probability that (a) any randomly selected man will deviate from 68 inches by 4 or more inches and (b) two randomly selected men *both* will deviate by this amount.

10. Referring to the standard normal table (Table A), find the probability that a randomly selected z score will be

   (a) above 1.96
   (b) below −1.96
   (c) either above 1.96 or below −1.96
   (d) between −1.96 and 0 or between 0 and 1.96
   (e) either above 2.57 or below −2.57
   (f) either above 3.30 or below −3.30

10
Sampling Distributions

10.1 –An Example
10.2 –A Closer Look at Sampling Distributions
10.3 –Some Important Symbols
10.4 –Mean of All Sample Means
10.5 –Standard Error of the Mean
10.6 –Shape of the Sampling Distribution
10.7 –Why the Central Limit Theorem Works
10.8 –Types of Sampling Distributions
10.9 –Summary
10.10–Exercises

Get ready for the most important concept in inferential statistics—the concept of a sampling distribution. Without sampling distributions, inferential statistics would be bogged down in the Stone Age of its development.

In practice, only one sample mean is actually observed. In order to make more refined generalizations to the population, however, this one sample mean is viewed within the context of its sampling distribution—the distribution of means for all possible samples that could have been taken from that population.

Most sampling distributions are impossible to construct from scratch because of their huge number of components. Statistical theory comes to our rescue and identifies some important properties shared by many of the sampling distributions spotlighted in this chapter.

10.1 An Example

There's a good chance that you've taken the Scholastic Aptitude Test (SAT), and you might even remember your scores for verbal and math aptitude. On a nationwide basis, the verbal scores for all college-bound students during a recent year were distributed around an average of 429 with a standard deviation of 110. An investigator at a university wishes to test the claim that, on the average, the SAT verbal scores for local freshmen equal the national average of 429. Since it's not feasible to obtain test scores for the entire freshman class, SAT verbal scores are obtained for a random sample of 100 freshmen. The mean score for this sample equals 462.

If each sample were an exact replica of the population, generalizations from the sample to the population would be most straightforward. Then, having observed a mean verbal score of 462 for a sample of 100 freshmen, we could have concluded, without even a pause, that the mean verbal score for the entire freshman class also equals 462 and, therefore, exceeds the national average.

Random samples rarely represent the underlying population exactly. Even a mean verbal score of 462 could originate, just by chance, from a population of freshmen whose mean equals the national average of 429. Accordingly, generalizations from a single sample to a population are much more tentative. Indeed, generalizations are based not merely on the single sample mean of 462, but also on its sampling distribution. The **sampling distribution of the mean** *refers to the probability distribution of means for all possible samples of a given size from some population.* In effect, a sampling distribution describes the entire spectrum of sample means that could occur just by chance, and thereby provides a frame of reference for generalizing from a single sample mean to a population mean. More specifically, if the sample mean of 462 can be viewed as a *probable outcome* in its sampling distribution (as described in the next chapter), then it's reasonable to conclude that the mean verbal score for the entire freshman class could equal the national average. Otherwise, if the sample mean of 462 must be viewed as an *improbable outcome* in its sampling distribution, then it's reasonable to conclude that the mean verbal score for the freshman class doesn't equal the national average.

Thus, when generalizing from a single sample mean to a population mean, we must consult the sampling distribution of the mean. In the present case, the sampling distribution is based on *all possible* samples, each of size 100, that could be taken from the local population of freshmen. All possible samples refers not to the number of samples of size 100 required to *completely survey* the local population of freshmen, but to the number of different ways in which a *single* sample of size 100 could be selected from this population.

All possible samples tends to be a huge number. For instance, if the local population contains at least 1000 freshmen, the total number of possible samples, each of size 100, becomes astronomical in size. The more than 300 digits in this number would, by comparison, dwarf the national debt. Even with the aid of a computer, it would be a horrendous task to construct this sampling distribution from scratch, itemizing each sample mean for all possible samples.

Fortunately, statistical theory comes to the rescue and supplies us with considerable information about the sampling distribution of the mean, as will be discussed in the remainder of this chapter. Armed with this information about sampling distributions, we'll return to the current example in the next chapter and test the claim that the mean verbal score for the local population of freshmen equals the national average of 429.

10.2 A Closer Look at Sampling Distributions

Let's establish precisely what constitutes a sampling distribution by creating one from scratch under highly simplified conditions. Imagine some ridiculously small population of four observations with values of 2, 3, 4, and 5, as shown in Figure 10.1. Next, itemize all possible samples, each of size two, that could be taken from this population. There are four possibilities on the first draw from the population and also four possibilities on the second draw from the population, as indicated in Table 10.1.* The two sets of possibilities combine to yield a total of 16 possible samples. At this point, remember, we're clarifying the notion of a sampling distribution. In practice, only one random sample, not 16 possible samples, would be taken from the population; a sample size would be very small relative to population size; and, of course, not all observations in the population would be known.

For each of the 16 possible samples, Table 10.1 also lists a sample mean (found by adding the two observations and dividing by 2) and its probability of occurrence (expressed as $\frac{1}{16}$, since each of the 16 possible samples are equally likely). When cast into a relative frequency or probability distribution, as in Table 10.2, the 16 sample means constitute the sampling distribution of the mean, previously defined as the

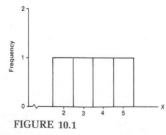

FIGURE 10.1

Graph of miniature population.

*Ordinarily, the same observation isn't sampled more than once. If employed with the present, highly simplified example, however, this restriction would magnify an unimportant technical adjustment.

SAMPLE	MEAN ($\bar{X}$)	PROBABILITY
(1) 2,2	2.0	$\frac{1}{16}$
(2) 2,3	2.5	$\frac{1}{16}$
(3) 2,4	3.0	$\frac{1}{16}$
(4) 2,5	3.5	$\frac{1}{16}$
(5) 3,2	2.5	$\frac{1}{16}$
(6) 3,3	3.0	$\frac{1}{16}$
(7) 3,4	3.5	$\frac{1}{16}$
(8) 3,5	4.0	$\frac{1}{16}$
(9) 4,2	3.0	$\frac{1}{16}$
(10) 4,3	3.5	$\frac{1}{16}$
(11) 4,4	4.0	$\frac{1}{16}$
(12) 4,5	4.5	$\frac{1}{16}$
(13) 5,2	3.5	$\frac{1}{16}$
(14) 5,3	4.0	$\frac{1}{16}$
(15) 5,4	4.5	$\frac{1}{16}$
(16) 5,5	5.0	$\frac{1}{16}$

**TABLE 10.1
ALL POSSIBLE SAMPLES
OF SIZE TWO FROM
MINIATURE POPULATION**

probability distribution of means for all possible samples of a given size from some population. Not all probabilities are equal in Table 10.2, you'll notice, since some values of the sample mean occur more than once among the 16 possible samples. For instance, a sample mean value of 3.5 appears among 4 of the 16 possibilities and has a probability of $\frac{4}{16}$.

The sampling distribution in Table 10.2 can be consulted to determine the probability of obtaining a particular sample mean or set of sample means. Thus, the probability is $\frac{1}{16}$ or .0625 that the value of a randomly selected sample mean equals 5.0. According to the addition rule for probabilities, described in the previous chapter, the probability is $\frac{1}{16} + \frac{1}{16} = \frac{2}{16}$ or .1250 that the value of a randomly selected sample mean equals either 5.0 or 2.0. Subsequent applications in inferential statistics will always entail, in some form or another, probability statements based on sampling distributions.

Figure 10.2 summarizes the previous discussion. It depicts the emergence of the sampling distribution of the mean from the set of all possible (16) samples of size two, based on the miniature population of four observations. Familiarize yourself with this figure. It will be referred to again in subsequent sections.

**TABLE 10.2
SAMPLING DISTRIBUTION
OF THE MEAN
(SAMPLES OF SIZE
TWO FROM MINIATURE
POPULATION)**

SAMPLE MEAN ($\bar{X}$)	PROBABILITY
5.0	$\frac{1}{16}$
4.5	$\frac{2}{16}$
4.0	$\frac{3}{16}$
3.5	$\frac{4}{16}$
3.0	$\frac{3}{16}$
2.5	$\frac{2}{16}$
2.0	$\frac{1}{16}$

10.3 Some Important Symbols

Having established precisely what constitutes a sampling distribution under highly simplified conditions, let's introduce the special symbols that identify the mean and standard deviation of the sampling distribution of the mean. Table 10.3 also lists the corresponding symbols for the sample and the population. To facilitate your understanding of the next few chapters, it would be wise to memorize these symbols.

TYPE OF DISTRIBUTION	MEAN	STANDARD DEVIATION
Sample	$\bar{X}$	S
Population	μ	σ
Sampling distribution of the mean	$\mu_{\bar{x}}$	$\sigma_{\bar{x}}$ $\left(\begin{array}{c}\text{standard error}\\ \text{of the mean}\end{array}\right)$

**TABLE 10.3
SYMBOLS FOR MEAN
AND STANDARD
DEVIATION OF THREE
TYPES OF
DISTRIBUTIONS**

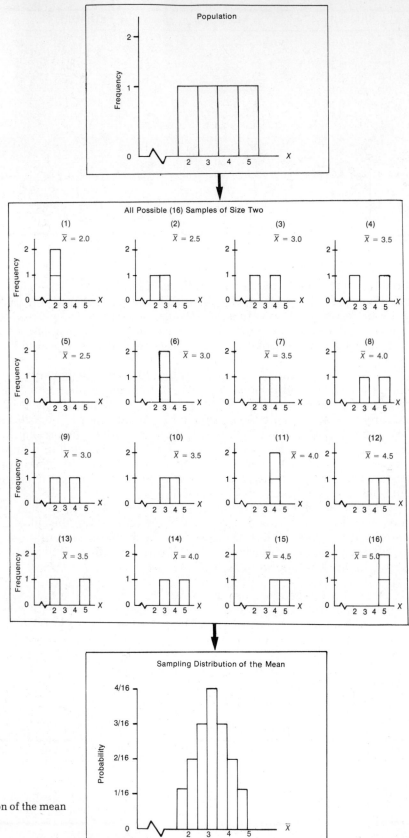

FIGURE 10.2

Emergence of the sampling distribution of the mean from all possible samples.

You're already acquainted with the English letters $\bar{X}$ and S, representing the mean and standard deviation of a sample. Totally new are the Greek letters μ (mu) and σ (sigma) representing the mean and standard deviation of the population. Also new are the Greek letters $\mu_{\bar{X}}$ and $\sigma_{\bar{X}}$, representing the mean of all sample means in the sampling distribution and the standard deviation of all sample means in the sampling distribution. To minimize confusion, the latter term, $\sigma_{\bar{X}}$, is often referred to as the standard error of the mean, or simply as the standard error.

Note that Greek letters are used to describe characteristics both of populations and of sampling distributions. This suggests a common feature. Both types of distribution deal with all possibilities, that is, with *all possible observations,* as in the case of the population, or with *all possible sample means,* as in the case of the sampling distribution of the mean.

With this background, we can identify the three important properties shared by many sampling distributions of the mean, including those encountered under more complex circumstances. In subsequent chapters, these three properties will form the basis for applied work in inferential statistics.

The distribution of sample means, itself, has a mean. *The mean of the sampling distribution always equals the mean of the population.* Expressed in symbols, we have

10.4 Mean of All Sample Means

$$\mu_{\bar{X}} = \mu \qquad (10.1)$$

where $\mu_{\bar{X}}$ represents the mean of the sampling distribution, and μ represents the mean of the population.

Since the mean of all sample means always equals the mean of the population, these two terms are used interchangeably in inferential statistics. Any claims about the population mean can be transferred directly to the mean of the sampling distribution, and vice versa. Thus, if the mean verbal score for the local population of freshmen equals the national average of 429, as claimed, then the mean of the sampling distribution also automatically equals 429. By the same token, it's permissible to view the one observed sample mean of 462 as a deviation either from the mean of the sampling distribution or from the mean of the population. It should be apparent, therefore, that whether an expression involves μ or $\mu_{\bar{X}}$ reflects, at most, a difference in emphasis on either the population or sampling distribution, respectively, rather than a difference in numerical value.

Although important, it's not particularly startling that the mean of all sample means equals the population mean. As can be seen in Figure 10.2, samples are not exact replicas of the population, and most sample means fail to coincide with the population mean (equal to 3.5 in Fig. 10.2). By taking the mean of all sample means, however, you effectively neutralize chance differences between sample means and retain a value equal to the population mean.

The distribution of sample means also has a standard deviation, referred to as the standard error of the mean. *The **standard error of the mean** equals the standard deviation of the population divided by the square root of the sample size.* Expressed in symbols,

10.5 Standard Error of the Mean

$$\sigma_{\bar{X}} = \frac{\sigma}{\sqrt{n}} \qquad (10.2)$$

where $\sigma_{\bar{X}}$ represents the standard error of the mean, σ represents the standard deviation of the population, and as usual, n represents the sample size.

The standard error of the mean serves as a special type of standard deviation that measures variability in the sampling distribution. It supplies us with a standard, much like a yardstick, that reflects the deviations of sample means about the mean of the sampling distribution. Because these deviations occur just by chance, whenever a sample fails to be an exact replica of the population, they are often referred to as "errors"—hence the term standard error of the mean. A majority of all sample means are within one standard error of the sampling distribution mean, and almost all sample means are within two standard errors of the sampling distribution mean.

A most important implication of Formula 10.2 is that whenever the sample size equals two or more, the variability of the sampling distribution is less than that in the population. A mild demonstration of this effect appears in Figure 10.2 where the means of all possible samples cluster closer to the population mean (equal to 3.5 in Fig. 10.2) than do the four observations in the population. A more dramatic demonstration occurs with larger sample sizes. Earlier in this chapter, for instance, 110 was given as the value of σ, the population standard deviation for verbal SAT scores. Much smaller is the variability in the sampling distribution among mean verbal scores, each based on samples of 100 freshmen. According to Formula 10.2,

$$\sigma_{\bar{X}} = \frac{\sigma}{\sqrt{n}} = \frac{110}{\sqrt{100}} = \frac{110}{10} = 11$$

In the present example, there is a tenfold reduction in variability from 110 to 11 when our focus shifts from the population to the sampling distribution.

According to Formula 10.2, any increase in sample size translates into a smaller standard error and, therefore, into a *new* sampling distribution with less variability. In other words, with a larger sample size, sample means cluster closer to the mean of the sampling distribution (and to the population mean), allowing more precise generalizations from samples to populations.

It's not surprising that variability should be less in sampling distributions than in populations. After all, the population standard deviation is directly affected by any relatively large or small observations within the population. The appearance of relatively large or small observations within a particular sample tends to affect the sample mean only slightly, because of the stabilizing presence in the same sample of other, more moderate observations or of extreme observations in the opposite direction. This stabilizing effect becomes even more pronounced with larger sample sizes.

10.6 Shape of the Sampling Distribution

A product of statistical theory known as the **central limit theorem** *states that the shape of the sampling distribution of the mean approximates a normal curve if sample size is sufficiently large.* According to this theorem, it doesn't matter whether the shape of the parent population is normal, positively skewed, negatively skewed, or some nameless, bizarre shape, as long as sample size is sufficiently large. What constitutes "sufficiently large" depends on the shape of the parent population. If the shape of the parent population is normal, then any sample size is sufficiently large. Otherwise, depending on the degree of non-normality in the parent population, a sample size between 25 and 100 is sufficiently large.

Given a population with a non-normal shape, as in Figure 10.2, the shape of the sampling distribution reveals a preliminary drift toward normality—that is, a shape having a peak in the middle with tapered flanks on either side—even for very small samples of size 2. Essentially this same drift describes the shapes of the sampling distributions, also based on samples of size 2, for two other populations with non-normal shapes, shown in Figure 10.3. When sample size equals 25, the shapes of these two sampling distributions closely approximate normality, as suggested in Figure 10.3.

Earlier in this chapter, 462 was given as the mean verbal score for a random sample of 100 freshmen. Because this sample size satisfies the requirements of the central limit theorem, we can view the sample mean of 462 as originating from a sampling distribution whose shape approximates a normal curve—even though we lack information about the shape of the population of verbal scores for the entire freshman class. It will be possible, therefore, to make precise statements about this sampling distribution, as described in the next chapter, by referring to the table for the standard normal curve.

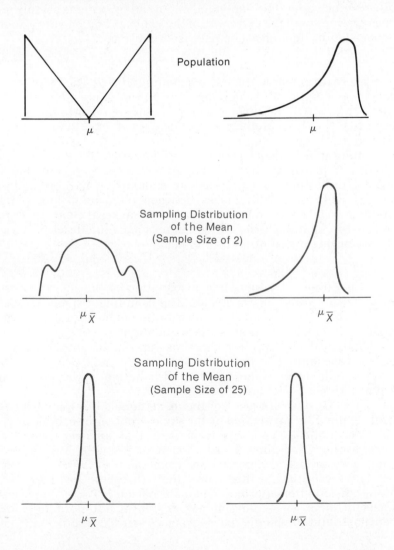

FIGURE 10.3

Effect of central limit theorem.

113

10.7 Why the Central Limit Theorem Works

In a normal curve, you'll recall, intermediate values are most prevalent while more extreme values, either larger or smaller, occupy the tapered flanks. Why, when sample size is large, does the distribution of sample means approximate a normal curve, even though the parent population might be non-normal?

When sample size is large, it's *most likely* that any single sample contains the full spectrum of small, intermediate, and large scores from the parent population *whatever its shape*. The calculation of a mean for this type of sample tends to neutralize or dilute the effects of any extreme scores, and the sample mean emerges with some intermediate value. Accordingly, intermediate values prevail in the distribution of sample means, and they cluster around a peak frequency representing the most common or modal value of the sample mean, as suggested in the bottom of Figure 10.3.

To account for the rarer sample mean values in the tails of the distribution, focus on those relatively infrequent samples that, just by chance, contain less than the full spectrum of scores from the parent population. Sometimes, because of the relatively large number of extreme scores in a particular direction, the calculation of a mean only slightly dilutes their effect, and the sample mean emerges with some more extreme value. The likelihood of obtaining extreme sample mean values declines with the extremity of the value, producing the smoothly tapered, slender tails that characterize a normal curve.

10.8 Types of Sampling Distributions

There are many different sampling distributions of means. Strictly speaking, a new sampling distribution is created either by a switch to a new population or to a new sample size. The population and sample size should, therefore, be specified for any particular sampling distribution of the mean.

Inferential statistics involves a wide variety of sampling distributions, not just those for means. For instance, there are sampling distributions for medians, proportions, standard deviations, and variances, as well as for differences between pairs of means, pairs of proportions, and so forth. Unless the context makes it clear, never refer simply to the sampling distribution but provide a full description, such as the sampling distribution of the mean.

10.9 Summary

The notion of a sampling distribution is the most important concept within inferential statistics. The sampling distribution of the mean is defined as the probability distribution of means for all possible samples of a given size from some population.

Statistical theory pinpoints three important properties of the sampling distribution of the mean:

(1) The mean of the sampling distribution equals the mean of the population.

(2) The standard error of the mean equals the standard deviation of the population, divided by the square root of the sample size. An important implication of this formula is that a larger sample size translates into a sampling distribution with a smaller variability, allowing more precise generalizations from samples to populations.

(3) According to the central limit theorem, the shape of the sampling distribution approximates a normal curve if the sample size is sufficiently large. Depending on the degree of non-normality in the parent population, a sample size between 25 and 100 is sufficiently large.

Any single sample mean can be viewed as originating from a sampling distribution whose mean equals the population mean (whatever its value); whose standard error equals the population standard deviation divided by the square root of the sample size; and whose shape approximates a normal curve (assuming that the sample size satisfies the requirements of the central limit theorem).

Shifts in either population or sample size create new sampling distributions. There are sampling distributions for a wide variety of measures, such as medians, proportions, and differences between pairs of sample means.

Important Terms & Symbols

Sampling distribution of the mean
Standard error of the mean ($\sigma_{\bar{X}}$)
Central limit theorem

1. Define the sampling distribution of the mean.

2. Imagine a very simple population consisting of only five observations: 2, 4, 6, 8, and 10.

    (a) List all possible samples of size two. (Hint: There are five possibilities on the first draw from the population and also five possibilities on the second draw from the population.)
    (b) Construct a relative frequency table showing the sampling distribution of the mean.
    (c) Given that the mean of this population equals 6, what is the mean of all sample means?
    (d) Given that the standard deviation of this population equals 2.83, what is the standard error of the mean?
    (e) Compare the shape of the sampling distribution with that of the population.
    (f) What's the probability of obtaining a sample mean larger than 7?

3. List the special symbol for the mean of the population __(a)__; mean of the sampling distribution of the mean __(b)__; mean of the sample __(c)__; standard error of the mean __(d)__; standard deviation of the sample __(e)__; standard deviation of the population __(f)__.

4. Someone claims that, since the mean of the sampling distribution equals the population mean, any single sample mean, therefore, must also equal the population mean. Any comments?

5. (a) A random sample of size 144 is taken from the local population of grade school children. Each child estimates the number of hours spent watching TV per week. The survey reveals a sample mean of 20 hours. At this point, what can be said about the sampling distribution of the mean?
    (b) Assume that, in fact, a standard deviation, σ, of 8 hours describes the TV estimates for the local population of school children. At this point, what can be said about the sampling distribution of the mean?
    (c) Assume that, in fact, a mean, μ, of 21 hours describes the TV estimates for the local population of school children. Now what can be said about the sampling distribution of the mean?

11
Introduction to Hypothesis Testing: the z Test

11.1 –An Example
11.2 –Hypothesis Testing: An Overview
11.3 –The z Test
11.4 –Statement of Problem
11.5 –Null Hypothesis (H_0)
11.6 –Alternative Hypothesis (H_1)
11.7 –Statistical Test
11.8 –Decision Rule
11.9 –Calculations
11.10–Decision
11.11–Interpretation
11.12–One-tailed and Two-tailed Tests
11.13–Choosing a Level of Significance (α)
11.14–Summary
11.15–Exercises

Want to test a hunch or hypothesis about some unknown population mean? Welcome to hypothesis testing, one of the two major areas in inferential statistics.

Under U.S. law, criminal suspects are presumed innocent until proven guilty. Under hypothesis testing procedures, the main hypothesis, often referred to as the null hypothesis, is presumed true until proven false. Once all evidence has been considered, a verdict is reached; the null hypothesis is either retained or rejected.

The evidence in a hypothesis test springs from the relationship between the one observed sample mean and the imaginary sampling distribution of the mean. If the observed sample mean is located inside predetermined boundaries (in the middle of the sampling distribution), it's judged to be a "probable outcome" under the hypothesis, and the hypothesis is retained. However, if the observed sample mean is located outside these predetermined boundaries, it's judged to be an "improbable outcome" under the hypothesis, and the hypothesis is rejected.

11.1   An Example

In the previous chapter, we postponed a test of the hypothesis that the mean SAT verbal score for all local freshmen equals the national average of 429. Now let's test this hypothesis, given a mean verbal score of 462 for a random sample of 100 freshmen.

If the hypothesis is true, then the sampling distribution of the mean—the distribution of means for all possible samples, each of size 100, from the local population of freshmen—is centered about the national average of 429. (Remember that the mean of the sampling distribution always equals the population mean.) In Figure 11.1, this sampling distribution is referred to as the *hypothesized* sampling distribution since its mean equals 429, the hypothesized mean verbal score for the local population of freshmen.

Anticipating the key role of the hypothesized sampling distribution in our hypothesis test, let's spotlight two more properties of this distribution. In Figure 11.1, vertical lines radiate out, at intervals of size 11, on either side of the hypothesized population mean of 429. These

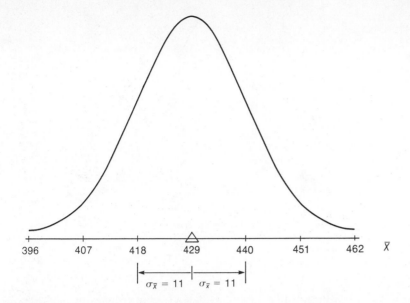

396   407   418   429   440   451   462 $\overline{X}$

$\sigma_{\overline{x}} = 11$ $\sigma_{\overline{x}} = 11$

FIGURE 11.1

Hypothesized sampling distribution of the mean ($\overline{X}$) centered about hypothesized population mean of 429.

intervals reflect the size of the standard error of the mean, $\sigma_{\overline{x}}$. To verify this fact, originally demonstrated in the previous chapter, substitute 110 for the population standard deviation, σ, and 100 for the sample size, n, in Formula 10.2 to obtain

$$\sigma_{\overline{X}} = \frac{\sigma}{\sqrt{n}} = \frac{110}{\sqrt{100}} = \frac{110}{10} = 11$$

Also notice that the shape of the hypothesized sampling distribution in Figure 11.1 approximates a normal curve since the sample size of 100 is large enough to satisfy the requirements of the central limit theorem. Eventually, with the aid of normal curve tables, we'll be able to construct boundaries for the range of sample mean values that represent probable (or improbable) outcomes under the hypothesis.

The hypothesis that the mean for the freshman class equals 429 tentatively is assumed to be true. It is tested by determining whether the one observed sample mean qualifies as a probable outcome or an improbable outcome in the hypothesized sampling distribution of Figure 11.1. *An observed sample mean qualifies as a probable outcome if the difference between its value and that of the hypothesized population mean is small enough to attribute to chance*—that is, if the sample mean doesn't deviate too far from the hypothesized population mean but appears to emerge from the dense concentration of possible sample means in the middle of the sampling distribution. Since, under these circumstances, there is no compelling reason for rejecting the hypothesis, the hypothesis is retained.

An observed sample mean qualifies as an improbable outcome if the difference between its value and the hypothesized value is too large to attribute to chance—that is, if the sample mean deviates too far from the hypothesized mean and appears to emerge from the sparse concentration of possible sample means in either "tail" of the sampling distribution. Since, under these circumstances, there are grounds for suspecting the hypothesis, the hypothesis is rejected.

Superimposed on the hypothesized sampling distribution in Figure 11.2 is one possible set of boundaries for the range of probable

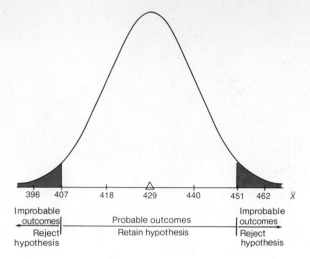

FIGURE 11.2

Hypothesized sampling distribution of the mean.

and improbable sample mean values. (Techniques for constructing these boundaries are described in Section 11.8.) If the one observed sample mean is located between 407 and 451, it qualifies as a probable outcome under the hypothesis, and the hypothesis is retained. If, however, the one observed sample mean is greater than or equal to 451 or less than or equal to 407, it qualifies as an improbable outcome under the hypothesis, and the hypothesis is rejected. Since, in fact, the observed sample mean of 462 exceeds 451, the hypothesis is rejected. On the basis of the present test, it's unreasonable to view the sample of 100 freshmen, with a mean verbal score of 462, as originating from a population whose mean equals the national average of 429. In other words, the investigator can conclude that the mean verbal score for the local population of freshmen doesn't equal the national average.

11.2 Hypothesis Testing: An Overview

Having rejected a hypothesis, we never know with absolute certainty whether that decision is correct or incorrect—unless, of course, we survey the entire population. Even if the hypothesis is true, there's a slight possibility that, just by chance, the one observed sample mean originates from the red rejection regions of the hypothesized sampling distribution, causing the true hypothesis to be rejected. This type of incorrect decision—rejecting a true hypothesis—is referred to as a type I error in Section 12.3.

On first impulse, it might seem desirable to abolish the red rejection regions in the hypothesized sampling distribution to insure that a true hypothesis is never rejected. A most unfortunate consequence of this strategy, however, is that no hypothesis, not even a radically false hypothesis, ever would be rejected. This type of incorrect decision—accepting a false hypothesis—is referred to as a type II error in Section 12.4.

Traditional hypothesis testing procedures, such as the one illustrated in Figure 11.2, tend to minimize both types of incorrect decisions. If the hypothesis is true, there is a high probability that the observed sample mean qualifies as a probable outcome under the hypothesis and that the true hypothesis will be retained. (This probability equals the proportion of white area in the hypothesized sampling distribution in Figure 11.2.) On the other hand, if the hypothesis is *seriously false,* because it differs considerably from the true population mean, there is also a high probability that the observed sample mean qualifies as an improbable outcome under the hypothesis and that the false hypothesis will be rejected. (This probability can't be determined from Figure 11.2 since, in this case, the hypothesized sampling distribution doesn't reflect the true sampling distribution. More about this in the next chapter.) Even though we never really know whether a particular decision is correct or incorrect, it's reassuring that, in the long run, most decisions will be correct—assuming the hypotheses are either true or seriously false.

11.3   The z Test

For the hypothesis test with SAT verbal scores, it's customary to base the test not on the hypothesized sampling distribution of $\bar{X}$ shown in Figure 11.2, but rather on its counterpart, the hypothesized sampling distribution of z shown in Figure 11.3. Now z represents a variation on the familiar standard score, and it displays all of the properties of distributions of standard scores described in Chapter 6. The conversion to z yields a sampling distribution that approximates the standard normal curve in Table A since, as indicated in Figure 11.3, the original hypothesized mean (429) emerges as 0 and the original standard error of the mean (11) emerges as 1. The shift from $\bar{X}$ to z eliminates the original units of measurement and standardizes the hypothesis testing procedure across all situations—without, however, affecting the results of the hypothesis test.

To convert a raw score into a standard score (also described in Chapter 6), express the raw score as a distance from its mean (by subtracting the mean from the raw score), and then split this distance into standard deviation units (by dividing with the standard deviation). Expressing this definition as a word formula, we have

$$\text{standard score} = \frac{\text{raw score} - \text{mean}}{\text{standard deviation}}$$

where, of course, the standard score indicates the deviation of the raw score, in standard deviation units, above or below the mean.

The z for the present situation emerges as a slight variation of this word formula. Replace the *raw score* with the one observed sample

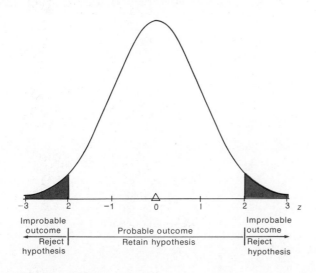

Improbable
outcome

Probable outcome

Improbable
outcome

Reject
hypothesis

Retain hypothesis

Reject
hypothesis

FIGURE 11.3

Hypothesized sampling distribution of z.

mean $\overline{X}$; replace the *mean* with the mean of the sampling distribution, that is, the hypothesized population mean μ_{hyp}; and replace the *standard deviation* with the standard error of the mean $\sigma_{\overline{X}}$. Now

$$z = \frac{\overline{X} - \mu_{hyp}}{\sigma_{\overline{X}}} \qquad (11.1)$$

where z indicates the deviation of the observed sample mean, in standard error units, above or below the hypothesized population mean.

To test the hypothesis for SAT verbal scores, we must determine the value of z from Formula 11.1. Given a sample mean of 462, a hypothesized population mean of 429, and a standard error of 11, we find

$$z = \frac{462 - 429}{11} = \frac{33}{11} = 3$$

The observed z of 3 exceeds the value of 2 specified in the hypothesized sampling distribution in Figure 11.3. Thus, the observed z qualifies as an improbable outcome under the hypothesis, and the hypothesis is rejected. The results of this test with z are the same as those for the original hypothesis test with $\overline{X}$.

When a hypothesis test is based on the sampling distribution of z, as in the present example, it's referred to as a **z test** *or,* more accurately, as a z test for a single population mean. The z test is accurate only when the population is normally distributed or the sample size is large enough to satisfy the requirements of the central limit theorem and the population standard deviation is known.

Having been exposed to some of the more important features of hypothesis testing, let's take a detailed look at the test for SAT verbal scores. The test procedure lends itself to a step-by-step description, beginning with a brief statement of the problem that inspired the test and ending with an interpretation of the test results. Whenever appropriate, this format will be used in the remainder of the book.

Hypothesis Test Summary:
z Test for a Population Mean
(SAT Verbal Scores)

Problem:
Does the mean SAT verbal score for all local freshmen equal the national average of 429?

Statistical Hypotheses:

$$H_0: \mu = 429$$
$$H_1: \mu \neq 429$$

Statistical Test:

$$z = \frac{\overline{X} - \mu_{hyp}}{\sigma_{\overline{X}}}$$

Decision Rule:
Reject H_0 at the .05 level of significance if $z \geq 1.96$ or $z \leq -1.96$ (where $\geq$ signifies greater than or equal to and $\leq$ signifies less than or equal to).

Calculations:

Given $\bar{X} = 462$; $\mu_{\text{hyp}} = 429$; $\sigma_{\bar{X}} = \dfrac{\sigma}{\sqrt{n}} = \dfrac{110}{\sqrt{100}} = 11$

$$z = \frac{462 - 429}{11} = 3$$

Decision:

Reject H_0.

Interpretation:

The mean SAT verbal score for all local freshmen does not equal—it exceeds—the national average of 429.

11.4 Statement of Problem

The formulation of a problem often represents the most crucial and exciting phase of an investigation. Indeed, the mark of a skillful investigator is to focus on an important problem that can be answered. Because of our emphasis on hypothesis testing, problems appear in this book as finished products, usually in the first one or two sentences of a new example.

11.5 Null Hypothesis (H_0)

Once the problem has been described, it must be translated into a statistical hypothesis about some population characteristic. Designated as the null hypothesis, this hypothesis becomes the focal point for the entire test procedure. In the test with SAT verbal scores, the null hypothesis asserts that the mean verbal score for the local population of freshmen equals the national average of 429. An equivalent statement, in symbols, reads

$$H_0\text{: } \mu = 429$$

where H_0 represents the null hypothesis and μ is the population mean for the local freshman class.

The **null hypothesis H_0** *supplies the value about which the hypothesized sampling distribution is centered.* It always makes a statement about a characteristic of the population, never about a characteristic of the sample. (The purpose of a hypothesis test, remember, is to determine the likelihood that a particular sample could have originated from a population with the hypothesized characteristic.) Furthermore, the null hypothesis always makes a claim about a single numerical value, never a range of values. In effect, the test procedure accommodates only one hypothesized sampling distribution centered about the single number in H_0—or its equivalent value of 0 along the z scale.

The number actually used in H_0 varies, of course, from problem to problem. Even for a given problem, this number could originate from any of several sources. For instance, it could be based on available information about some relevant population other than the target population, as in the present example, where 429 reflects the mean SAT verbal scores for all college-bound students during a recent year. It also could be based on some existing standard or theory—for example, that the current mean verbal score of local freshmen should equal 475 because that was the mean score achieved by local freshmen during the last five years.

If, as sometimes happens, it's impossible to identify a meaningful null hypothesis, don't try to salvage the situation with arbitrary numbers. Instead use another entirely different technique known as estimation, which is described in Chapter 16.

11.6 Alternative Hypothesis (H_1)

In the present example, the alternative hypothesis asserts that the mean verbal score for the local population of freshmen does not equal the national average of 429. An equivalent statement, in symbols, reads

$$H_1: \mu \neq 429$$

where H_1 represents the alternative hypothesis, μ is the population mean for the local freshmen class, and $\neq$ indicates not equal to.

Generally speaking, the **alternative hypothesis H_1** *asserts the opposite of the null hypothesis.* A decision to retain the null hypothesis implies a lack of support for the alternative hypothesis, while a decision to reject the null hypothesis implies support for the alternative hypothesis.

As will be described in Section 11.12, the alternative hypothesis may assume any one of three different forms, depending on the perspective of the investigator. Regardless of its particular form, however, H_1 always specifies a range of possible values about the single number (429) in H_0.

11.7 Statistical Test

To determine whether the observed sample mean (462) qualifies as a probable or improbable outcome under H_0 (429), we must consult an appropriate statistical test. In the present case, given that the sample size (100) is large enough to satisfy the requirements of the *central limit theorem, and that the population standard deviation* (110) is known, the z test is appropriate. The one observed sample mean $\bar{X}$ can be expressed as z with the aid of Formula 11.1:

$$z = \frac{\bar{X} - \mu_{\text{hyp}}}{\sigma_{\bar{X}}}$$

where all terms have been defined in previous sections. If H_0 is true, the resulting z originates from a sampling distribution that approximates the standard normal curve in Table A.

11.8 Decision Rule

A **decision rule** *specifies precisely when H_0 should be retained* (because the observed z qualifies as a probable outcome), *and when H_0 should be rejected* (because the observed z qualifies as an improbable outcome). There are many possible decision rules. One very common decision rule specifies that H_0 should be rejected if the observed z equals or is more positive than 1.96, or if the observed z equals or is more negative than -1.96. In other words, H_0 should be retained if the observed z falls between -1.96 and 1.96. Heretofore, for the sake of simplicity, this decision rule only has been approximated. Thus, -1.96 and 1.96 should replace -2 and 2, respectively, in Figure 11.3.

Figure 11.4 indicates that this pair of z scores, -1.96 and 1.96, are boundaries for the middle .95 of the total area (1.00) under the hypothesized sampling distribution for z. Derived from the normal curve table, as you can verify by checking Table A, these two z scores *separate probable from improbable outcomes* and hence dictate whether H_0 should be retained or rejected. Because of their vital role in the decision about H_0, these scores are referred to as **critical z scores.**

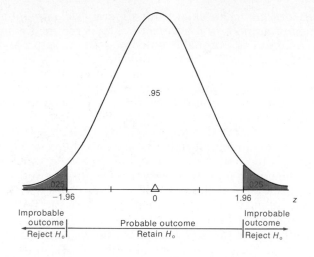

FIGURE 11.4

Hypothesized sampling distribution of z ($\alpha = .05$)

Figure 11.4 also indicates *the proportion* (.025 + .025 = .05) *of total area under the sampling distribution that is identified with improbable outcomes.* This proportion is often referred to as the **level of significance** of the statistical test and symbolized by the Greek letter **α** (alpha). In the present example, the level of significance, α, equals .05.

The level of significance indicates the degree of rarity that, if achieved by an observed z, triggers the rejection of H_0. For instance, the .05 level of significance indicates that H_0 should be rejected if the observed z could have occurred just by chance, with a probability of only .05 (one chance out of twenty) *or less.*

11.9 Calculations

Having specified both the statistical test and the decision rule, we can use information from the sample to calculate a value for z. As has been noted previously, use Formula 11.1 to convert the observed sample mean of 462 to a z of 3.

11.10 Decision

Either retain or reject H_0, depending on the location of the observed z value relative to the critical z values specified in the decision rule. According to the present rule, H_0 should be rejected if the observed z equals or exceeds a critical z of 1.96. The observed z is 3. Therefore, we reject H_0 since it's unlikely that the observed z originates from the hypothesized sampling distribution.

11.11 Interpretation

Finally, interpret the decision in terms of the original problem. In the present example, it can be concluded that the mean SAT verbal score for the local freshman class doesn't equal the national average of 429.

Although not a strict consequence of the present test, a more specific conclusion is possible. Since the sample mean of 462 (or its equivalent z of 3) falls in the upper rejection region of the hypothesized sampling distribution, it can be concluded that the mean SAT verbal score of the local freshman class *exceeds* the national average of 429. By the same token, if the observed sample mean or its equivalent z had fallen in the lower rejection region of the hypothesized sampling distribution, it could have been concluded that the mean for all local freshmen is exceeded by the national average.

There are subtle, but important, differences in the interpretation of decisions to retain H_0 and to reject H_0. H_0 is retained whenever the observed z qualifies as a probable outcome—on the assumption that H_0 is true. Therefore, H_0 *could* be true. But the same sample result would also qualify as a probable outcome when the original value in H_0 (429) is replaced with a similar value. Thus, retention of H_0 must be viewed as a relatively weak decision. Many statisticians prefer to describe this decision as simply a *failure to reject H_0,* rather than as the retention of H_0. If H_0 had been retained in the present example, it would have been appropriate to conclude, not that the mean verbal score for all local freshmen equals the national average, but that the mean verbal score *could* equal the national average, as well as many other possible values in the general vicinity of the national average.

On the other hand, H_0 is rejected whenever the observed z qualifies as an improbable outcome—one that could have occurred just by chance with a probability of .05 or less—on the assumption that H_0 is true. This suspiciously rare outcome implies that H_0 must not be true. Rejection of H_0 can be viewed as a strong decision. When H_0 was rejected in the present example, it was appropriate to report a definitive conclusion that the mean verbal score for all local freshmen does not equal the national average.

Having identified the basic format for hypothesis testing, let's consider some techniques that can make the test more responsive to special conditions. These techniques should be selected before the collection of data.

11.12 One-tailed and Two-tailed Tests

Generally speaking, the alternative hypothesis, H_1, is the complement of the null hypothesis, H_0. Under typical conditions, the form of H_1 resembles that shown for the previous hypothesis test, namely,

$$H_1: \mu \neq 429$$

This alternative hypothesis says that the null hypothesis should be rejected if the mean verbal score for all local freshmen differs in either direction from the national average of 429. An observed sample mean qualifies as an improbable outcome if it deviates either too far below or too far above the national average. Figure 11.5A shows rejection regions that are associated with both tails of the hypothesized sampling distribution. The corresponding decision rule, with its pair of critical z scores of -1.96 and 1.96, is referred to as a **two-tailed** or **non-directional test.**

Now let's assume that the investigation of SAT verbal scores was prompted by complaints from instructors about the poor preparation of local freshmen. Let's assume also that if the investigation supports these complaints, a remedial program will be instituted. Under these circumstances, the investigator might prefer a hypothesis test that is specially designed to detect only whether the mean verbal score for all local freshmen *is less than* the nation average.

This alternative hypothesis reads:

$$H_1: \mu < 429$$

It reflects a concern that the null hypothesis be rejected only if the mean verbal score for all local freshmen is less than the national average of 429. Accordingly, an observed sample mean triggers the decision to reject H_0 only if it deviates too far below the national average. Figure 11.5B illustrates a rejection region that is associated with only the lower tail of the hypothesized sampling distribution. The corresponding

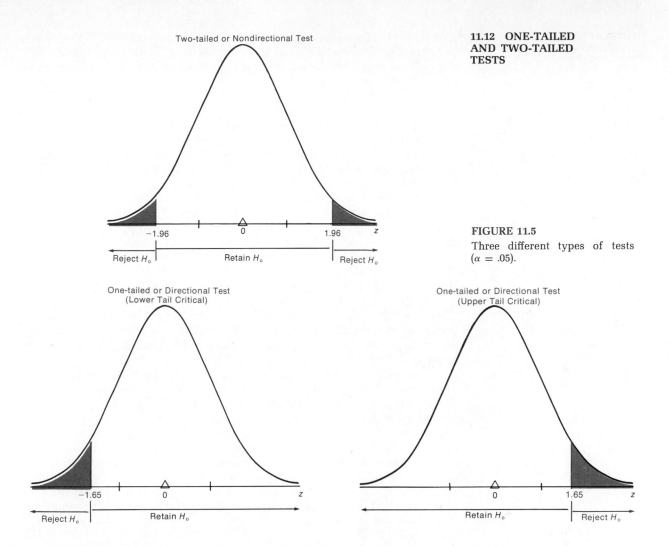

FIGURE 11.5
Three different types of tests ($\alpha = .05$).

decision rule, with its critical z of -1.65, is referred to as a **one-tailed** or **directional test** *with the lower tail critical.* Use Table A to verify that, if the critical z equals -1.65, then .05 of the total area under the sampling distribution has been allocated to the lower rejection region. Notice that the level of significance, α, equals .05 for this one-tailed test and also for the original two-tailed test.

This new one-tailed test is extra sensitive to any drop in the mean for all local freshmen below the national average. If, in fact, H_0 is false because a drop has occurred, then the observed sample mean is more likely to deviate below the national average. As can be seen in Figure 11.5A and B, an observed deviation in the direction of concern—below the national average—is more likely to penetrate the broader rejection region for the one-tailed test than that for the two-tailed test. Therefore, the decision to reject a *false* H_0 is more likely to occur in the one-tailed test than in the two-tailed test.

Figure 11.5C illustrates a **one-tailed** *or* **directional test** *with the upper tail critical.* This one-tailed test is the mirror image of the previous test. Now the alternative hypothesis reads:

$$H_1: \mu > 429$$

and its critical z equals 1.65. This test is specially designed to detect only whether the mean verbal score for all local freshmen *exceeds* the national average. For example, the investigation might have been prompted by an alumnus who will donate a large sum of money if it can be demonstrated that, on the average, the SAT verbal scores of all local freshmen exceed the national average.

If, prior to a hypothesis test, there's an exclusive concern that the true population mean differs from the hypothesized population mean in a particular direction, use the appropriate one-tailed or directional test for extra sensitivity. Otherwise, use the more customary two-tailed or non-directional test.

Having committed yourself to a one-tailed test, with its single rejection region, you must retain H_0 regardless of how far the observed sample mean deviates from the hypothesized sample mean in the direction of "no concern." For instance, if a one-tailed test with the lower tail critical had been used with the data for 100 freshmen from the previous example, H_0 would have been retained because, even though the observed z equals an impressive value of 3, it deviates in the direction of no concern—in this case, above the national average. Clearly, a one-tailed test should be adopted only when there is absolutely no concern about deviations, even very large deviations, in one direction. If there is the slightest concern about these deviations, use a two-tailed test.

Ideally, the selection of a one- or two-tailed test should be made before the collection of data, but in any event, it shouldn't be influenced by the value of the observed z. Never "peek" at the value of the observed z to determine whether to locate the rejection region for a one-tailed test in the upper or lower tail of the sampling distribution. To qualify as a one-tailed test, the location of the rejection region must reflect the investigator's exclusive concern, *prior to any inspection of the data,* about deviations in a particular direction. Indeed, the investigator should be able to muster a compelling reason—based on an understanding of the problem being tested—to support the direction of the one-tailed test.

11.13 Choosing a Level of Significance (α)

The level of significance indicates how rare an observed z must be before H_0 can be rejected. To reject H_0 at the .05 level of significance implies that the observed z could have occurred just by chance with a probability of only .05 (one chance out of twenty) *or less.*

The level of significance also spotlights an inherent risk in hypothesis testing—the risk of rejecting a true H_0. When the level of significance equals .05, there is a probability of .05 that the observed z will stray into the rejection region and cause a true H_0 to be rejected.

When the rejection of a true H_0 is particularly serious, some smaller level of significance can be selected. For example, the .01 level of significance implies that, before H_0 can be rejected, the observed z must achieve a degree of rarity equal to .01 (one chance out of one hundred) or less; it also limits, to a probability of .01, the risk of rejecting a true H_0. For example, the .01 level might be used in a hypothesis test where the rejection of a true H_0 would cause the introduction of a costly new remedial education program even though, in fact, the mean verbal score for all local freshmen equals the national average. An even smaller level of significance, such as the .001 level, might be used when the rejection of a true H_0 has horrendous consequences—for instance, the public distribution of a new drug even though human consumption is sometimes fatal.

Although many different levels of significance are possible, most tables for hypothesis tests are geared to the .05 and .01 levels. In this book, the level of significance will be specified for you. In real life applications, you, as investigator, will have to select a level of significance. *Unless there are obvious reasons for selecting either a larger or smaller level of significance, use the customary .05 level*—the largest level of significance reported in most professional journals.

When testing hypotheses with the z test, you may find it helpful to refer to Table 11.1, which lists the critical z values for one- and two-tailed tests at the .05 and .01 levels of significance. These z values were obtained from Table A.

11.14   Summary

To test a hypothesis about the population mean, a single observed sample mean is viewed within the context of a hypothesized sampling distribution, which is centered about the hypothesized population mean. If the sample mean appears to emerge from the dense concentration of possible sample means in the hypothesized sampling distribution, it qualifies as a probable outcome, and the hypothesis is retained. On the other hand, if the sample mean appears to emerge from the sparse concentration of possible sample means in the hypothesized sampling distribution, it qualifies as an improbable outcome, and the hypothesis is rejected.

Even though we never know whether a particular decision is correct or incorrect, it's reassuring that, in the long run, most decisions will be correct—assuming the hypotheses are either true or seriously false.

Hypothesis tests are based not on the sampling distribution of $\bar{X}$, expressed in original units of measurement, but on its counterpart, the sampling distribution of z.

When testing a hypothesis, adopt a step-by-step procedure, beginning with a statement of the problem and ending with an interpretation of the test results. Intermediate steps are as follows:

Identify the Statistical Hypotheses.   First specify the null hypothesis, expressed symbolically as

$$H_0: \mu = \text{some number}$$

The null hypothesis supplies the value about which the hypothesized sampling distribution is centered. Depending on the outcome of the hypothesis test, H_0 is either retained or rejected.

TYPE OF TEST	LEVEL OF SIGNIFICANCE (α)	
	.05	.01
Two-tailed or non-directional test	± 1.96	± 2.58
($H_1: \mu \neq$ some number)		
One-tailed or directional test with lower tail critical	-1.65	-2.33
($H_1: \mu <$ some number)		
One-tailed or directional test with upper tail critical	$+1.65$	$+2.33$
($H_1: \mu >$ some number)		

TABLE 11.1
CRITICAL z VALUES

Then select the alternative hypothesis from among the following three possibilities:

$$(1) \ H_1: \mu \neq \text{some number}$$

This expression reflects a concern that the true population mean doesn't equal some hypothesized number, and it translates into a decision rule for a two-tailed or non-directional test.

$$(2) \ H_1: \mu < \text{some number}$$

This expression reflects an exclusive concern that the true population mean *falls below* the hypothesized number, and it translates into a decision rule for a one-tailed or directional test with the lower tail critical.

$$(3) \ H_1: \mu > \text{some number}$$

This expression reflects an exclusive concern that the true population mean *exceeds* the hypothesized number, and it translates into a decision rule for a one-tailed or directional test with the upper tail critical.

Use the more sensitive one-tailed test only when, prior to an investigation, there's an exclusive concern about deviations in a particular direction. Otherwise, use a two-tailed test.

Specify the Statistical Test. In the present chapter, the z test is appropriate. This test assumes that the sample size is large enough to invoke the central limit theorem and that the population standard deviation is known.

Specify a Decision Rule. This rule indicates precisely when H_0 should be rejected. The exact form of the decision rule depends on the type of test (whether two-tailed, one-tailed with the lower tail critical, or one-tailed with the upper tail critical) and on the level of significance (whether .05 or .01). In any event, H_0 is rejected whenever the observed z deviates further from 0 than the critical z.

The level of significance indicates how rare an observed z must be (on the assumption that H_0 is true) before H_0 can be rejected. Unless there are obvious reasons for selecting either a larger or smaller level of significance, use the customary .05 level.

Calculate the Value of the Observed z. Express the one observed sample mean as an observed z, using Formula 11.1.

Make a Decision. Either retain or reject H_0 depending on the location of the observed z relative to the critical z scores.

Important Terms & Symbols ━━━━━━━━━━━━━━━━━━━━

z test
Null hypothesis (H_0)
Alternative hypothesis (H_1)
Decision rule
Critical z scores
Level of significance (α)
Two-tailed (or non-directional) test
One-tailed (or directional) test

11.15 Exercises

1. According to the American Psychological Association, members with a Ph.D. degree and a full-time teaching appointment earn, on the average, $21,500 per year, with a standard deviation of $3,000. An investigator wishes to determine whether $21,500 also describes the mean salary for all female

members with a Ph.D. degree and a full-time teaching appointment. Salaries are obtained for a random sample of 100 women from this population, and the mean salary equals $20,300. Using the step-by-step procedure described in this chapter, test the null hypothesis at the .05 level of significance.

2. For the population at large, the Wechsler Intelligence Scale is designed to yield a normal distribution of test scores with a mean of 100 and a standard deviation of 15. School district officials wonder whether, on the average, an IQ score of 100 describes the intellectual aptitudes of all students within their district. Accordingly, Wechsler IQ scores are obtained for a random sample of 25 of their students, and the mean IQ is found to equal 105. Using the customary procedure, test the null hypothesis at the .05 level of significance.

3. An assembly line at a candy plant is designed to yield two-pound boxes of assorted candies whose weights, in fact, follow a normal distribution with a mean of 33 ounces and a standard deviation of .30 ounce. A random sample of 36 boxes from the production of the most recent shift reveals a mean weight of 33.09 ounces. Test the null hypothesis at the .05 level of significance.

4. According to a recent government survey, the daily one-way commuting distance of U.S. workers averages 13 miles with a standard deviation of 13 miles. An investigator wishes to determine whether the national average describes the mean commuting distance for all workers in the Chicago area. Commuting distances are obtained for a random sample of 169 workers from this area, and the mean distance is found to be 14.5 miles. Test the null hypothesis at the .05 level of significance.

5. It has been suggested that the null hypothesis always states a claim about a population, *never about a sample*. Someone insists, nevertheless, on testing a null hypothesis that the population mean equals the observed value of the sample mean. What's wrong with this procedure?

6. Why not conclude that the null hypothesis is true when—as a result of a hypothesis test—the null hypothesis is retained?

7. Each of the following statements could represent the point of departure for a hypothesis test. Given only the information in each statement, would you use a two-tailed (or non-directional) test; one-tailed (or directional) test with the lower tail critical; or a one-tailed (or directional) test with the upper tail critical? Indicate your decision by specifying the appropriate H_1.

   (a) To increase rainfall, extensive cloud-seeding experiments are to be conducted, and the results are to be compared with a baseline figure of 0.54 inch of rainfall (for comparible periods when cloud seeding wasn't done).
   (b) Public health statistics indicate, we'll assume, that American males gain an average of 23 pounds during the twenty-year period after age forty. An ambitious weight-reduction program, spanning twenty years, is being tested with a random sample of forty-year-old men.
   (c) When untreated during their lifetimes, cancer-susceptible mice have an average lifespan of 134 days. To determine the effects of a potentially life-prolonging (and cancer-retarding) drug, the average lifespan is determined for a randomly selected group of mice that receives this drug.

8. Supply the missing word(s) in the following statements.

   If the one observed sample mean can be viewed as a(n) __(a)__ outcome under the hypothesis, H_0 is __(b)__. Otherwise, if the one observed sample mean can be viewed as a(n) __(c)__ outcome under the hypothesis, H_0 is __(d)__.

The pair of z scores that separate probable and improbable outcomes are referred to as __(e)__ z scores. Within the hypothesized sampling distribution, the proportion of area allocated to improbable outcomes is referred to as the _/(f)/_ and symbolized with the Greek letter __(g)__.

When based on the sampling distribution of z, the hypothesis test is referred to as a __(h)__ test. This test is appropriate only if sample size is sufficiently large to satisfy the _/(i)/_ and if the _/(j)/_ is known.

If prior to a hypothesis test, there's an exclusive concern about deviations in a particular direction, use a _/(k)_ or __(l)__ test. Otherwise, use a __(m)/__ or __(n)__ test.

If H_0 is rejected at the .05 level of significance, this implies that, if H_0 were __(o)__, the observed z could have occurred just by chance with a probability of only __(p)__ or less.

When the rejection of a true H_0 is particularly serious, use the __(q)__ level of significance, or even the __(r)__ level of significance.

12

Two Types of Error and Sample Size Selection

12.1 –Vitamin C Experiment
12.2 –Four Possible Outcomes
12.3 –H_0 is True (& the Type I Error)
12.4 –H_0 is False (& the Type II Error)
12.5 –Influence of Sample Size
12.6 –Selection of Sample Size
12.7 –Small, Medium, and Large Effects
12.8 –Additional Comments about the Sample Size Table
12.9 –Summary
12.10–Exercises

When testing a null hypothesis, our object is to retain a true hypothesis. We also hope to reject a false hypothesis—particularly a seriously false one. In well-designed hypothesis tests, two types of errors are controlled. You can consult a table to answer the common question: "How large should my sample size be?"

Personal experience, loosely conducted experiments, appeals to authority—all have been used to justify the daily ingestion of vitamin C. An investigator wishes to test a hunch that vitamin C increases the intellectual aptitude of high school students. After being randomly selected from some large school district, each of 36 students takes a daily dose of 50 milligrams of vitamin C (the commonly recommended dosage) for a period of two months, prior to being tested on the Wechsler Intelligence Scale.

12.1 Vitamin C Experiment

Ordinarily, Wechsler IQ scores for all students in this school district approximate a normal distribution with a mean of 100 and a standard deviation of 15. According to the null hypothesis, a mean of 100 still would describe the distribution of IQ scores even if all students in the district were to receive the vitamin C treatment, that is,

$$H_0: \mu = 100$$

Rejection of H_0 would support the investigator's hunch that vitamin C increases intellectual aptitude.

To determine whether the sample mean IQ for the 36 students qualifies as a probable or improbable outcome under the null hypothesis, a z test will be used. A z test is appropriate since the population standard deviation is known to equal 15 and the sample size of 36 is sufficiently large—indeed, when the population approximates a normal distribution, as in the present case, any sample size is sufficiently large—to satisfy the central limit theorem.

Although poorly designed, the present experiment supplies a perspective that will be most useful in later chapters. A better designed experiment would contrast the IQ scores of experimental subjects, who receive vitamin C, with the IQ scores of control subjects, who receive fake vitamin C. Hypothesis tests for this type of experiment are described in Chapter 14.

Summarized in the box below are those features of the hypothesis test that can be identified before the collection of any data for the vitamin C experiment. The directional H_1 reflects an exclusive concern that H_0 be rejected only if the true population mean exceeds the hypothesized population mean of 100. In effect, H_0 is to be rejected only if there is evidence that vitamin C increases the mean IQ for all students in the district.

Hypothesis Test Summary:
z Test for a Population Mean
(Prior to Vitamin C Experiment)

Problem:

Does the daily ingestion of vitamin C cause an increase, on the average, among Wechsler IQ scores of all students in the school district?

Statistical Hypotheses:

$$H_0: \mu = 100$$
$$H_1: \mu > 100$$

Statistical Test:

$$z = \frac{\bar{X} - \mu_{\text{hyp}}}{\sigma_{\bar{X}}}$$

Decision Rule:

Reject H_0 at the .05 level of significance if $z \geq 1.65$.

Calculations:

$$\sigma_{\bar{X}} = \frac{\sigma}{\sqrt{n}} = \frac{15}{\sqrt{36}} = \frac{15}{6} = 2.5$$

When conducting a hypothesis test, we hope to make a correct decision. If vitamin C increases IQ, we hope to reject H_0; otherwise, we hope to retain H_0. It's been suggested that, generally speaking, hypothesis tests tend to produce correct decisions when the hypothesis is either true or seriously false. This claim will be examined in the context of the vitamin C experiment, as also will be the claim, mentioned briefly in Chapter 11, that there are two types of incorrect decisions—the type I and type II errors. First, however, let's consider, one at a time, each of the four possible outcomes for any hypothesis test, as summarized in Table 12.1.

**TABLE 12.1
POSSIBLE
OUTCOMES OF
HYPOTHESIS TEST**

	STATUS OF H_0	
DECISION	True H_0	False H_0
Retain H_0	Correct decision	Type II error (Miss)
Reject H_0	Type I error (False alarm)	Correct decision

(1) If H_0 is true (because vitamin C doesn't cause an increase in IQ), then *it's a correct decision to retain the true* H_0 (and conclude that there's no evidence that vitamin C increases IQ).

(2) If H_0 is true, then *it's a* **type I error** *to reject the true* H_0 (and conclude that vitamin C increases IQ when, in fact, it doesn't). Type I errors are sometimes called "false alarms" because, as with their firehouse counterparts, they trigger wild goose chases after something that doesn't exist. For instance, a type I error might encourage a batch of worthless experimental efforts to discover precisely what dosage level of vitamin C maximizes the "increase" in IQ.

(3) If H_0 is false (because vitamin C causes an increase in IQ), then *it's a* **type II error** *to retain the false* H_0 (and conclude that there is a lack of evidence that vitamin C increases IQ when, in fact, it does). Type II errors are sometimes called "misses" because they fail to detect potentially important relationships (such as between vitamin C and IQ).

(4) If H_0 is false, then *it's a correct decision to reject the false* H_0 (and conclude that vitamin C increases IQ).

Specification of these four possible outcomes presumes that we know whether H_0 is true or false. In practice, of course, we never know whether we're testing a true H_0 or a false H_0—that's the reason for the hypothesis test. Well-designed hypothesis tests must be concerned with both the probability of a type I error, in the event that H_0 is true, and the probability of a type II error, in the event that H_0 is false. Subsequent sections describe how this is accomplished—assuming first that H_0 is true and then that H_0 is false. Although you might view this approach as hopelessly theoretical, read the next few sections carefully, for they have important practical implications for any hypothesis test.

I HOPE THIS ISN'T A TYPE I ERROR

Pretend we have special knowledge that, in fact, vitamin C doesn't increase IQ, that is, H_0 is true. Nevertheless, as a theoretical exercise, we wish to check the properties of the projected one-tailed hypothesis test described in Section 12.1. Under these special circumstances, the hypothesized sampling distribution shown in Figure 12.1 also qualifies as the one true sampling distribution from which will originate the one sample mean (or z) observed in the experiment.

When a randomly selected sample mean originates from the small red portion of the sampling distribution, its z value equals or exceeds 1.65, and hence H_0 is rejected. Since H_0 is true, this is a type I

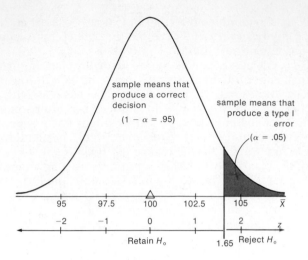

FIGURE 12.1

H_0 is true (vitamin C causes no increase in IQ).

error—a false alarm that vitamin C increases IQ even though, in fact, it doesn't. The probability of a type I error simply equals α, the level of significance. (The level of significance, remember, indicates the proportion of area under the sampling distribution in the rejection region for H_0.) In the present case, the probability of a type I error equals .05, as indicated in Figure 12.1.

When a sample mean originates from the large white portion of the sampling distribution, its z value is less than 1.65, and H_0 is retained. Since H_0 is true, this is a correct decision—announced as a lack of evidence that vitamin C increases IQ. The probability of a correct decision equals $1 - \alpha$, that is, .95.

Thus, if H_0 is true, the present test will produce a correct decision with a probability .95 and a type I error with probability .05. If a false alarm has serious consequences, the probability of a type I error can be reduced to .01 or even to .001 simply by using the .01 or .001 levels of significance, respectively. One of these levels of significance might be preferred for the vitamin C test if, for instance, a false alarm could cause the adoption of an expensive program to supply vitamin C to all students in the district.

To summarize, if H_0 is true, correct decisions tend to occur much more frequently than type I errors. A type I error occurs only when, just by chance, a sample mean strays into the rejection region for the true H_0. The **probability of a type I error, α,** *always equals the level of significance.* Therefore, the level of significance indicates the degree of control over the type I error.

12.4 H_0 is False (& the Type II Error)

Pretend we have special knowledge that H_0 is false because, in fact, vitamin C increases IQ. Furthermore, by most standards H_0 is "seriously false" because vitamin C produces a ten-point increase in the mean IQ for all students in the district. (Remember, we're describing the mean for an entire population, not just a sample.) As before, we wish to check the properties of the projected one-tailed hypothesis test.

Under these special circumstances, as shown in Figure 12.2, there are two different sampling distributions—the *hypothesized* sampling distribution centered about the hypothesized population mean of 100 and the *true* sampling distribution centered about the true population mean of 110. The hypothesized sampling distribution fades into the

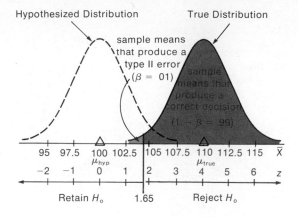

FIGURE 12.2

H_o is "seriously false" (vitamin C causes large increase in IQ).

background after serving as the basis for the familiar decision rule with a critical z of 1.65. Henceforth, the true sampling distribution attracts total attention because it serves as the source for the one sample mean (or z) that will be observed in the experiment.

When a randomly selected sample mean originates from the very small black portion of the true sampling distribution, its z value is less than 1.65, and H_0 is retained. Since H_0 is false, this is an incorrect decision or type II error: it implies a lack of evidence that vitamin C increases IQ. With the aid of tables for the normal curve, it can be demonstrated that the probability of a type II error, symbolized by the Greek letter β (beta), equals .01.

(The present argument doesn't require that you know how to calculate this probability of .01 or those given in the remainder of the chapter. In brief, these probabilities represent areas under the true sampling distribution found by expressing the critical z as a deviation from the true population mean (110) rather than from the hypothesized population mean (100). This issue is pursued in Exercise 4 at the end of the chapter.)

When a sample mean originates from the large red portion of the true sampling distribution, its z value equals or exceeds 1.65, and H_0 is rejected. Since H_0 is false, this is a correct decision: it tells us that vitamin C does, in fact, increase IQ. The probability of a correct decision, symbolized as $1 - \beta$, equals .99.

If H_0 is seriously false, because vitamin C increases the population mean IQ by 10 points, the projected one-tailed test does quite well. There's a high probability of .99 that a correct decision will be made, and the probability of a type II error, β, equals only .01 for this same test. Certainly these error rates justify the earlier claim that hypothesis tests tend to produce correct decisions when the hypothesis is either true or seriously false.

The projected hypothesis test doesn't fare nearly as well if H_0 is only "mildly false"—if, for instance, vitamin C increases the population mean IQ by only three points. Once again, as indicated in Figure 12.3, there are two different sampling distributions—the *hypothesized* sampling distribution centered about the hypothesized population mean of 100 and the *true* sampling distribution centered about the true population mean of 103. After the customary decision rule has been constructed with the aid of the hypothesized sampling distribution, attention is shifted to the true sampling distribution from which the one observed sample mean actually will originate.

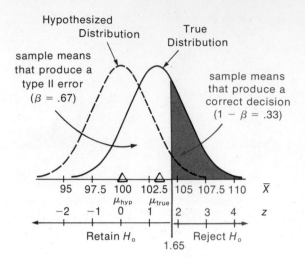

FIGURE 12.3

H_0 is "mildly false" (vitamin C causes small increase in IQ).

As before, when a randomly selected sample mean originates from the black portion of the true sampling distribution, its z value is less than 1.65, and H_0 is retained. Since H_0 is false, this is a type II error, and it occurs with probability .67. On the other hand, when a sample mean originates from the red portion of the true sampling distribution, its z value equals or exceeds 1.65, and H_0 is rejected. Since H_0 is false, this is a correct decision, and it occurs with probability .33. Clearly, when H_0 is only mildly false, the probability of a type II error is quite high. The small difference of three points between the hypothesized and true population means is missed two out of three times.

To summarize, the **probability of a type II error, β,** *depends on the size of the difference between true and hypothesized population means.* If H_0 is false, a type II error is committed whenever, just by chance, a sample mean strays into the retention region for the false H_0. This is less likely to occur—and, therefore, β is small—when H_0 is seriously false (because of a considerable difference between true and hypothesized population means). It is more likely to occur when H_0 is only mildly false (because of a slight difference between true and hypothesized population means). As will be described in the next section, however, if it's important to reject even a mildly false H_0, then β can be reduced to any desired value merely by increasing the size of the sample.

12.5 Influence of Sample Size

Ordinarily, the investigator might not be too concerned about missing, with probability .67, the fact that vitamin C adds only 3 points to the population mean IQ. But, under special circumstances, this error rate might be excessive. For example, previous experimentation might have established that vitamin C has many positive effects, including the reduction of common colds and no apparent negative side effects. Furthermore, huge quantities of vitamin C might be available at no cost to the school district. The establishment of one more positive effect, even a fairly mild one, such as a small increase in the population mean IQ, might clinch the case for supplying vitamin C to all students in the district. Therefore, the investigator might wish to use a test procedure for which, when H_0 is mildly false, β is substantially less than .67.

To reduce β, the probability of a type II error, increase the sample size. For instance, the number of subjects in the vitamin C experiment could be increased from 36 to 100. Given that H_0 is only mildly false, as before, Figure 12.4 shows the properties of the new hypothesis test. Most important, the value of β has been reduced from .67 to .36. Now, because of the increase in sample size, a three-point difference between the true and hypothesized population means will be missed only about one time in three.

It's instructive to compare the sampling distribution of the mean for 36 students, as shown in Figure 12.3, with that of the mean for 100 students, as shown in Figure 12.4. In the latter case, the true sampling distribution is more compactly distributed about its mean of 103, and therefore, a sample mean is less likely to stray, just by chance, into the retention region for the false H_0. Essentially, it's this reduction in variability that produces the drop in β.

The precise reduction in variability can, of course, be determined from the formula for the standard error of the mean, $\sigma_{\bar{X}}$, namely

$$\sigma_{\bar{X}} = \frac{\sigma}{\sqrt{n}}$$

For a population standard deviation of 15, when the sample size equals 36, as in the original experimental design, the standard error equals 2.5. When the sample size equals 100, as in the new experimental design, the standard error shrinks to 1.5. An obvious implication is that the standard error term (and indirectly, β) can be reduced to any desired value merely by increasing sample size. To cite an extreme case, when the sample size equals 10,000 subjects (!), the standard error shrinks to a tiny 0.15 and the mildly false H_0 always would be rejected since, for all practical purposes, β equals 0.

12.6 Selection of Sample Size

You needn't be content with a crude trial-and-error approach to the selection of sample size. Much like a pilot's pre-flight check before attempting flight, *there exists a table that, prior to the actual investigation, can be consulted to determine the appropriate sample size for a hypothesis test with specified values for both α and β.* To use this table, you must identify the smallest difference between true and hypothesized population means for which the value of β should be specified.

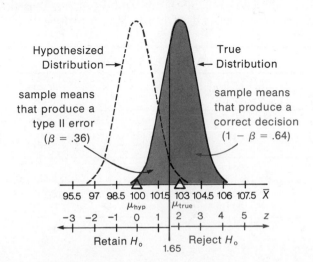

FIGURE 12.4

H_0 is "mildly false" (sample size = 100).

In experiments, *any difference between the true and hypothe-sized population mean is referred to as an* **effect.** Prior to the vitamin C experiment, the investigator might identify the smallest effect (of vitamin C on IQ) for which the probability of a type II error, β, should be specified. Under the particular circumstances described in the previous section, a three-point difference in IQ might qualify as the smallest effect for which β is to be specified.

Access to the table for sample size—much as access to the table for the normal curve—requires that a raw score of three IQ points be converted to a type of standard score, symbolized as d. The conversion to d requires merely that the smallest important difference between true and hypothesized population means be divided by the population standard deviation, as indicated by the following formula:

$$d = \frac{\mu_{\text{true}} - \mu_{\text{hyp}}}{\sigma} \qquad (12.1)$$

When the population standard deviation equals 15, as in the current example, we have

$$d = \frac{103 - 100}{15} = \frac{3}{15} = 0.2$$

Expressed as a pure number, free of the original units of measurement in IQ points, d indicates that the effect size corresponds to 0.2 of the population standard deviation. Thus, **d** *describes the size of an effect in units of the population standard deviation.*

Armed with a d of 0.2, we can consult Table 12.2 to determine what sample size will insure that the one-tailed hypothesis test for vitamin C complies with pre-established error rates. Let's say that for α and β this rate should equal .05 each. Refer to the bottom half of Table 12.2 (for one-tailed tests) and locate the cell intersected by a d value of 0.2 and by the desired α and β values of .05 each. The entry for the intersected cell is 272, the required sample size. If 272 subjects are used in the vitamin C experiment, the hypothesis test will comply with the pre-established error rates of .05 for α and β.

The value of β equals .05 only when the effect actually equals three IQ points. A smaller, presumably unimportant effect would be missed more often (with a probability greater than .05), while a larger, more important effect would be missed less often (with a probability smaller than .05). The present test procedure guarantees that any important effect will be missed with, *at most*, a probability of .05.

TABLE 12.2

**APPROXIMATE SAMPLE
SIZE FOR HYPOTHESIS
TEST: SINGLE SAMPLE**

	TWO-TAILED TEST			
Effect size d	.05 (α) .20 (β)	.01 (α) .20 (β)	.05 (α) .05 (β)	.01 (α) .05 (β)
0.2 (small)	198	294	326	447
0.5 (medium)	34	51	54	75
0.8 (large)	15	22	23	32
	ONE-TAILED TEST			
Effect size d	.05 (α) .20 (β)	.01 (α) .20 (β)	.05 (α) .05 (β)	.01 (α) .05 (β)
0.2 (small)	156	253	272	396
0.5 (medium)	27	43	45	66
0.8 (large)	12	19	19	28

There are, of course, an unlimited number of *d* values, each corresponding to a different effect size. The three values of *d* shown in Table 12.2 correspond to small, medium, and large effects, according to one rule of thumb.* With practice, you may choose to modify this scheme or even ignore it. It is based on the notion that a *d* value of 0.2 generally corresponds to a small effect—that is, a difference equivalent to 0.2 of the population standard deviation (3 IQ points in the vitamin C experiment). A *d* value of 0.5 generally corresponds to a medium effect (7.5 IQ points in the vitamin C experiment), and a *d* value of 0.8 generally corresponds to a large effect (12 IQ points in the vitamin C experiment).

Notice in Table 12.2 that the required sample size drops when attention is shifted to medium and large effects. For instance, maybe the district-wide vitamin C program would involve considerably more effort than presumed heretofore, and this would not be worth the trouble unless it produced an IQ increase of at least 7.5 points. In this case, the investigator might wish to concentrate on a medium effect (with a *d* of 0.5). According to Table 12.2, only 45 subjects are required to guarantee the desired error rates of .05 for α and β. If, in addition, a district-wide vitamin C program would exclude other potentially promising programs, the investigator might wish to concentrate on a large effect (with a *d* of 0.8). Now, according to Table 12.2, only 19 subjects are required to guarantee the desired error rates of .05.

The decision to concentrate on small, medium, or large effects requires a certain tolerance for ambiguity, particularly in uncharted areas of research where not even an estimate of the population standard deviation is available. (Under these circumstances, incidentally, it's probably best to think directly in terms of *d* values rather than original units of measurement.) Don't let this scare you away from using Table 12.2. Prior to conducting an experiment, you'll doubtless familiarize yourself with any relevant background information, and you might conduct several small pilot studies. In the process, you should develop a fairly clear notion about what constitutes the smallest important difference—if you force yourself to confront this problem.

A case can be made for always consulting Table 12.2 to determine the appropriate sample size prior to any hypothesis test. This would represent a distinct improvement over the arbitrary determination of sample size that often results in either an excessively large or an unduly small sample size. An excessively large sample size translates into an extra sensitive test procedure that detects even a very small effect that lacks practical importance. An unduly small sample size translates into an insensitive test procedure that misses even a very large, important effect. Judicious use of Table 12.2 yields a sample size that, being neither too large nor too small, translates into a test procedure that detects all important effects within some tolerable error rate for α and β.

Several additional features of Table 12.2 deserve comment. For any given effect size, sample size increases with movement from the leftmost column, where the total error rate for α and β is highest, to the rightmost column, where total error rate is lowest. In other words, a lower overall error rate requires a larger sample size.

12.7 Small, Medium, and Large Effects

12.8 Additional Comments about the Sample Size Table

---

* See Cohen, J.: *Statistical Power Analysis for the Behavioral Sciences*, Rev. Ed., Academic Press, New York, 1977. This book also supplies tables for many more *d* values.

You might be struck by the fact that, although values of .05 and .01 are specified for α, only values of .20 and .05 are specified for β in Table 12.2. The relatively smaller α values reflect a view, fairly prevalent in the behavioral sciences, that type I errors are more serious than type II errors. According to this perspective, a type I error or false alarm, which stimulates a batch of worthless experiments, is more costly than a type II error or miss, which, if important, will tend to be detected by subsequent experiments. Given this perspective, you should usually consult the leftmost column of Table 12.2, where α equals .05 and β equals .20, *unless* either the type I or type II error appears to be particularly serious.

A comparison of corresponding cells in Table 12.2 reveals that the required sample size is consistently smaller for one-tailed than for two-tailed tests. When used appropriately, as described in Section 11.12, a one-tailed test always is preferable to a two-tailed test.

Until now, Table 12.2 has been used to determine the appropriate sample size *prior* to an experiment. Table 12.2 also can be used with existing studies to evaluate, *in retrospect,* the properties of completed hypothesis tests. For example, pretend that a two-tailed test has been conducted at the .05 level of significance with a sample size of 30. Letting β equal .20, refer to the leftmost column in the upper half of Table 12.2 and locate 34, the table entry closest to the sample size of 30. Given this test procedure, a medium effect has a probability of only about .20 of being missed—and a large effect has an even smaller, unspecified probability of being missed. The sample size of 30 is adequate to detect, within a tolerable margin of error, medium and large effects.

Table 12.2 reappears as one part of Table H in the Appendix. Occasionally, during subsequent chapters, references will be made to Table H and the control of α and β through sample size selection. Whenever possible, Table H should be consulted prior to a hypothesis test. Although not listed in this book, sample size tables also are available for many other types of hypothesis tests.

12.9 Summary

There are four possible outcomes for any hypothesis test:
(1) If H_0 is true, it's a correct decision to retain the true H_0.
(2) If H_0 is true, it's a type I error to reject the true H_0.
(3) If H_0 is false, it's a type II error to retain the false H_0.
(4) If H_0 is false, it's a correct decision to reject the false H_0.
Well-designed hypothesis tests must control both the probability of a type I error, α, in the event that H_0 is true, and the probability of a type II error, β, in the event that H_0 is false.

The probability of a type I error, α, always equals the level of significance.

The probability of a type II error, β, depends on the size of the difference between the true and hypothesized population means. When this difference is large, reflecting a seriously false H_0, the value of β tends to be low, but when this difference is small, reflecting a mildly false H_0, the value of β tends to be high.

To reduce β, increase the sample size. Prior to an investigation, a table can be consulted to identify the sample size that guarantees a hypothesis test will not exceed pre-established error rates of α and β. Access to this table requires that a value be assigned to d, the smallest important difference between true and hypothesized population means divided by the population standard deviation. The sample size table in this book is limited to only three *d* values that correspond to small, medium, and large effects.

The sample size table has a number of implications, including the traditional view that the type I error is more serious than the type II error, and the preference for one-tailed tests rather than two-tailed tests, whenever appropriate. Furthermore, it can be used to evaluate, in retrospect, the properties of completed hypothesis tests.

Important Terms & Symbols

Type I error
Type II error
Probability of a type I error (α)
Probability of a type II error (β)
Effect
Effect size in units of σ (d)

12.10 Exercises

1. (a) List the four possible outcomes for any hypothesis test.
   (b) Under the U.S. criminal code, a defendant is presumed innocent unless proven guilty. Viewing a criminal trial as a hypothesis test (with H_0 specifying that the defendant is innocent), describe each of the four possible outcomes.

2. Define the type I error and the type II error.

3. Supply the missing word(s) in the following statements.

   The probability of a type I error, α, is fixed by the __/(a)/__. The probability of a type II error, β, is determined primarily by the difference between the __(b)__ and __(c)__ population means and by the __/(d)/__. An increase in sample size always produces a(n) __(e)__ in the probability of a type __(f)__ error. To use the sample size table, you must express the __(g)__ between true and hypothesized population means relative to the population __/(h)/__. Values for d are classified as __(i)__, _____, and _____ effects. The appropriate sample size guarantees that the hypothesis test will comply with specified values of __(j)__ and __(k)__ for a given __(l)__ size.

4. (Optional) In Section 12.4 it was suggested that β, the probability of a type II error, can be found when the true population mean is known.
   (a) For the one-tailed test described in this chapter, verify that β equals .01 when the true population mean is 110.
   (b) For the same test, verify that β equals .67 when the true population mean is 103. (Before referring to Table A, express the critical z as a deviation from the true population mean rather than from the hypothesized population mean.)

5. Prior to a hypothesis test, we must be concerned about both a type I and type II error.
   (a) Is this still true, assuming that the test has been conducted, and the hypothesis already has been retained?
   (b) Assuming that the hypothesis already has been rejected?

6. What four bits of information are required to use Table 12.2 to select sample size?

7. Chapter 11 focused on a two-tailed hypothesis test, with $\alpha = .05$, that the mean SAT verbal score for the local population of freshmen equals the national average of 429.

(a) In words, describe the meaning of a type I error for this investigation.
(b) In words, describe the meaning of a type II error for this investigation.
(c) Given a population standard deviation of 110, what difference between true and hypothesized population means corresponds to a small effect?
(d) If β should have a value of .20 for a small effect, what would be an appropriate sample size?

8. For a given hypothesis test, check the one characteristic in each pair that requires the smaller sample size.

(a_1) one-tailed test (a_2) two-tailed test
(b_1) $\alpha = .05$ (b_2) $\alpha = .01$
(c_1) $\beta = .05$ (c_2) $\beta = .20$
(d_1) small effect (d_2) medium effect
(e_1) medium effect (e_2) large effect

9. Specify the appropriate sample size for the following hypothesis tests:

(a) Two-tailed test, with $\alpha = .05$ and $\beta = .05$ (for d of 0.5)
(b) One-tailed test, with $\alpha = .01$ and $\beta = .05$ (for d of 0.8)
(c) Two-tailed test, with $\alpha = .01$ and $\beta = .20$ (for d of 0.2)

10. Comment critically on the following experimental reports.

(a) Using a group of 6 subjects, an investigator announces that H_0 was retained at the .05 level of significance.
(b) Using a sample of 800 subjects, an experimenter reports that H_0 was rejected at the .05 level of significance.

13

t Test for One Sample

t TEST FOR A
POPULATION MEAN

13.1 –A Gas Mileage
     Investigation
13.2 –Estimating the Population
     Standard Deviation
13.3 –t Ratio
13.4 –t Distribution
13.5 –t Tables
13.6 –t Test for Gas Mileage
     Investigation
13.7 –Assumptions
13.8 –Degrees of Freedom
13.9 –Hypothesis Tests: An
     Overview

t TEST FOR A
POPULATION
CORRELATION
COEFFICIENT

13.10–t Test for the Greeting
     Card Exchange
13.11–Assumptions and a
     Limitation
13.12–Summary
13.13–Exercises

When information from a sample is used to estimate the unknown population standard deviation, as is typically the case, the z test must be replaced by a new test—the t test. The discovery of the t test was prompted by the quality control program at Guinness Brewery.

As will become apparent in the next few chapters, the t test is used in a wide variety of situations. Indeed, it's one of the most prevalent tests in statistics.

An important notion in statistics, known as degrees of freedom, is introduced in this chapter and reappears throughout the remainder of the book.

t TEST FOR A
POPULATION MEAN

13.1 A Gas Mileage Investigation

Federal law may eventually specify that new automobiles must average, for instance, 45 miles per gallon of gasoline. Since, of course, it would be impossible to test all new cars, tests are based on random samples from the entire production of each car model. If a hypothesis test indicates sub-standard performance, the manufacturer is penalized $50 per car for the entire production, but if the test indicates standard performance (or better), the manufacturer is given a tax break equivalent to $10 per car.

In these tests, the null hypothesis states that the population mean for some car model equals 45 miles per gallon. The alternative hypothesis reflects an exclusive concern that the population mean is less than 45 miles per gallon. Symbolically, the two statistical hypotheses read:

$$H_0: \mu = 45$$
$$H_1: \mu < 45$$

From most perspectives, including the manufacturer's, a type I error (stiff penalty even though car complies with standard) is more serious than a type II error (mild tax break even though car fails to comply with standard). It's appropriate, therefore, to let α, the probability of a type I error, equal .01 and not to be too concerned about β, the probability of a type II error, unless H_0 is seriously false. Given a one-tailed test with $\alpha = .01$ and $\beta = .20$ for $d = 0.8$, the sample size should equal 19, according to Table 12.2 or Table H. To simplify computations for the present example, however, this recommendation is ignored in favor of a very small sample size of 6. Even with this small sample size, the value of α can be maintained at .01, but the value of β is larger than that specified above.

Often, as in the present example, the population standard deviation is unknown, and it must be estimated from the sample. An estimate of the population standard deviation is described in the next section, and then this estimate is used in the formula for a new hypothesis test—the *t* test.

13.2 Estimating the Population Standard Deviation

One possible estimate of the unknown population standard deviation, σ, is the sample standard deviation, S, as defined in Formula 5.1:

$$S = \sqrt{\frac{\Sigma(X - \bar{X})^2}{n}}$$

This formula appears to provide a most straightforward solution to our problem. Simply calculate S from Formula 5.1 and use that value to estimate σ. Unfortunately, Formula 5.1 is designed for descriptive purposes, as in Part I of this book, rather than for estimation purposes, as in the present situation.

A better estimate of σ can be obtained by replacing n with $n - 1$ in the denominator of the formula for S. Denoted by the small letter s, this estimate of σ is defined as

$$s = \sqrt{\frac{\Sigma(X - \bar{X})^2}{n - 1}} \tag{13.1}$$

Always use this version of the **sample standard deviation, *s*,** *to estimate* σ. The value of s can be determined either from the deviation formula, as defined above, or from the following raw score formula:

$$s = \sqrt{\frac{n\Sigma X^2 - (\Sigma X)^2}{n(n - 1)}} \tag{13.2}$$

This formula can be derived from Formula 5.2, the original raw score formula for S, by replacing one n term in the denominator with $n - 1$.

Now s replaces σ in the formula for the standard error of the mean. Instead of

$$\sigma_{\bar{X}} = \frac{\sigma}{\sqrt{n}}$$

we have

$$s_{\bar{X}} = \frac{s}{\sqrt{n}} \tag{13.3}$$

where $s_{\bar{X}}$ represents the **estimated standard error of the mean,** s is defined in Formulas 13.1 or 13.2, and n equals the sample size. The expression for the estimated standard error appears in hypothesis tests about population means whenever, as in the current example, the population standard deviation is unknown.

13.3 *t* Ratio

The shift from the standard error of the mean, $\sigma_{\bar{X}}$, to its estimate, $s_{\bar{X}}$, has an important effect on the entire hypothesis test for a population mean. The familiar z test, which presumes that the ratio

$$z = \frac{\bar{X} - \mu_{\text{hyp}}}{\sigma_{\bar{X}}}$$

is normally distributed, must be replaced by a new t test, which presumes that the ratio

$$t = \frac{\overline{X} - \mu_{\text{hyp}}}{s_{\overline{X}}} \qquad (13.4)$$

is distributed according to "Student's" distribution (or simply the "t" distribution). We'll explain the origin of this equation in the following section.

To appreciate the difference between z and t, imagine a theoretical exercise where, even though the population standard deviation is known, both z and t are calculated for a series of random samples from the same population. Both ratios share a common source of chance sampling variability—the chance differences among sample means, $\overline{X}$, that appear in their numerators. Both ratios also share the hypothesized population mean, μ_{hyp}, which remains the same for all samples. However, t possesses a second source of chance sampling variability—the chance differences among estimated standard errors, $s_{\overline{X}}$, that appear (as products of each new sample standard deviation) in the denominator of t. When expressed as t ratios, these two independent sources of chance sampling variability—one in the numerator, the other in the denominator—generate extra variability. The tails of the sampling distribution for t are more inflated than the tails of the sampling distribution for z, particularly when sample size is small.

While doing quality control work for Guinness Brewery during the early 1900s, William Gosset discovered the sampling distribution of t and subsequently reported his achievement under the pen name Student. Actually, Gosset discovered not just one but an entire family of t distributions. Each t distribution is associated with a special number referred to as "degrees of freedom" (discussed in more detail in Section 13.8). When testing a hypothesis about the population mean, as in the current example, the number of degrees of freedom, abbreviated df, always equals the sample size minus one. Symbolically,

13.4 t Distribution

$$df = n - 1 \qquad (13.5)$$

For instance, if the investigation of gas mileage involves 6 cars, the corresponding t test is based on a sampling distribution with 5 degrees of freedom (from $df = 6 - 1$).

Figure 13.1 shows three t distributions. In the limit, when there are an infinite (∞) number of degrees of freedom, the distribution of t is the same as the standard normal distribution of z. Notice that, even

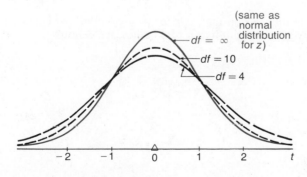

FIGURE 13.1

Various t distributions.

with only 4 or 10 degrees of freedom, a *t* distribution shares a number of properties with the normal distribution. All *t* distributions are symmetrical, unimodal, and bell-shaped with a dense concentration that peaks in the middle (when *t* equals 0) and tapers off both to the right of the middle (as *t* becomes more positive) and to the left of the middle (as *t* becomes more negative). *The inflated tails of the* t *distribution, particularly apparent with small values of* df, *constitute the most important difference between* t *and* z *distributions.*

13.5   *t* Tables

To save space, tables for *t* distributions concentrate only on critical values of *t* that correspond to the more common levels of significance. Table B of the Appendix lists critical *t* values for either one- or two-tailed hypothesis tests at the .05, .01, and .001 levels of significance. All listed critical *t* values are positive and originate from the upper half of each distribution. Because of the symmetry of the *t* distribution, you can obtain the corresponding critical *t* values for the lower half of each distribution merely by placing a negative sign in front of any entry in the table.

To find a critical *t* in Table B, read the entry in the cell intersected by the row for the correct number of degrees of freedom and the column for the test specifications. For example, in the gas mileage investigation, first locate the row corresponding to 5 degrees of freedom, and then locate the column for a one-tailed test at the .01 level of significance. The intersected cell specifies 3.365. A negative sign must be placed in front of 3.365 since the hypothesis test requires the lower tail to be critical. Thus, −3.365 is the critical *t* for the gas mileage investigation, and the corresponding decision rule is illustrated in Figure 13.2.

If the gas mileage investigation had involved a two-tailed test (still at the .01 level with 5 degrees of freedom), then the intersected cell would have specified 4.032. A positive sign and a negative sign would have to be placed in front of 4.032 since both tails are critical. In this case, ±4.032 would have been the pair of critical *t* values.

If the desired number of degrees of freedom doesn't appear in the *df* column of Table B, use the row with the next smallest number of degrees of freedom. For example, when 36 degrees of freedom are specified, use information from the row for 30 degrees of freedom.

13.6   *t* Test for Gas Mileage Investigation

Hypothesis Test Summary:
t Test for a Population Mean
(Gas Mileage Investigation)

Problem:

Does the mean gas mileage for some population of cars drop below the legally required minimum of 45 miles per gallon?

Statistical Hypotheses:

$$H_0: \mu = 45$$
$$H_1: \mu < 45$$

Statistical Test:

$$t = \frac{\bar{X} - \mu_{\text{hyp}}}{s_{\bar{X}}}$$

Decision Rule:
    Reject H_0 at the .01 level of significance if $t \leq -3.365$ (from Table B, given $df = n - 1 = 6 - 1 = 5$).

Calculations:
    Given $\bar{X} = 43$, $s_{\bar{X}} = 0.89$ (See Table 13.1 for computations.)
$$t = \frac{43 - 45}{0.89} = -2.25$$

Decision:
    Retain H_0

Interpretation:
    The population mean gas mileage *could* equal 45 miles per gallon; the manufacturer should receive a tax break of $10 per car.

The present *t* test is based on gas mileage values of 40, 44, 46, 41, 43, and 44 miles per gallon for six randomly selected cars. Table 13.1 lists the various computations that produce a *t* of -2.25.

13.7 Assumptions

When testing hypotheses about population means, use *t* rather than *z* if, as is typically the case, the population standard deviation is unknown. Strictly speaking, when using the *t* test, you must assume that the underlying population is normally distributed. Even when this normality assumption is violated, the *t* test retains much of its accuracy as long as sample size is not too small. If a very small sample size (less than about 10) is being used and you believe that the sample originates from a non-normal population—possibly because of a pronounced positive or negative skew among observations in the sample—it would be wise to increase the sample size before attempting a hypothesis test.

13.8 Degrees of Freedom

The notion of degrees of freedom will be used throughout the remainder of this book. **Degrees of freedom (df)** *refer to the number of values, within a given set of values, that are free to vary.* Typically, less than all values within the set are free to vary. For example, even though the gas mileage data consist of six values—40, 44, 46, 41, 43, and 44—the *t* test for these data has only five degrees of freedom. In other words, only five of these six values are free to vary. Let's look at this more closely.
    The loss of one degree of freedom occurs, without fanfare, when the unknown population standard deviation, σ, is estimated from the

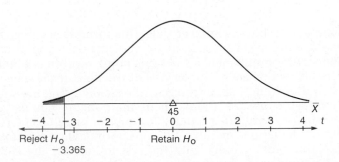

FIGURE 13.2

Hypothesized sampling distribution of *t* (gas mileage investigation).

TABLE 13.1
CALCULATIONS FOR
t TEST
(GAS MILEAGE
INVESTIGATION)

I. FINDING $\bar{X}$ AND s

(a) Computational sequence:

Assign a value to n ①.

Sum all X scores ②.

Substitute numbers in formula ③ and solve for $\bar{X}$.

Square each X score ④, one at a time, and then add all squared X scores ⑤.

Substitute numbers into the formula ⑥ and solve for s.

(b) Data and computations:

X	④ X²
40	1600
44	1936
46	2116
41	1681
43	1849
44	1936

① $n = 6$ ② $\Sigma X = 258$ ⑤ $\Sigma X^2 = 11118$

③ $\bar{X} = \dfrac{\Sigma X}{n} = \dfrac{258}{6} = 43$

⑥ $s = \sqrt{\dfrac{n(\Sigma X^2) - (\Sigma X)^2}{n(n-1)}} = \sqrt{\dfrac{6(11118) - (258)^2}{6(6-1)}}$

$= \sqrt{\dfrac{66708 - 66564}{30}} = \sqrt{\dfrac{144}{30}} = \sqrt{4.8} = 2.19$

II. FINDING $s_{\bar{X}}$

(a) Computational sequence:

Substitute numbers obtained above in formula ⑦ and solve for $s_{\bar{X}}$.

(b) Computations:

⑦ $s_{\bar{X}} = \dfrac{s}{\sqrt{n}} = \dfrac{2.19}{\sqrt{6}} = \dfrac{2.19}{2.45} = 0.89$

III. FINDING THE OBSERVED t

(a) Computational sequence:

Assign value to μ_{hyp} ⑧, the hypothesized population mean.

Substitute numbers obtained above in formula ⑨ and solve for t.

(b) Computations:

⑧ $\mu_{hyp} = 45$

⑨ $t = \dfrac{\bar{X} - \mu_{hyp}}{s_{\bar{X}}} = \dfrac{43 - 45}{0.89} = \dfrac{-2}{0.89} = -2.25$

sample standard deviation, s. The missing degree of freedom can be detected most easily by referring to Formula 13.1 for s, that is

$$s = \sqrt{\dfrac{\Sigma(X - \bar{X})^2}{n - 1}}$$

To estimate σ from the gas mileage data, this formula specifies that the six original gas mileage values, X, be expressed, one at a time, as deviations about their sample mean, $\bar{X}$. At this point, a subtle mathematical restriction causes the loss of one degree of freedom. It's always true, as demonstrated in Table 13.2 for the gas mileage data, that *the sum of all values, expressed as deviations about their mean, equals zero.* (If you're skeptical, verify this property for any set of values.) Given *any* five of the six deviations in Table 13.2, the sixth deviation is not free to vary—its value must comply with the restriction that the sum of all deviations equals zero. For instance, given the first five deviations in

X	$X - \bar{X}$	POSITIVE DEVIATIONS	NEGATIVE DEVIATIONS	TABLE 13.2 DEMONSTRATION: $\Sigma(X - \bar{X}) = 0$ (GAS MILEAGE DATA)
40	40 — 43		— 3	
44	44 — 43	1		
46	46 — 43	3		
41	41 — 43		— 2	
43	43 — 43	0		
44	44 — 43	1	——	
		5	— 5	

Therefore, $\Sigma(X - \bar{X}) = 5 + (-5) = 5 - 5 = 0$

Table 13.2, the sixth deviation must equal 1, and therefore, the sixth observation must equal 44, that is, the mean of 43 plus the deviation of 1.

A slightly different perspective may further clarify the notion of degrees of freedom. The population standard deviation reflects the deviations of observations about the population mean, $X - \mu$. To estimate the unknown population standard deviation, it would be most efficient to take a random sample of these deviations, but this is usually impossible because the population mean is unknown. Typically, therefore, the unknown population mean, μ, must be replaced with the sample mean, $\bar{X}$, and a random sample of these deviations, $X - \bar{X}$, must be used to estimate the population deviations, $X - \mu$. Even though there are n deviations in the sample, only $n - 1$ of these deviations supply a true picture of the population deviations—because only $n - 1$ of the sample deviations are free to vary, given that the sum of the n deviations about *their own sample mean* always equals zero.

"HE KEEPS MUMBLING ABOUT NOT HAVING ANY DEGREES OF FREEDOM LEFT."

The original version of the sample standard deviation, S, tends to underestimate the population standard deviation, σ. This occurs because, even though there are only $n - 1$ independent deviations in the numerator, n is retained in the denominator. A better estimate of σ is obtained when the denominator term reflects the number of degrees of freedom in the numerator, as in the formula for s, where the denominator term equals $n - 1$.

In subsequent sections, other mathematical restrictions will be encountered, and sometimes more than one degree of freedom will be lost. In any event, however, degrees of freedom always indicate the number of values free to vary, given one or more mathematical restrictions.

13.9 Hypothesis Tests: An Overview

The remainder of this chapter deals with a t test for a population correlation coefficient. Furthermore, much of the remainder of this book concerns itself with an alphabet-soup variety of tests—F, U, T, H, as well as z and t—for an assortment of situations. Notwithstanding the new formulas with their special symbols, *all of these tests represent variations on a common theme: if some observed characteristic, such as the mean for one or more random samples, qualifies as an improbable outcome under the null hypothesis, the hypothesis is rejected. Otherwise, the hypothesis is retained.* To determine whether an outcome is improbable, the observed characteristic is converted to some new value, such as t, and compared with critical values from the appropriate sampling distribution. Generally speaking, if the observed value equals or exceeds a positive critical value (or if it equals or is more negative than a negative critical value) the outcome is viewed as improbable, and the hypothesis is rejected.

13.10 t Test for the Greeting Card Exchange

In Chapter 7, .80 describes the sample correlation coefficient, r, between cards given and cards received by five friends. Any conclusions about the correlation coefficient in the underlying population—for instance, the population of all friends—must consider chance sampling variability, as described by the sampling distribution of r.

Let's view the greeting card data for the five friends as a random sample from the population of all friends. Then it's possible to test the null hypothesis that the **population correlation coefficient,** symbolized by the Greek letter ρ (rho), equals zero. In other words, it's possible to test the hypothesis that, in the population of all friends, there is no correlation between cards given and cards received.

A t test can be used to determine whether an r of .80 qualifies as a probable or improbable outcome under the null hypothesis. To obtain a value for t, use the following formula:

$$t = \frac{r(\sqrt{n-2})}{\sqrt{1-r^2}} \tag{13.5}$$

where **r** refers to the **sample correlation coefficient** (Formula 7.1 or 7.2), and n refers to the number of pairs of observations. As implied by the term in the numerator, this ratio is distributed with $n-2$ degrees of freedom. When pairs of observations are represented as points in a scatterplot, r presumes that the cluster of points approximates a straight line. Two degrees of freedom are lost because points are free to vary only about some straight line that, itself, always depends on two points.

Hypothesis Test Summary:
t Test for a Population Correlation Coefficient
(Greeting Card Exchange)

Problem:

Could there be a correlation between cards given and cards received for the population of all friends?

Statistical Hypotheses:

$$H_0: \rho = 0$$
$$H_1: \rho \neq 0$$

Statistical Test:

$$t = \frac{r(\sqrt{n-2})}{\sqrt{1-r^2}}$$

Decision Rule:

Reject H_0 at the .05 level of significance if $t \leq -2.571$ or if $t \geq 2.571$ (from Table B, given $df = n - 2 = 5 - 2 = 3$).

Calculations:

Given $r = 0.80$ and $n = 5$

$$t^* = \frac{(.80)(\sqrt{5-2})}{\sqrt{1-(.80)^2}} = \frac{(.80)(\sqrt{3})}{\sqrt{1-.64}} = \frac{(.80)(1.73)}{\sqrt{.36}}$$

$$= \frac{1.38}{.6} = 2.30$$

Decision:

Retain H_0.

Interpretation:

The population correlation coefficient *could* equal zero; there might not be any relationship between cards given and cards received in the population of friends.

According to the present hypothesis test, the population coefficient *could* equal zero. This conclusion might seem surprising, given that an r of .80 was observed for the greeting card exchange. When the value of r is based on only 5 pairs of observations, as in the present example, its sampling variability is huge, and in fact, an r of .88 would have been required to reject the null hypothesis. Ordinarily, a serious investigation would use a larger sample size—preferably the sample size for a hypothesis test with specified values of α and β for some effect (or value of ρ).

13.11 Assumptions and a Limitation

When using the t test for the population correlation coefficient, you must assume that the sample originates from a *normal bivariate population*. This term means that the separate population distributions for each variable (X and Y) should be normal and that the relationship between the two variables should be described by a straight line. When these assumptions are suspect—when, for instance, the observed distribution for one variable appears to be extremely non-normal—test results are only approximate and should be interpreted accordingly.

The present t test can't be used to test the hypothesis that the population correlation coefficient ρ equals some number other than zero. When ρ is not equal to zero, the sampling distribution of r is skewed and can't be adequately approximated by the symmetrical t distribution. A better approximation is supplied by Fisher's r to z transformation, as described in more advanced statistics books.

When the population standard deviation, σ, is unknown, it must be estimated with the sample standard deviation, s, defined in Formula 13.1. By the same token, the standard error of the mean, $\sigma_{\bar{x}}$, then must be estimated with $s_{\bar{x}}$, defined in Formula 13.3.

Under these circumstances, t rather than z should be used to test a hypothesis about the population mean. The t ratio, as defined in Formula 13.4, is distributed with $n - 1$ degrees of freedom, and critical t values are obtained from Table B.

When using the t test, you must assume that the underlying population is normally distributed. Violations of this assumption are important only when sample size is less than about 10.

Degrees of freedom (df) refer to the number of values, within a set of values, that are free to vary. When σ is estimated with s, only $n - 1$ of the deviations in the numerator are free to vary. The loss of one degree of freedom is reflected in the selection of $n - 1$ as the denominator term for s.

To test the hypothesis that the population correlation coefficient equals zero, use the t test defined in Formula 13.5. This t ratio is distributed with $n - 2$ degrees of freedom, and critical values of t can be obtained from Table B.

When using the t test for a population correlation coefficient, you must assume that the population distributions for X and for Y are normally distributed and that the relationship between X and Y is linear.

Important Terms & Symbols ━━━━━━━━━━━━━━━━

Sample standard deviation (s)
Estimated standard error of the mean ($s_{\bar{x}}$)
t test for a population mean
Degrees of freedom (df)
Sample correlation coefficient (r)
Population correlation coefficient (ρ)
t test for a population correlation coefficient

━━━━━━━━━━━━━━━━━━━━━━━━━━━━━━━━━━━━━

13.13 Exercises

1. A consumers' group suspects that a large food chain makes extra money by supplying less than the specified weight of 16 ounces in their standard packages of ground beef. A random sample of ten packages reveals the following weights in ounces: 16, 15, 14, 15, 14, 15, 16, 14, 14, 14. Test the null hypothesis with t, using the .05 level of significance.

2. A library system lends books for periods of 21 days. This policy is being re-evaluated in view of the possibility of a new loan period that could be either longer or shorter than 21 days. To aid in this decision, book lending records are consulted to determine the loan periods actually used by patrons. A random sample of eight records reveals the following loan periods in days: 21, 15, 12, 24, 20, 21, 13, 16. Test the null hypothesis with t, using the .05 level of significance.

3. It's a well-established fact that lab rats require, on the average, 32 trials before reaching a criterion of three consecutive errorless trials in a complex water maze. To determine whether a mildly adversive stimulus has any effect on performance, a sample of seven lab rats are given a mild electrical shock just prior to each trial. They require the following number of trials before reaching criterion: 35, 38, 39, 33, 31, 32, 36. Test the null hypothesis with t, using the .05 level of significance.

4. A tire manufacturer wishes to determine whether, on the average, a brand of steel-belted radial tires provides more than 50,000 miles of wear. A random sample of 36 tires yields a mean, $\overline{X}$, of 52,100 miles and a standard deviation, s, of 2500 miles. Use t to test the null hypothesis at the .01 level of significance.

5. Assume that, on the average, healthy young adults dream 90 minutes each night, as inferred from a number of measures, including rapid eye movement. An investigator wishes to determine whether the consumption of alcohol just prior to sleep affects the amount of dream time. After consuming a standard amount of alcohol, dream time is monitored for each of 28 healthy young adults in a random sample. Results show a mean, $\overline{X}$, of 88 minutes and a standard deviation, s, of 9 minutes. Use t to test the null hypothesis at the .05 level of significance.

6. (Optional) Even though the population standard deviation is unknown, an investigator uses z rather than the more appropriate t to test a hypothesis at the .05 level of significance. Does .05 describe the true level of significance? If not, is the true level of significance larger or smaller than .05?

7. (Optional) When discussing degrees of freedom, it was suggested that the most efficient way to estimate the population standard deviation, σ, would be to take a random sample of the deviations $X - \mu$. Ordinarily, this is impossible because the value of the population mean, μ, is unknown. If the value of μ were known, how many degrees of freedom would be associated with an estimate of σ based on n of these deviations $(X - \mu)$?

8. A random sample of 27 California taxpayers reveals an r of .43 between years of education and annual income. Use t to test the null hypothesis at the .05 level of significance that there is no relationship between educational level and annual income among the population of California taxpayers.,

9. A random sample of 38 statistics students from a large statistics class reveals an r of $-.24$ between test score on a statistics exam and amount of time spent taking the exam. Test the null hypothesis with t, using the .01 level of significance.

14

t Test for Two Independent Samples

14.1 –Blood-doping Experiment
14.2 –Two Populations
14.3 –Sampling Distribution of $\bar{X}_1 - \bar{X}_2$
14.4 –Mean of the Sampling Distribution
14.5 –Standard Error of the Sampling Distribution
14.6 –z Test
14.7 –Estimating the Population Variance
14.8 –Estimating the Standard Error
14.9 –t Ratio
14.10–t Test for the Blood-doping Experiment
14.11–Selection of Sample Size
14.12–Assumptions

MORE GENERAL
COMMENTS
ABOUT
HYPOTHESIS
TESTS

14.13–Secondary Status of the Null Hypothesis
14.14–Statistical Significance vs. Practical Importance
14.15–A Note on Usage
14.16–Suspending Judgment about the Null Hypothesis
14.17–Summary
14.18–Exercises

Well-designed experiments usually consist of at least two groups—an experimental group and a control group. A t test can be used to determine whether the difference between the two groups qualifies as a probable or an improbable outcome under the null hypothesis.

Typically, an investigation is inspired by an informal hunch or research hypothesis, not by the null hypothesis. Nevertheless, there are several reasons why the null hypothesis, rather than the research hypothesis, is tested directly.

Don't confuse statistical significance with practical importance; sometimes a statistically significant result might lack practical importance.

14.1 Blood-doping Experiment

During a recent session of the summer Olympics, there was speculation that some long distance runners might have increased their endurance through "blood-doping." In this procedure, extra oxygen-carrying red blood cells are injected into the athlete's blood stream just prior to the competitive event. An investigator wants to determine whether, in fact, blood doping increases the endurance of athletes under controlled laboratory conditions. (We'll ignore the very real ethical and legal issues raised by this type of experimentation.) Volunteer athletes from the local track team are randomly assigned to one of two groups: an experimental group (X_1), which is blood doped, or a control group (X_2), which is not blood doped but receives an injection of harmless fluid. After being injected, each athlete runs on a rapid treadmill until exhausted. Total time on the treadmill is used as the measure of endurance.

In previous chapters, when the null hypothesis for a single population mean was tested, considerable attention was devoted to the single sample mean, $\bar{X}$, and its sampling distribution. In the present chapter, the null hypothesis for the difference between two population means will be tested, and we will focus on the **difference between sample means** for the experimental and control groups, $\bar{X}_1 - \bar{X}_2$, respectively, and its sampling distribution. Many similarities between one- and two-sample hypothesis tests will become apparent in this chapter.

The subjects in the blood-doping experiment originate from a very limited real population—all volunteer athletes from the local track team. This is hardly an inspiring target for statistical inference. A standard remedy is to characterize the sample of athletes *as if* it were a random sample from a much larger hypothetical population, loosely defined as "all similar volunteer athletes who could conceivably participate in the experiment." Strictly speaking, there are two hypothetical populations, one defined for the endurance scores of athletes who are blood doped and the other for the endurance scores of athletes who are not blood doped. These two populations are cited in the null hypothesis, and as noted previously, any generalizations to hypothetical populations must be viewed as provisional conclusions. Only additional experimentation can resolve whether a given experimental finding merits the generality assigned to it by the investigator.

14.2 Two Populations

In practice, of course, there's only one observed difference between sample means, $\bar{X}_1 - \bar{X}_2$. To determine whether this difference qualifies as a probable or an improbable outcome under the null hypothesis, it is viewed as originating from the sampling distribution of the difference between sample means, that is, the distribution of differences between sample means for all possible pairs of random samples—of given sizes—from the two underlying populations. Because of the huge number of possibilities, the sampling distribution of $\bar{X}_1 - \bar{X}_2$ is not constructed from scratch. Instead, as with the sampling distribution of $\bar{X}$ described in Chapter 10, statistical theory must be relied on for information about its mean and standard error.

14.3 Sampling Distribution of $\bar{X}_1 - \bar{X}_2$

In the one-sample case, the mean of the sampling distribution equals the population mean. Likewise, in the two-sample case, *the mean of the sampling distribution equals the difference between population means.* Expressed in symbols,

$$\mu_{\bar{X}_1 - \bar{X}_2} = \mu_1 - \mu_2 \tag{14.1}$$

where $\mu_{\bar{X}_1 - \bar{X}_2}$ represents the mean of the sampling distribution of $\bar{X}_1 - \bar{X}_2$ and $\mu_1 - \mu_2$ represents the **difference between population means.**

14.4 Mean of the Sampling Distribution

Formula 14.1 is not particularly startling. Because of sampling variability, it's unlikely that the one observed difference between sample means equals the difference between population means. However, since not just one, but all possible differences between sample means contribute to the mean of the sampling distribution, the effects of the sampling variability are neutralized, and the mean of the sampling distribution equals the difference between population means. Accordingly, these two terms are used interchangeably in statistical inference. Any claims about the difference between population means can be transfered directly to the mean of the sampling distribution of $\bar{X}_1 - \bar{X}_2$.

The difference between population means reflects the effect of blood doping on endurance. If blood doping has little or no effect on endurance, then endurance scores will tend to be about the same for both the experimental and control populations of athletes, and the difference between population means will hover close to zero. If blood doping facilitates endurance, scores for the experimental population will tend to exceed those for the control population, and the difference between population means will be positive; the stronger the facilitative

155

effect of blood doping on endurance, the larger the positive difference between population means. Finally, if blood doping hinders endurance, endurance scores for the experimental population will tend to be exceeded by those for the control population, and the difference between population means will be negative.

The hypothesis test for blood doping is geared to the difference between population means. The null hypothesis will assume that the difference between population means equals zero, and the decision about the null hypothesis will reflect whether the one observed difference between sample means qualifies as a probable or an improbable outcome under the null hypothesis.

14.5 Standard Error of the Sampling Distribution

The sampling distribution of $\bar{X}_1 - \bar{X}_2$ also has a standard deviation, referred to as the standard error of the difference between sample means. Symbolized as $\sigma_{\bar{X}_1 - \bar{X}_2}$, the formula for this standard error reads:

$$\sigma_{\bar{X}_1 - \bar{X}_2} = \sqrt{\frac{\sigma_1^2}{n_1} + \frac{\sigma_2^2}{n_2}} \qquad (14.2)$$

where σ_1^2 and σ_2^2 represent the two population variances—that is, the squares of the two population standard deviations—and n_1 and n_2 represent the two sample sizes.

The standard error $\sigma_{\bar{X}_1 - \bar{X}_2}$ shares a number of common properties with the standard error $\sigma_{\bar{X}}$, originally described in Section 10.5. For instance, with increases in both sample sizes, the size of $\sigma_{\bar{X}_1 - \bar{X}_2}$ becomes much smaller than either of the population standard deviations. As a result, values of $\bar{X}_1 - \bar{X}_2$ tend to cluster relatively close to the sampling distribution mean (and the difference between population means), allowing more precise generalizations from samples to populations.

14.6 *z* Test

Eventually, we want to test the null hypothesis that blood doping has no effect on endurance. This test is based not on the hypothesized sampling distribution of $\bar{X}_1 - \bar{X}_2$, but on its counterpart, the hypothesized sampling distribution of *t*. There is also a *z* test for two population means, but its use requires that both population standard deviations be known. In practice, this information is rarely available, and therefore, the *z* test is hardly ever appropriate. No further description of the *z* test will be given in this or the next chapter.

14.7 Estimating the Population Variance

Use of the *t* test requires that the unknown population standard deviations (or their squares, the population variances) be estimated from the sample standard deviations (or their squares, the sample variances). It's customary to assume that the two population variances are equal:

$$\sigma_1^2 = \sigma_2^2 = \sigma^2$$

This premise, officially known as the homogeneity of variance assumption, redefines the estimation problem. Now, σ_1^2 and σ_2^2 needn't be estimated separately with their respective sample variances s_1^2 and s_2^2. Instead, the unknown population variance σ^2, presumed to be common to both populations, can be estimated with a single combination of both sample variances.

As noted in Section 13.8, a more refined estimate of the unknown population standard deviation occurs when the sample standard deviation is adjusted for degrees of freedom. The same is true for the sample variance. The best estimate of the common population variance is produced when the sample variances s_1^2 and s_2^2 are combined on the basis of their degrees of freedom, $n_1 - 1$ and $n_2 - 1$, respectively, as follows:

$$s_p^2 = \frac{(n_1 - 1)s_1^2 + (n_2 - 1)s_2^2}{(n_1 - 1) + (n_2 - 1)} \qquad (14.3)$$

The symbol s_p^2 designates the common or **pooled variance estimate;** s_1^2 and s_2^2 are the variances for the two samples; and n_1 and n_2 are the two sample sizes. Values of s_1^2 and s_2^2 can be found by squaring each of the values obtained for s_1 and s_2 in Formula 13.1.

Multiplying each sample variance by its degrees of freedom, as in Formula 14.3, insures that the contributions of s_1^2 and s_2^2 are proportionate to their degrees of freedom. For instance, if s_1^2 contains twice as many degrees of freedom as s_2^2, then the value of s_1^2 also counts twice as much as s_2^2 in determining the final value of s_p^2.

The term in the denominator of Formula 14.3 implies that the degrees of freedom for the pooled variance estimate equals the sum of the two sample sizes minus two, that is, $n_1 + n_2 - 2$. Two degrees of freedom are lost because observations in each of the two samples are expressed as deviations about their respective sample means.

When Formula 14.3 is viewed in its entirety, s_p^2 can be characterized as the mean of s_1^2 and s_2^2, once these estimates have been adjusted for their degrees of freedom. Accordingly, if the values of s_1^2 and s_2^2 are different, s_p^2 will always assume some intermediate value.

14.8 Estimating the Standard Error

On the assumption that the two population variances are equal, substitute the new variance estimate, s_p^2, for both σ_1^2 and σ_2^2 in Formula 14.2 for the standard error. The new formula reads:

$$s_{\bar{X}_1 - \bar{X}_2} = \sqrt{\frac{s_p^2}{n_1} + \frac{s_p^2}{n_2}} \qquad (14.4)$$

where $s_{\bar{X}_1 - \bar{X}_2}$ represents the **estimated standard error of the difference between means,** s_p^2 is the pooled variance estimate, defined in Formula 14.3, and n_1 and n_2 are the two sample sizes. Now when the estimated standard error appears in the denominator of Formula 14.5, as shown below, the resulting t ratio can be used to test a hypothesis about the difference between two population means.

14.9 *t* Ratio

In the blood-doping experiment, the null hypothesis can be tested with the following ratio:

$$t = \frac{(\bar{X}_1 - \bar{X}_2) - (\mu_1 - \mu_2)_{\text{hyp}}}{s_{\bar{X}_1 - \bar{X}_2}} \qquad (14.5)$$

which complies with a t distribution having $n_1 + n_2 - 2$ degrees of freedom. In Formula 14.5, $\bar{X}_1 - \bar{X}_2$ represents the one observed difference between sample means, $(\mu_1 - \mu_2)_{\text{hyp}}$ represents the hypothesized difference (of zero) between population means, and $s_{\bar{X}_1 - \bar{X}_2}$ represents the estimated standard error, as defined in Formula 14.4.

According to the null hypothesis for the blood-doping experiment, the difference between the mean endurance scores for the experimental population and the control population equals zero, that is, blood doping has no effect—or a null effect—on endurance scores. An equivalent statement, in symbols, reads:

$$H_0: \mu_1 - \mu_2 = 0$$

where H_0 represents the null hypothesis, and μ_1 and μ_2 represent the mean endurance scores for experimental and control populations, respectively. Although, strictly speaking, the hypothesized difference between population means could equal any number, this difference almost invariably (always in this book) equals zero.

The investigator probably wants to reject the null hypothesis only if there's evidence that blood doping increases endurance scores. Given this perspective, the alternative hypothesis should specify that the difference between population means exceeds zero in favor of blood doping. An equivalent statement, in symbols, reads:

$$H_1: \mu_1 - \mu_2 > 0$$

where H_1 represents the alternative hypothesis and, as above, μ_1 and μ_2 represent the mean endurance scores for the experimental and control populations, respectively. This directional alternative hypothesis translates into a one-tailed test with the upper tail critical.

There are two other possible alternative hypotheses. Another directional hypothesis, expressed as

$$H_1: \mu_1 - \mu_2 < 0$$

translates into a one-tailed test with the lower tail critical, while a nondirectional hypothesis, expressed as

$$H_1: \mu_1 - \mu_2 \neq 0$$

translates into a two-tailed test. As emphasized in Section 11.12, directional alternative hypotheses should be used only when there's an exclusive concern about differences in a particular direction.

Ideally, at this point, the investigator should consult Table H to determine the sample size required for a hypothesis test with specified error rates of α and β for some effect size d. However, let's ignore this issue until the next section and merely assume that the one-tailed hypothesis test for blood doping is to be conducted at the .05 level of significance for very small samples of only five athletes per group. Figure 14.1 illustrates the decision rule for this test.

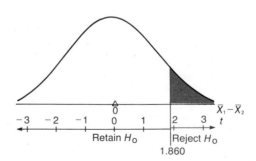

FIGURE 14.1

Hypothesized sampling distribution of *t* (blood-doping experiment).

Problem:

Does the population mean endurance score for athletes who are blood doped exceed that for athletes who are not blood doped?

Statistical Hypotheses:

$$H_0: \mu_1 - \mu_2 = 0$$
$$H_1: \mu_1 - \mu_2 > 0$$

Statistical Test:

$$t = \frac{(\bar{X}_1 - \bar{X}_2) - (\mu_1 - \mu_2)_{\text{hyp}}}{s_{\bar{X}_1 - \bar{X}_2}}$$

Decision Rule:

Reject H_0 at the .05 level of significance if $t \geq 1.860$ (from Table B, given $df = n_1 + n_2 - 2 = 5 + 5 - 2 = 8$).

Calculations:

$$t = \frac{(8 - 6) - (0)}{1.10}$$
$$= 1.82 \text{ (See Table 14.1 for all computations.)}$$

Decision:

Retain H_0.

Interpretation:

The difference between population means *could* equal zero; there is no evidence that blood doping increases the endurance scores of athletes.

The present test is based on five endurance scores for the experimental athletes and five endurance scores for the control athletes shown in Table 14.1. For computational convenience, these scores have been rounded to the nearest minute. (In practice, they would reflect more precise measurement.) Table 14.1 also lists the various computations that produce a t of 1.82.

14.11 Selection of Sample Size

Table H can be consulted to determine a more appropriate sample size for each group in the blood-doping experiment. Procedures for using the bottom half of Table H for two-sample tests are much the same as those for using the top half of Table H for one-sample tests, as described in Section 12.6. The first step is to specify, prior to the experiment, tolerable error rates, α and β, for some effect size d. In the two-sample case, the effect size d is defined as follows:

$$d = \frac{(\mu_1 - \mu_2)_{\text{true}}}{\sigma} \tag{14.6}$$

where $(\mu_1 - \mu_2)_{\text{true}}$ refers to the true difference between population means, and σ refers to the population standard deviation, which is

TABLE 14.1
CALCULATIONS FOR
t TEST: TWO
INDEPENDENT SAMPLES
(BLOOD-DOPING
EXPERIMENT)

I. FINDING $\bar{X}_1$, $\bar{X}_2$, s_1^2, AND s_2^2

(a) Computational sequence:

Assign a value to n_1 **①**.

Sum all X_1 scores **②**.

Substitute numbers in formula **③** and solve for $\bar{X}$.

Square each X_1 score **④**, one at a time, and then add all squared X_1 scores **⑤**.

Substitute numbers into formula **⑥** and solve for s_1^2.

Repeat this entire computational sequence for n_2 and X_2, and solve for $\bar{X}_2$ and s_2^2.

(b) Data and computations:

Endurance Scores (Minutes)

Experimental		Control	
X_1	X_1^2 ④	X_2	X_2^2
5	25	5	25
8	64	4	16
9	81	8	64
8	64	6	36
10	100	7	49

① $n_1 = 5$ **②** $\Sigma X_1 = 40$ **⑤** $\Sigma X_1^2 = 334$ $n_2 = 5$ $\Sigma X_2 = 30$ $\Sigma X_2^2 = 190$

③ $\bar{X}_1 = \dfrac{\Sigma X_1}{n_1} = \dfrac{40}{5} = 8$

$\bar{X}_2 = \dfrac{\Sigma X_2}{n_2} = \dfrac{30}{5} = 6$

⑥ $s_1^2 = \dfrac{n_1(\Sigma X_1^2) - (\Sigma X_1)^2}{n_1(n_1 - 1)}$

$= \dfrac{5(334) - (40)^2}{5(5 - 1)}$

$= \dfrac{1670 - 1600}{5(4)}$

$= \dfrac{70}{20} = 3.5$

$s_2^2 = \dfrac{n_2(\Sigma X_2^2) - (\Sigma X_2)^2}{n_2(n_2 - 1)}$

$= \dfrac{5(190) - (30)^2}{5(5 - 1)}$

$= \dfrac{950 - 900}{5(4)}$

$= \dfrac{50}{20} = 2.5$

II. FINDING s_p^2

(a) Computational sequence:

Substitute numbers obtained above in formula **⑦** and solve for s_p^2.

(b) Computations:

⑦ $s_p^2 = \dfrac{(n - 1)s_1^2 + (n_2 - 1)s_2^2}{(n_1 - 1) + (n_2 - 1)} = \dfrac{(5 - 1)3.5 + (5 - 1)2.5}{(5 - 1) + (5 - 1)}$

$= \dfrac{(4)3.5 + (4)2.5}{(4) + (4)} = \dfrac{14 + 10}{8} = \dfrac{24}{8} = 3$

III. FINDING $s_{\bar{X}_1 - \bar{X}_2}$

(a) Computational sequence:

Substitute numbers obtained above in formula **⑧** and solve for $s_{\bar{X}_1 - \bar{X}_2}$.

(b) Computations:

⑧ $s_{\bar{X}_1 - \bar{X}_2} = \sqrt{\dfrac{s_p^2}{n_1} + \dfrac{s_p^2}{n_2}} = \sqrt{\dfrac{3}{5} + \dfrac{3}{5}}$

$= \sqrt{\dfrac{6}{5}} = \sqrt{1.2} = 1.10$

IV. FINDING THE OBSERVED t

(a) Computational sequence:

Substitute numbers obtained above in formula **⑨**, as well as a value of 0 for the expression $(\mu_1 - \mu_2)_{\text{hyp}}$, and solve for t.

(b) Computations:

⑨ $t = \dfrac{(\bar{X}_1 - \bar{X}_2) - (\mu_1 - \mu_2)_{\text{hyp}}}{s_{\bar{X}_1 - \bar{X}_2}}$

$= \dfrac{(8 - 6) - 0}{1.10} = \dfrac{2}{1.10} = 1.82$

assumed to be the same for both populations. Essentially, d is a type of standard score. It reflects the smallest difference in population means, expressed relative to the population standard deviation, for which β should have some specified value. As before, only three values of d are listed in the bottom half of Table H, and these reflect the rule of thumb for small, medium, and large effects.

In the blood-doping experiment, the consequences of a type I error (an extensive blood-doping program even though it has no effect on endurance) are probably more serious than those of a type II error (no blood-doping program even though it increases endurance). It's appropriate, therefore, to let α, the probability of a type I error, equal .01 rather than .05, and not to be too concerned about β, the probability of a type II error. Given a one-tailed test, with $\alpha = .01$ and $\beta = .20$ for a $d = 0.8$, *each* group in the blood-doping experiment should consist of 33 athletes, according to Table H. The recommended sample size is considerably larger than the five athletes per group used in the previous example.

14.12 Assumptions

Use of the present test for a difference between population means presupposes that both samples are independent. *Two* **independent samples** *occur if observations in one sample are not paired with those in the other sample.* The current t test for two independent samples should be contrasted with the t test for two dependent samples described in the next chapter.

Furthermore, the present t test assumes that both underlying populations are normally distributed with equal variances. This test can be used with confidence as long as both sample sizes are equal and each is fairly large (greater than about 10). Otherwise, watch out for any conspicuous departures from either normality or equality of variances in the two sets of sample observations. In those situations where either of these assumptions seems to be seriously violated, increase the sample size before attempting a hypothesis test, or use a less sensitive test, such as the Mann-Whitney U test described in Chapter 20.

MORE GENERAL COMMENTS ABOUT HYPOTHESIS TESTS

14.13 Secondary Status of the Null Hypothesis

Even though the null hypothesis is the focus of a statistical test, it's usually of secondary concern to the investigator. In the blood-doping experiment, for instance, the investigator probably hopes to muster support for the alternative hypothesis (that the mean endurance score for the population of experimental athletes exceeds the mean endurance score for the population of control athletes). Indeed, the alternative hypothesis often corresponds to the investigator's **research hypothesis**—the informal hypothesis or hunch that inspired the entire investigation.

There are several reasons why the research hypothesis is not tested directly. To be tested, a hypothesis must specify a single number about which the hypothesized sampling distribution can be constructed. The null hypothesis satisfies this requirement; it specifies that the difference between population means equals zero. Typically, the research hypothesis violates this requirement. For instance, in the blood-doping experiment, it merely specifies that, insofar as blood doping increases endurance, the difference between population means should be *some* positive number. *The research hypothesis lacks the necessary precision to be tested directly.*

Logical considerations also argue against the direct testing of the research hypothesis. As has been mentioned previously, the decision to retain the null hypothesis is weaker than the decision to reject it. Retention of the null hypothesis signifies merely that it could be true. If tested with the same data, other similar hypotheses—for instance, that the population mean difference is in the *vicinity* of zero—also would have been retained, and therefore, they also could be true. Clearly, retention of the null hypothesis provides, at most, only weak support for that hypothesis.

On the other hand, rejection of the null hypothesis signifies that it's almost surely false. If the null hypothesis is almost surely false, then the alternative hypothesis is almost surely true.

It makes sense, therefore, to use the research hypothesis as the alternative hypothesis. If, as hoped, the data favor the research hypothesis, the test will generate strong support for your "hunch"—it's almost surely true. If the data don't favor the research hypothesis, the hypothesis test will generate, at most, weak support for the null hypothesis—it could be true. Weak support for the null hypothesis is of little consequence since this hypothesis usually serves only as a convenient testing device.

14.14 Statistical Significance vs. Practical Importance

Hypothesis tests are often referred to as significance tests. When the null hypothesis is rejected, **statistical significance** has been established.

Don't confuse statistical significance with practical importance. Statistical significance merely indicates that a test result is improbable, given that the null hypothesis is true. It doesn't indicate whether the null hypothesis is seriously false (because of a huge difference between population means) or mildly false (because of a slight difference between population means).

Having encountered a statistically significant test, always check the size of the observed effect. In the present context, always check the size of the difference between sample means. For instance, imagine a report that blood doping has a statistically significant effect on endurance scores. However, a little detective work reveals that, on the average, blood doping actually increases endurance by only three seconds—a small fraction of the sample standard deviation. From most perspectives, this result lacks practical importance, and blood doping should not be employed even though the test results were statistically significant.

Statistical significance that lacks practical importance is often caused by the use of excessively large sample sizes. With large sample sizes—for instance, with 1000 athletes per group in the blood-doping experiment—even a very small, unimportant difference between population means will be detected (because of the small standard error), and the test will be statistically significant. To avoid excessively large sample sizes in your own investigations, select sample size with the aid of tables, such as Table H. When used appropriately these tables specify a sample size that will detect, *with a high probability,* only those differences that you, the investigator, judge to be practically important.

14.15 A Note on Usage

Published reports of hypothesis tests usually are brief. Often they consist of only an interpretative comment, plus a parenthetical statement that summarizes the statistical analysis. A published report of the previous hypothesis test might read as follows:

"There is a lack of evidence that, on the average, blood doping increases endurance scores ($t(8) = 1.82$, $p > .05$)."

The parenthetical statement indicates that a t based on 8 degrees of freedom was found to equal 1.82. Furthermore, p, the probability of the observed t—on the assumption that H_0 is true—exceeds .05.* Since p exceeds .05, the observed t qualifies as a probable outcome, and therefore, the null hypothesis is retained, as implied in the interpretative statement.

If the null hypothesis had been rejected in the previous hypothesis test—for instance, if an observed t of 2.50 exceeds the critical t of 1.860—a published report might read as follows:

"There is evidence that, on the average, blood doping increases endurance scores ($t(8) = 2.50$, $p < .05$)."

Now, since p, the probability of the observed t, is less than .05, the observed t qualifies as an improbable outcome, and the null hypothesis is rejected.

Notice the importance of the *direction of the inequality symbol.* When $p > .05$ the null hypothesis is retained, but when $p < .05$ the null hypothesis is rejected.

14.16 Suspending Judgment about the Null Hypothesis

Some investigators adopt a less structured approach to hypothesis testing. The null hypothesis is neither retained nor rejected, but viewed with *degrees of suspicion,* depending on the size of p. According to this view, the more improbable the test result (as when, for instance, $p < .001$ or, at least, $p < .01$), the more suspect is the null hypothesis—and the more credible is the research hypothesis. A single research report might describe a batch of tests where some have $p < .05$, others have $p < .01$, and still others have $p < .001$.

This approach has merit. Having eliminated the requirement that the null hypothesis either be retained or be rejected, you can postpone a decision until sufficient evidence has been mustered, possibly from a series of investigations. This perspective is very attractive when test results are borderline as, for instance, in the previous hypothesis test where the null hypothesis had to be retained even though the observed t of 1.82 was only slightly less than the critical t of 1.860.

One weakness of this less structured approach is that, in the absence of a firm commitment either to retain or to reject the null hypothesis according to some predetermined level of significance, it's impossible to consider the important notions of type I and type II errors, as well as the selection of an appropriate sample size. For these reasons, a more structured approach to hypothesis testing is emphasized throughout this book.

14.17 Summary

To test the null hypothesis for the difference between two population means, you must deal with the difference between sample means, $\bar{X}_1 - \bar{X}_2$. Statistical theory pinpoints two important properties of the sampling distribution of $\bar{X}_1 - \bar{X}_2$:

---

*Strictly speaking, p refers to the probability of not just the one observed t but to the probability of all ts as extreme as, or *more extreme than,* the observed t—on the assumption that H_0 is true.

(1) The *mean* of the sampling distribution equals the difference between population means.

(2) The *standard error* of the difference between sample means is defined in Formula 14.2.

The null hypothesis for the difference between two population means can be tested with t, as defined in Formula 14.5. The t ratio is distributed with $n_1 + n_2 - 2$ degrees of freedom, as described in Table B.

The null hypothesis takes the form:

$$H_0: \mu_1 - \mu_2 = 0$$

The alternative hypothesis must be selected from among the following three possibilities:

(1) Non-directional:

$$H_1: \mu_1 - \mu_2 \neq 0$$

(2) Directional, lower tail critical:

$$H_1: \mu_1 - \mu_2 < 0$$

(3) Directional, upper tail critical:

$$H_1: \mu_1 - \mu_2 > 0$$

Table H can be consulted to determine, prior to an investigation, the required sample sizes in order for α and β to equal a preselected value for some effect size d.

The t test for a difference between population means presupposes that the two samples are independent. Furthermore, it assumes that both underlying populations are normally distributed with equal variances. Except under the most extreme circumstances, violations of the latter assumptions only slightly affect the accuracy of the t test.

Although the research hypothesis, rather than the null hypothesis, is of primary concern, the research hypothesis is not tested directly because it lacks the necessary precision. Logical considerations also argue against the direct testing of the research hypothesis.

Don't confuse statistical significance with practical importance. To gauge the practical importance of a statistically significant result, check the size of the observed effect, that is, the size of the difference between sample means.

Important Terms & Symbols

Difference between sample means $(\overline{X}_1 - \overline{X}_2)$
Difference between population means $(\mu_1 - \mu_2)$
Pooled variance estimate (s_p^2)
Estimated standard error of the difference between means $(s_{\overline{X}_1 - \overline{X}_2})$
t test for two population means: independent samples
Research hypothesis
Statistical significance

14.18 Exercises

1. To test compliance with authority, a classical experiment in social psychology requires subjects to administer increasingly painful electric shocks to seemingly helpless victims, who agonize in an adjacent room. Each

subject earns a score between 0 and 25, depending on the point at which the subject refuses to comply with authority—an experimenter who orders the administration of increasingly intense shocks. A score of 0 signifies the subject's unwillingness to comply at the very outset, while a score of 25 signifies the subject's willingness to comply completely with orders.

You are curious about the effect of a "committee atmosphere" on compliance with authority. In the experimental condition, shocks are administered only after an affirmative decision by the committee, consisting of one "real" subject and two "stooges," who simply go along with the decision of the real subject. In the control condition, shocks are administered only after an affirmative decision by a solitary "real" subject.

Six subjects are randomly assigned to the committee condition and six subjects are randomly assigned to the solitary condition. A compliance score is obtained for each subject. Use t to test the null hypothesis at the .05 level of significance.

COMPLIANCE SCORES

Committee	Solitary
2	3
5	8
20	7
15	10
4	14
10	0

2. A psychologist wants to determine the effect of instructions on the time required to solve a mechanical puzzle. Each of twenty volunteers is given the same mechanical puzzle to be solved as rapidly as possible. Prior to the task, subjects are randomly assigned, in equal numbers, to receive two different sets of instructions. One group is told that their task is difficult, while the other group is told that their task is easy. The score for each subject reflects the time in minutes required to solve the puzzle.

SOLUTION TIMES

"Difficult" Task	"Easy" Task
5	13
20	6
7	6
23	5
30	3
4	6
9	10
8	20
20	9
12	12

(a) Use t to test the null hypothesis at the .05 level of significance.
(b) If the investigator had wanted $\alpha = .05$ and $\beta = .20$ for a medium effect, what should have been the sample size?

3. Let's return to the investigator, first described in Chapter 12, who wants to determine whether daily doses of vitamin C increase intellectual aptitude. Now a total of seventy high school students are randomly designated as either experimental subjects, who receive daily doses of 50 milligrams of vitamin C, or control subjects, who receive daily doses of fake vitamin C.

After two months of daily doses, IQ scores are obtained. The mean IQ for the experimental subjects ($\bar{X}_1$) equals 110, the mean IQ for the control subjects ($\bar{X}_2$) equals 108, and the estimated standard error equals 1.80.

(a) Using t, test the null hypothesis at the .01 level of significance.
(b) How might the results of this test appear in a published report?
(c) Is the present sample size of 35 per group adequate to detect a small, medium, or large effect, according to Table H?

4. Is the performance of college students affected by grading policy? Within the same introductory biology class, a total of eighty student volunteers are randomly assigned, in equal numbers, to either take the class for letter grades or for a simple pass/fail. At the end of the academic term, the mean achievement score for the letter grade students equals 86.2, while the mean achievement score for pass/fail students equals 81.6. The estimated standard error is 1.50.

(a) Use t to test the null hypothesis at the .05 level of significance.
(b) How might the results of this test appear in a published report?
(c) Most students would doubtless prefer to select their favorite grading policy rather than be randomly assigned to a particular grading policy. Why not, therefore, replace random assignment with self-selection?

5. An investigator wishes to determine whether alcoholic consumption causes a deterioration in the performance of automobile drivers. Prior to the driving test, subjects drink a glass of orange juice which, in the case of experimental subjects, is laced with two ounces of vodka. Performance is measured by the number of errors made on a driving simulator, such as those found in many amusement park arcades. A total of one hundred twenty volunteer subjects are randomly assigned, in equal numbers, to the two groups. The mean number of errors for the experimental subjects equals 26.4, while that for the control subjects equals 18.6. The estimated standard error equals 2.4.

(a) Use t to test the null hypothesis at the .05 level of significance.
(b) How might the results of this test appear in a published report?
(c) Is the present sample size of 60 per group adequate to detect a small, medium, or large effect, according to Table H?

6. Statistically significant results are reported for experiment A and for experiment B. Although both use the .05 level of significance and are similar in every other respect, experiment A employed 20 subjects per group, while experiment B employed 100 subjects per group.

(a) Which experiment most likely reflects an effect that has practical importance?
(b) Which experiment is most likely to detect a medium effect?
(c) Which experiment is most likely to commit a type I error?
(d) Given the above information, is it possible to characterize one experimental result as being better than the other? If so, which?

15

t Test for Two Dependent Samples

15.1 –Matching Pairs of Athletes in the Blood-doping Experiment
15.2 –Sampling Distribution of $\bar{X}_1 - \bar{X}_2$ for Two Dependent Samples
15.3 –t Ratio
15.4 –t Test for the Blood-doping Experiment (Two Dependent Samples)
15.5 –Selection of Sample Size
15.6 –Assumptions
15.7 –To Match or Not to Match?
15.8 –Using the Same Subject in Both Groups
15.9 –Hypothesis Tests for Population Means: An Overview
15.10–Summary
15.11–Exercises

You can view the t test for two dependent samples as being closely related to the t test for two independent samples, or from a slightly different perspective, as being closely related to the t test for a single sample.

When used appropriately, matching reduces the standard error. Otherwise, matching can waste valuable degrees of freedom.

It's often tempting to use subjects twice—once in the experimental condition and once in the control condition. This technique can be hazardous to your research career.

Preliminary studies of blood doping, as described in Chapter 14, might reveal an unexpected phenomenon: lightweight athletes in both the blood-doped and control groups have better endurance scores than heavier athletes. This factor complicates the search for the effect of blood doping on endurance scores.

In subsequent experiments, therefore, it might be advantageous to pair athletes with similar body weights. Prior to the collection of data, athletes could be matched for body weight, beginning with the two lightest and ending with the two heaviest athletes. Once a member of a given pair has been randomly assigned either to the experimental or control group, the other member of that pair is automatically assigned to the remaining group. Thereafter, athletes are treated as in the original blood-doping experiment; that is, endurance scores are obtained for both experimental athletes (who are blood doped) and control athletes (who are injected with a harmless fluid).

Because athletes within each pair are matched for body weight, pairs of endurance scores will tend to be similar, and the statistical test must be altered to reflect this new dependency between the scores for *pairs* of experimental and control athletes. This new test for two dependent samples is relatively straightforward, and under appropriate circumstances, it's preferred to the test for two independent samples described in the previous chapter.

15.1 Matching Pairs of Athletes in the Blood-doping Experiment

15.2 Sampling Distribution of $\overline{X}_1 - \overline{X}_2$ for Two Dependent Samples

The mean of the sampling distribution for two dependent samples is the same as the mean of the sampling distribution for two independent samples. In both cases, the mean of the sampling distribution equals the difference between population means.

However, the formulas for the standard error differ for the two sampling distributions. According to statistical theory, when samples are dependent, the formula for the standard error reads

$$\sigma_{\overline{X}_1-\overline{X}_2} = \sqrt{\frac{\sigma_1^2}{n_1} + \frac{\sigma_2^2}{n_2} - 2\rho\left(\frac{\sigma_1}{\sqrt{n_1}}\right)\left(\frac{\sigma_2}{\sqrt{n_2}}\right)} \qquad (15.1)$$

where σ_1^2 and σ_2^2 are the population variances, n_1 and n_2 are the sample sizes, and ρ is the Pearson correlation coefficient for the population of all pairs of scores X_1 and X_2. In the current example, ρ represents the correlation coefficient for pairs of endurance scores when all athletes in the experimental and control populations are matched for body weight.

The negative term in Formula 15.1 distinguishes the standard error for two dependent samples from that for two independent samples (Formula 14.2). This term reflects the degree of dependency between paired observations, and when combined with the remaining terms, causes a reduction in the standard error—a most desirable consequence. If this reduction can be anticipated by the investigator prior to the collection of data, it can be used to justify a smaller sample size, as will be described in Section 15.6. Otherwise, it automatically reduces β, the probability of a type II error and, therefore, makes the hypothesis test more sensitive.

15.3 t Ratio

When using t to test the null hypothesis for the current blood-doping experiment, computations can be simplified by working directly with the difference between pairs of endurance scores, that is, by working directly with

$$D = X_1 - X_2 \qquad (15.2)$$

where **D** is the **difference score,** and X_1 and X_2 are the endurance scores for pairs of experimental and control athletes, respectively. The use of difference scores converts a two-sample problem into a one-sample problem and thus eliminates the need to calculate a correlation coefficient.

The null hypothesis can be tested with the following ratio:

$$t = \frac{\overline{D} - \mu_{D_{\text{hyp}}}}{s_{\overline{D}}} \qquad (15.3)$$

which complies with a t distribution having $n - 1$ degrees of freedom, where n *equals the number of difference scores.* In Formula 15.3, $\overline{D}$ represents the **sample mean of the difference scores;** $\mu_{D_{\text{hyp}}}$ represents the hypothesized mean (of zero) for all difference scores in the population; and $s_{\overline{D}}$ represents the **estimated standard error** of the mean of the difference scores, as defined in Formula 15.4.

The estimated standard error, $s_{\overline{D}}$, can be obtained from the following expression:

$$s_{\overline{D}} = \frac{s_D}{\sqrt{n}} \qquad (15.4)$$

where s_D represents the sample standard deviation for the observed difference scores, as defined next in Formula 15.5, and n equals the number of difference scores.

Finally, the sample standard deviation, s_D, can be obtained from the following expression:

$$s_D = \sqrt{\frac{n\Sigma D^2 - (\Sigma D)^2}{n(n-1)}} \qquad (15.5)$$

where D represents the difference scores, and n equals the number of difference scores.

Except for a change in notation from X to D, the above formulas for two dependent samples are exactly the same as their counterparts for one sample in Chapter 13. Once difference scores have been obtained, the computational procedures for the current test should seem familiar since they already have been encountered in Chapter 13.

The use of difference scores transforms the original pair of populations—one for scores of experimental athletes, the other for scores of control athletes—into a single population of difference scores, and the null hypothesis can be expressed in terms of this new population. If blood doping has no consistent effect on endurance scores when athletes are paired for body weight, the **population mean of all difference scores, μ_D,** should equal zero. In symbols, an equivalent statement reads:

15.4 t **Test for the Blood-doping Experiment (Two Dependent Samples)**

$$H_0: \mu_D = 0$$

As before, the investigator probably wants to reject the null hypothesis only if there is evidence that blood doping increases endurance scores. An equivalent statement, in symbols, reads:

$$H_1: \mu_D > 0$$

This directional alternative hypothesis translates into a one-tailed test with the upper tail critical.

There are two other possible alternative hypotheses. Another directional hypothesis, expressed as

$$H_1: \mu_D < 0$$

translates into a one-tailed test with the lower tail critical, while a non-directional hypothesis, expressed as

$$H_1: \mu_D \neq 0$$

translates into a two-tailed test.

Hypothesis Test Summary:
t Test for Two Population Means: Dependent Samples
(Blood-doping Experiment)

Problem:

Does the population mean endurance score for athletes who are blood doped exceed that for athletes who aren't blood doped, given that athletes are matched for body weight?

Statistical Hypotheses:

$$H_0: \mu_D = 0$$
$$H_1: \mu_D > 0$$

Statistical Test:

$$t = \frac{\bar{D} - \mu_{D_{hyp}}}{s_{\bar{D}}}$$

Decision Rule:
   Reject H_0 at the .05 level of significance if $t \geq 2.015$ (from Table B, given $df = n - 1 = 6 - 1 = 5$).

Calculations:

$$t = \frac{2 - 0}{0.68} = 2.94 \text{ (See Table 15.1 for computations.)}$$

Decision:
   Reject H_0.

Interpretation:
   There is evidence that, when athletes are matched for body weight, blood doping tends to increase endurance.

The present test is based on endurance scores for six pairs of athletes (matched for body weight) shown in Table 15.1. This table also shows the various computations that produce a *t* of 2.94.

According to the present test, the null hypothesis can be rejected, and there is evidence that, when athletes are matched for body weight, blood doping increases endurance. It's important to mention the matching procedure in any conclusion. When matching was absent, as in the previous test of blood doping with two independent samples, the null hypothesis was retained, not rejected. In effect, the matching procedure eliminates one source of variability among endurance scores—the variability due to differences in body weight—that otherwise inflates the standard error term and causes an increase in β, the probability of a type II error.

15.5 Selection of Sample Size

To determine the required sample size for two dependent samples, follow the same procedure as described in Section 14.11 for two independent samples. Given the specifications for the previous test with two independent samples—that is, given a one-tailed test with $\alpha = .01$ and $\beta = .20$ for $d = 0.8$—the bottom half of Table H reveals that 33 *pairs* of athletes should be employed in the blood-doping experiment.

When samples are dependent, the entry from Table H can be reduced without sacrificing test specifications, because of the extra precision due to matching. Simply multiply the entry from Table H by $1 - \rho$, where ρ represents the population correlation coefficient for all pairs of endurance scores. Ordinarily, ρ is not known, and it must be estimated. For instance, one estimate of ρ is 0.66, the value of the sample correlation coefficient, *r*, obtained (from Formula 7.2) for the six pairs of endurance scores in Table 15.1. To determine the appropriate sample size, proceed as follows:

i) Locate the appropriate entry in Table H:

33

ii) Multiply this entry by $1 - r$, that is, in the present case, by $1 - .66$ or .34, and round up to the next highest whole number:

$$(33)(.34) = 11.22 \text{ or } 12$$

TABLE 15.1
CALCULATIONS FOR
t TEST: TWO
DEPENDENT SAMPLES
(BLOOD-DOPING
EXPERIMENT)

I. FINDING $\bar{D}$ AND s_D

(a) Computational sequence:

Assign a value to n, the number of paired scores **①**.
Subtract X_2 from X_1 to obtain D **②**.
Sum all D scores **③**.
Substitute numbers in formula **④** and solve for $\bar{D}$.
Square each D score **⑤**, one at a time, and then add all squared D scores **⑥**.
Substitute numbers into formula **⑦** and solve for s_D.

(b) Data and computations:

Endurance Scores (Minutes)

Pair	Experimental Athletes X_1	Control Athletes X_2	Difference Scores ② D	⑤ D^2
1	9	7	2	4
2	4	5	−1	1
3	8	4	4	16
4	8	6	2	4
5	10	7	3	9
6	10	8	2	4
① $n = 6$			③ $\Sigma D = 12$	⑥ $\Sigma D^2 = 38$

$$④\quad \bar{D} = \frac{\Sigma D}{n} = \frac{12}{6} = 2$$

$$⑦\quad s_D = \sqrt{\frac{n(\Sigma D^2) - (\Sigma D)^2}{n(n-1)}} = \sqrt{\frac{6(38) - (12)^2}{6(6-1)}}$$

$$= \sqrt{\frac{228 - 144}{6(5)}} = \sqrt{\frac{84}{30}}$$

$$= \sqrt{2.8} \qquad = 1.67$$

II. FINDING $s_{\bar{D}}$

(a) Computational sequence:

Substitute numbers obtained above in formula **⑧** and solve for $s_{\bar{D}}$.

(b) Computations:

$$⑧\quad s_{\bar{D}} = \frac{s_D}{\sqrt{n}} = \frac{1.67}{\sqrt{6}} = \frac{1.67}{2.45} = 0.68$$

III. FINDING THE OBSERVED t

(a) Computational sequence:

Substitute numbers obtained above in formula **⑨**, as well as a value of 0 for $\mu_{D_{hyp}}$, and solve for t.

(b) Computations:

$$⑨\quad t = \frac{\bar{D} - \mu_{D_{hyp}}}{s_{\bar{D}}} = \frac{2 - 0}{0.68} = 2.94$$

It would be wise to increase this number by one since the above procedure tends to underestimate required sample sizes (particularly small required sample sizes of 15 or less). A cautious investigator might increase this number even more, possibly to about 20, because of the considerable sampling variability of an r based on only six pairs of observations. In any event, when matching involves a relevant variable, such as body weight in the blood-doping experiment, sample size can be reduced considerably—in the present case, from 33 to 20 or less—without sacrificing test specifications.

15.6 Assumptions

The present test for a difference between population means presupposes that both samples are dependent. **Dependent samples** *occur if observations in one sample are paired with those in the other sample.*

Furthermore, the present *t* test assumes that the population of difference scores is normally distributed. This test can be used with confidence as long as sample size is fairly large (greater than about 10 pairs). Otherwise, watch out for conspicuous departures from normality. In those situations where this assumption seems to be seriously violated, either increase the sample size before attempting a hypothesis test, or use a less sensitive test, such as the Wilcoxon *T* test described in Chapter 20.

15.7 To Match or Not to Match?

Inappropriate matching is costly and should be avoided. Matching usually takes extra effort, such as the preliminary weighing of athletes, and wastes a few subjects who can't be matched because, for instance, they weigh too much or too little.

A shift in the unit of analysis from original observations, as in two independent samples, to the differences between *pairs* of original observations, as in two dependent samples, causes the degrees of freedom to be reduced by a factor of one half. For example, when a total of 30 subjects are used in an experiment with two independent samples, the *t* test has 28 degrees of freedom (from $15 + 15 - 2$), but when the same total number of subjects are sorted into 15 pairs in an experiment with two dependent samples, the *t* test has only half as many degrees of freedom, that is, 14 (from $15 - 1$). In the absence of effective matching, the net effect might be a *less* sensitive hypothesis test because of the wasted degrees of freedom.

When samples are dependent, any conclusion applies only to a population with matching restrictions. In the most recent hypothesis test, there is evidence that blood doping increases endurance scores *only* in a population of athletes *who are matched for body weight.* There is no basis for assuming that blood doping also will produce a demonstrable increase in endurance scores under less controlled circumstances—when athletes aren't matched for body weight.

Matching is desirable only when some uncontrolled variable appears to have a considerable impact on the variable being measured. Appropriate matching reduces variability—and the estimated standard error—and, therefore, produces a more sensitive hypothesis test (or a test that requires a smaller sample size).

To identify a variable worthy of matching, you should familiarize yourself with all previous research in the area and conduct pilot studies. Use matching only if, prior to the full-fledged investigation, you're able to detect an uncontrolled variable that, when identified, aids your interpretation of preliminary findings.

15.8 Using the Same Subject in Both Groups

As a special case, two dependent samples might involve the use of the same subjects in both samples. If, for instance, the same athletes are used in both the experimental and control groups of the blood-doping experiment, any variability due to "individual differences" would be eliminated. This technique controls the groups not only for body weight but also for any other possibly important characteristic of subjects, such as physical strength, age, sex, experience, attitude, and so forth. Ideally, any differences between pairs of endurance scores then would be primarily due to blood doping—a most desirable consequence.

Unfortunately, the attractiveness of this design often fades with closer inspection. Since each athlete performs twice, once in the experimental condition and once in the control condition, there's a possibility that performance in one condition might be "contaminated" by the subject's prior experience with the other condition. For instance, if some athletes are tested first under the experimental condition and then under the control condition, sufficient time must elapse between these two conditions to eliminate any lingering effects due to blood doping. If there is any concern that these effects can't be eliminated, use each subject in only one condition.

Otherwise, when subjects do perform double duty in both conditions, *it's customary to randomly assign half of the subjects to experience the two conditions in a particular order*—say, first the experimental and then the control condition—*while the other half of the subjects experience the two conditions in the reverse order.* Known as **counter-balancing,** this adjustment eliminates a potential bias in favor of one condition merely because most subjects happen to experience it first (or second).

15.9 Hypothesis Tests for Population Means: An Overview

Previous chapters have described a variety of t tests for population means. Ordinarily, it's fairly easy to decide whether to use a t test for one sample, two independent samples, or two dependent samples. When the test is to be based on a single set of observations, use the t test for one sample. When the test is to be based on two sets of observations that aren't paired, use the t test for two independent samples. When the test is to be based on two sets of observations that are paired, because subjects are matched or because the same subject is measured twice, use the t test for two dependent samples. The more distinctive features of each of these three tests are summarized in Table 15.2.

15.10 Summary

When two samples are dependent—that is, when pairs of observations are matched—the sampling distribution of $\bar{X}_1 - \bar{X}_2$ is very similar to its counterpart for two independent samples. A major difference is the negative term, reflecting the degree of dependency between paired observations, that appears in the standard error formula for two dependent samples.

TABLE 15.2
SUMMARY OF t TESTS FOR POPULATION MEANS

TYPE OF SAMPLE	SAMPLE MEAN	NULL HYPOTHESIS	STANDARD ERROR	t RATIO	DEGREES OF FREEDOM
One sample	$\bar{X}$	$H_0: \mu =$ some number	$s_{\bar{X}}$ (Formula 13.3)	$\dfrac{\bar{X} - \mu_{hyp}}{s_{\bar{X}}}$	$n - 1$
Two Independent Samples (No pairing)	$\bar{X}_1 - \bar{X}_2$	$H_0: \mu_1 - \mu_2 = 0$	$s_{\bar{X}_1 - \bar{X}_2}$ (Formula 14.4)	$\dfrac{(\bar{X}_1 - \bar{X}_2) - (\mu_1 - \mu_2)_{hyp}}{s_{\bar{X}_1 - \bar{X}_2}}$	$n_1 + n_2 - 2$
Two Dependent Samples (Pairing)	$\bar{D}$	$H_0: \mu_D = 0$	$s_{\bar{D}}$ (Formula 15.4)	$\dfrac{\bar{D} - \mu_{D_{hyp}}}{s_{\bar{D}}}$	$n - 1$ (where n refers to pairs of observations)

In the two dependent sample case, the *t* test is defined in Formula 15.3. This *t* ratio is distributed with $n - 1$ degrees of freedom (in Table B), given that n equals the number of paired observations.

The formula for *t* is based on the difference, D, between pairs of observations. The use of difference scores transforms the original pair of populations into a single population of difference scores. The null hypothesis takes the form:

$$H_0: \mu_D = 0$$

where μ_D represents the population mean for all difference scores. The alternative hypothesis must be selected from among the following three possibilities:

(1) Non-directional:

$$H_1: \mu_D \neq 0$$

(2) Directional, lower tail critical:

$$H_1: \mu_D < 0$$

(3) Directional, upper tail critical:

$$H_1: \mu_D > 0$$

To determine the required sample size for two dependent samples, consult Table H, following the same procedure as would have been used for two independent samples. When samples are dependent, however, the entry from Table H can be reduced by $(1 - \rho)$, where this factor reflects the extra precision gained through matching.

When using the *t* test for two dependent samples, you must assume that the population of difference scores is normally distributed. Violations of this assumption are relatively unimportant as long as sample sizes are not too small.

Inappropriate matching is costly and should be avoided. When matching is appropriate, it reduces the size of the estimated standard error—a most desirable consequence. Only match when, prior to the full-fledged investigation, you're able to detect a heretofore uncontrolled variable that aids the interpretation of preliminary findings.

As a special case, two dependent samples may involve the use of the same subjects in both samples. This technique eliminates any variability due to individual differences. There is the possibility, however, that performance in one condition might be contaminated by the subject's prior experience with the other condition. If this possibility can't be eliminated, use each subject in only one condition.

Important Terms & Symbols

Difference score (D)
Sample mean of the difference scores ($\bar{D}$)
Population mean of the difference scores (μ_D)
Estimated standard error ($s_{\bar{D}}$)
t test for two population means, dependent samples
Counterbalancing

15.11 Exercises

1. An investigator wants to test still another claim for vitamin C, namely, that it reduces the frequency of common colds. To eliminate the variability due to different family environments, pairs of children from the same family are

randomly assigned either to the experimental group, which receives vitamin C, or to the control group, which receives fake vitamin C. At the end of the study, after an entire school year, each child has a score that reflects the total number of days ill due to colds, as determined from daily inspections by the school nurse. The following scores were obtained for ten pairs of children:

Days Ill Due to Colds

PAIR NO.	EXPERIMENTAL (X_1)	CONTROL (X_2)
1	2	3
2	5	4
3	7	9
4	0	3
5	3	5
6	7	7
7	4	6
8	5	8
9	1	2
10	3	5

(a) Using t, test the null hypothesis at the .05 level of significance.

(b) Assume that ρ is estimated as .60. Is the present sample of 10 pairs adequate to detect a small, medium, or large effect, according to Table H?

(c) Still assuming that ρ is estimated as .60, how many pairs should have been used in the present experiment in order to satisfy the following test specifications: $\alpha = .05$ and $\beta = .05$ for $d = 0.5$?

2. An educational psychologist wants to check the claims of some spiritualists that the daily practice of "ABC" meditation will improve the academic achievement of practitioners. To control the experiment for academic aptitude, pairs of college students with similar grade point averages (GPA) are randomly assigned either to the experimental group, which receives daily training in ABC meditation, or to the control group, which doesn't receive training in meditation. At the end of the experiment, which spans one semester, the following GPAs are reported for the seven pairs of participants:

GPAs

PAIR NO.	EXPERIMENTAL (X_1)	CONTROL (X_2)
1	4.00	3.75
2	2.67	2.74
3	3.65	3.42
4	2.11	1.67
5	3.21	3.00
6	3.60	3.25
7	2.80	2.65

(a) Using t, test the null hypothesis at the .01 level of significance.

(b) Assuming $\rho = .50$, is the present sample size of 7 pairs adequate to detect a small, medium, or large effect?

3. A public health investigator wishes to determine whether a new anti-smoking film actually reduces the daily consumption of cigarettes by heavy smokers. The mean daily cigarette consumption is calculated for each of

eight heavy smokers during the month *before* the film presentation and also during the month *after* the film presentation, with the following results:

MEAN DAILY CIGARETTE CONSUMPTION

SMOKER NO.	BEFORE FILM (X_1)	AFTER FILM (X_2)
1	28	26
2	29	27
3	31	32
4	44	40
5	35	35
6	20	16
7	50	47
8	25	23

(a) Using t, test the null hypothesis at the .05 level of significance. (This illustrates another variation on the two dependent sample design, where the experimental condition—in this case, the film presentation—is sandwiched in time between repeated measurements for the same subject.)

(b) What might be done to improve the design of this experiment?

4. A manufacturer of a gas additive claims that it improves gas mileage under virtually any kind of driving conditions. A random sample of 30 drivers test this claim by determining their gas mileage for a full tank of gas that contains the additive (X_1) and for a full tank of gas that doesn't contain the additive (X_2). The sample mean difference, $\bar{D}$, equals 2.12 miles (in favor of the additive) and the estimated standard error equals 1.50 miles.

(a) Using t, test the null hypothesis at the .05 level of significance.
(b) Assuming ρ is estimated as .20, is the present sample of 30 pairs sufficient to detect a small, medium, or large effect?
(c) Are there any special precautions that should be taken with the present experimental design?

5. Although samples are actually dependent, an investigator ignores this fact in the statistical analysis and uses a t test for two independent samples. How will this mistake affect the probability of a type II error?

6. (a) When done properly, what's accomplished by matching?
(b) What's the cost of matching?
(c) What makes a variable worthy of matching?
(d) What serious complication can negate any advantages gained by using the same subjects in both conditions?

16
Estimation

ESTIMATING ONE POPULATION MEAN

16.1 – Investigation of SAT Verbal Scores, Revisited
16.2 – Point Estimates for μ
16.3 – Confidence Intervals for μ
16.4 – Why Confidence Intervals Work
16.5 – Confidence Intervals for μ Based on z
16.6 – Interpretation of a Confidence Interval
16.7 – Level of Confidence
16.8 – Effect of Sample Size
16.9 – Confidence Intervals for μ Based on t

ESTIMATING THE DIFFERENCE BETWEEN POPULATION MEANS

16.10 – Blood-doping Experiment, Revisited
16.11 – Point Estimates for $\mu_1 - \mu_2$
16.12 – Confidence Intervals for $\mu_1 - \mu_2$
16.13 – Confidence Intervals for $\mu_1 - \mu_2$ Based on t (Two Independent Samples)
16.14 – Confidence Intervals for $\mu_1 - \mu_2$ Based on t (Two Dependent Samples)

GENERAL COMMENTS ABOUT CONFIDENCE INTERVALS

16.15 – Hypothesis Tests or Confidence Intervals?
16.16 – Other Types of Confidence Intervals
16.17 – Summary
16.18 – Exercises

When asked to guess someone's age, you probably would use one of two types of estimates. For instance, you might estimate a person's age to be either "23 years" or "between 20 and 25 years." These estimates represent crude models of point estimates and confidence intervals, two types of estimates in statistics.

"Confidence" has a well-defined meaning in statistics. In fact, we routinely claim to be 95 percent confident or 99 percent confident that some range of values includes an unknown characteristic of the population.

Use confidence intervals whenever possible—they tend to be more informative than hypothesis tests.

In Chapter 11, an investigator was concerned about detecting any difference between the mean SAT verbal score for all local freshmen and the national average. As has been seen, this concern translates into a hypothesis test, and with the aid of a z test, it was concluded that the local population exceeds the national average. Given a concern about the national average, this conclusion is most informative—it might even create some joy among local university officials. However, the same SAT investigation could have been prompted by a wish merely *to estimate* the value of the local population mean rather than *to test a hypothesis* based on the national average. This new concern translates into an estimation problem, and with the aid of point estimates and confidence intervals, known sample characteristics can be used to estimate the mean SAT verbal score for all local freshmen.

ESTIMATING ONE POPULATION MEAN

16.1 Investigation of SAT Verbal Scores, Revisited

16.2 Point Estimates for μ

*A **point estimate** for μ specifies a single value that represents the unknown population mean.* This is the most straightforward type of estimation procedure. If a random sample of 100 local freshmen reveals a mean SAT verbal score of 462, then 462 is the point estimate of the unknown population mean for all local freshmen. The best single point estimate for the unknown population mean is simply the observed value of the sample mean.

Although straightforward, simple, and precise, point estimates suffer from a basic deficiency—they tend to be inaccurate. It's unlikely that a single sample mean equals the population mean (because of sampling variability). By the same token, it's unlikely that a point estimate, such as 462, coincides with the mean SAT verbal score for all local freshmen. Furthermore, point estimates convey no indication of this inaccuracy, causing statisticians to prefer another, more realistic type of estimate.

16.3 Confidence Intervals for μ

*A **confidence interval** for μ specifies a range of values that includes the unknown population mean a certain percent of the time.* For instance, using techniques described below, the SAT investigator might claim, *with 95 percent confidence,* that the interval between 440.44 and 483.56 includes the value of the unknown mean SAT verbal score for all local freshmen. To be 95 percent confident signifies that, if these intervals were constructed for a long series of samples, approximately 95 percent would include the mean verbal score for all local freshmen. In the long run, 95 percent of these confidence intervals are true because they include the unknown population mean. The remaining 5 percent are false because they fail to include the unknown population mean.

16.4 Why Confidence Intervals Work

Essentially, confidence intervals are based on some well-established properties of sampling distributions. First, let's briefly identify these properties—encountered previously under slightly different circumstances—and then show how they lead to the emergence of confidence intervals.

Focus on the sampling distribution from which the sample mean of 462 originates. As indicated in Figure 16.1, this sampling distribution

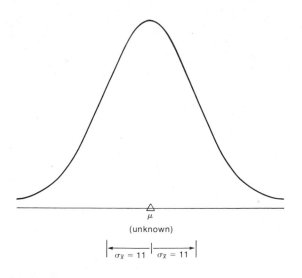

FIGURE 16.1
Sampling distribution of the mean (SAT verbal scores).

is centered about the unknown population mean for all local freshmen (whatever its value) because the mean of the sampling distribution always equals the population mean. The standard error of the sampling distribution equals 11, a value obtained from the formula for the standard error by dividing 110 (the population standard deviation) by the square root of 100 (the sample size). We know that the shape of the sampling distribution approximates a normal distribution, since the sample size of 100 satisfies the requirements of the central limit theorem.

In practice, only one sample mean is actually taken from the sampling distribution and used to construct a single 95 percent confidence interval. However, imagine taking not just one but a series of randomly selected sample means from this sampling distribution. For each of these sample means, construct a 95 percent confidence interval by adding 1.96 standard errors to the sample mean and by subtracting 1.96 standard errors from the sample mean, that is, use the expression

$$\bar{X} \pm 1.96\,\sigma_{\bar{X}}$$

to obtain a 95 percent confidence interval for each sample mean.

Why, according to statistical theory, do 95 percent of these confidence intervals include the unknown population mean? As indicated in Figure 16.2, since the sampling distribution is normal, 95 percent of all sample means are within 1.96 standard errors of the unknown population mean. Therefore, when sample means are expanded into confidence intervals—by adding 1.96 standard errors to the sample mean and subtracting 1.96 standard errors from the sample mean—95 percent of all possible confidence intervals are true because they include the unknown population mean. To illustrate this point, fifteen of the sixteen sample means shown in Figure 16.2 are within 1.96 standard errors of the unknown population mean. These fifteen confidence intervals, shown in black, include the value of the unknown population mean.

It's also the case that 5 percent of all confidence intervals fail to include the unknown population mean. As indicated in Figure 16.2, 5 percent of all sample means are not within 1.96 standard errors of the unknown population mean. Therefore, when sample means are expanded into confidence intervals—by adding 1.96 standard errors to the sample mean and subtracting 1.96 standard errors from the sample mean—5 percent of all possible confidence intervals are false because they fail to include the unknown population mean. To illustrate this point, only one of the sixteen sample means shown in Figure 16.2 is not within 1.96 standard errors of the unknown population mean. Only one confidence interval, shown in red, fails to include the value of the unknown population mean.

16.5 Confidence Intervals for μ Based on z

To determine the previously reported confidence interval (from 440.44 to 483.56) for the unknown mean SAT verbal score of all local freshmen, use the following general expression:

$$\bar{X} \pm (z_{\text{conf}})(\sigma_{\bar{X}}) \qquad (16.1)$$

where $\bar{X}$ represents the sample mean; z_{conf} represents a number from the normal tables that satisfies the confidence specifications for the confidence interval; and $\sigma_{\bar{X}}$ represents the standard error of the mean.

Given that the sample mean SAT verbal score, $\bar{X}$, equals 462, that

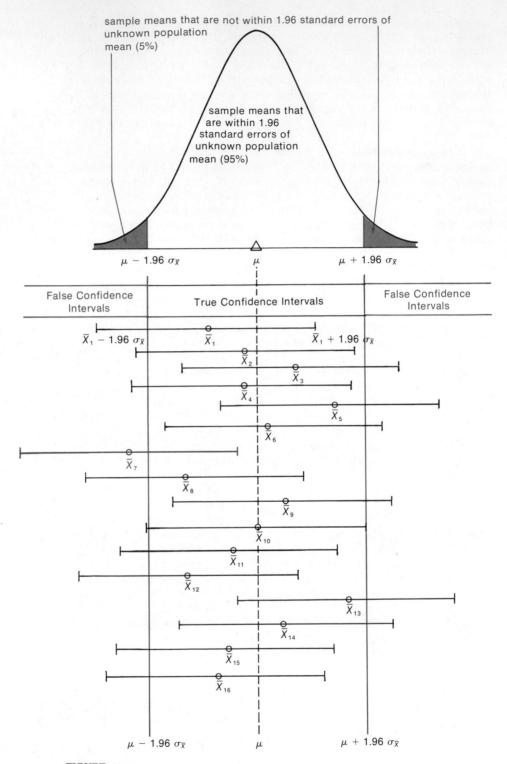

FIGURE 16.2

A series of 95 percent confidence intervals (emerging from sampling distribution).

z_{conf} equals 1.96, and that the standard error, $\sigma_{\bar{X}}$, equals 11, Formula 16.1 becomes

$$462 \pm (1.96)(11)$$

This expression, in turn, splits into two expressions, one for the lower limit of the confidence interval,

$$\begin{aligned} \text{lower limit} &= 462 - (1.96)(11) \\ &= 462 - 21.56 \\ &= 440.44 \end{aligned}$$

and the other for the upper limit of the confidence interval,

$$\begin{aligned} \text{upper limit} &= 462 + (1.96)(11) \\ &= 462 + 21.56 \\ &= 483.56 \end{aligned}$$

Now it can be claimed, with 95 percent confidence, that the interval between 440.44 and 483.56 includes the value of the unknown mean SAT verbal score for all local freshmen.

The use of Formula 16.1 to construct confidence intervals assumes that the population standard deviation is known, and that the population is normal or the sample size is sufficiently large—at least 25—to satisfy the requirements of the central limit theorem.

A 95 percent confidence claim reflects a long-term performance rating for an extended series of confidence intervals. If a series of confidence intervals is constructed to estimate the same population mean, as in Figure 16.2, approximately 95 percent of these intervals should include the population mean. In practice, only one confidence interval—not a series of intervals—is constructed, and that one interval is either true or false. Of course, *we never really know whether a particular confidence interval is true or false* unless the entire population is surveyed. When the level of confidence equals 95 percent (or more), however, we can be "reasonably confident" that the one observed confidence interval includes the true population mean. For instance, we can be reasonably confident that the interval between 440.44 and 483.56 includes the true mean SAT verbal score for all local freshmen.

16.6 Interpretation of a Confidence Interval

The **level of confidence** *indicates the percent of time that a series of confidence intervals includes the unknown population mean.* Any level of confidence may be assigned to a confidence interval merely by substituting an appropriate value for z_{conf} in Formula 16.1. For instance, to construct a 99 percent confidence interval from the data for SAT verbal scores, first consult Table A to verify that z_{conf} values of ± 2.58 define the middle 99 percent of the total area under the normal curve. Then substitute numbers for symbols in Formula 16.1 to obtain

16.7 Level of Confidence

$$462 \pm (2.58)(11)$$

This expression splits into

$$\begin{aligned} \text{lower limit} &= 462 - (2.58)(11) \\ &= 462 - 28.38 \\ &= 433.62 \end{aligned}$$

and

$$\text{upper limit} = 462 + (2.58)(11)$$
$$= 462 + 28.38$$
$$= 490.38$$

It can be claimed, *with 99 percent confidence,* that the interval between 433.62 and 490.38 includes the value of the unknown mean SAT verbal score for all local freshmen. This implies that, in the long run, 99 percent of these confidence intervals will include the unknown population mean.

Notice that the 99 percent confidence interval (from 433.62 to 490.38) for the above data is wider and, therefore, less precise than the corresponding 95 percent confidence interval (from 440.44 to 483.65). The shift from a 95 percent to a 99 percent level of confidence requires an increase in the value of z_{conf} from ± 1.96 to ± 2.58. This increase, in turn, causes a wider, less precise confidence interval. Any shift to a higher level of confidence always produces a wider, less precise confidence interval—unless offset by an increase in sample size, as mentioned in the next section.

Although many different levels of confidence have been used, 95 *percent and* 99 *percent are the most prevalent.* Generally speaking, a larger level of confidence, such as 99 percent, should be reserved for those situations where a false interval might have particularly serious consequences—for example, the widely publicized failure of national opinion pollsters to predict the election of President Truman in 1948.

16.8 Effect of Sample Size

The larger the sample size, the smaller the standard error, and hence, the more precise (narrower) the confidence interval. Indeed, as sample size grows larger, the standard error approaches zero, and the confidence interval shrinks to a point estimate.

As with hypothesis tests, sample size can be selected according to specifications established prior to the investigation. To generate a confidence interval that possesses some desired width, yet complies with the desired level of confidence, refer to formulas for sample size in other statistics books.* Valid use of these formulas requires that, prior to the investigation, the population standard deviation be either known or estimated.

16.9 Confidence Intervals for μ Based on t

In Chapter 13, an investigator was concerned about detecting any drop in the mean gas mileage for some population of cars below the legally required minimum of 45 miles per gallon. As has been seen, this concern translates into a hypothesis test, and with the aid of a t test, it was concluded that the population mean *could* equal 45 miles per gallon and that the manufacturer should receive a tax break of $10 per car.

Under slightly different circumstances, the investigator might wish to estimate the unknown mean gas mileage for the population of cars, rather than test a hypothesis based on 45 miles per gallon. For

---

*For instance, see Minium, E.W., *Statistical Reasoning in Psychology and Education,* 2nd Ed., Wiley, New York, 1978, p. 318.

example, there might be no legally required minimum of 45 miles per gallon, but merely a desire on the part of the manufacturer to estimate the mean gas mileage for a population of cars—possibly as a first step toward the design of a new, improved version of the current model.

When the population standard deviation is unknown and, therefore, must be estimated, as in the present case, t replaces z in Formula 16.1, and the new formula for a confidence interval reads as follows:

$$\bar{X} \pm (t_{\text{conf}})(s_{\bar{X}}) \qquad (16.2)$$

where $\bar{X}$ represents the sample mean; t_{conf} represents a number (distributed with $n - 1$ degrees of freedom) from the t tables, which satisfies the confidence specifications for the confidence interval; and $s_{\bar{X}}$ represents the estimated standard error of the mean, defined in Formulas 13.2 and 13.3.

To find the appropriate value for t_{conf} in Formula 16.2, refer to Table B. Read the entry from the cell intersected by the row for the correct number of degrees of freedom and the column for the confidence specifications. In the current problem, for example, if a 95 percent confidence interval is desired, first locate the row corresponding to 5 degrees of freedom (from $df = n - 1 = 6 - 1 = 5$), and then locate the column for the 95 percent level of confidence, that is, the column heading identified with a single asterisk. (A double asterisk identifies the column for the 99 percent level of confidence.) The intersected cell specifies that a value of 2.571 be entered in Formula 16.2.

Given this value for t_{conf}, as well as (from Table 13.1) values of 43 for $\bar{X}$, the sample mean gas mileage, and 0.89 for $s_{\bar{X}}$, the estimated standard error, Formula 16.2 becomes

$$43 \pm (2.571)(0.89)$$

As usual, this expression splits into

$$\text{lower limit} = 43 - 2.29$$
$$= 40.71$$

and

$$\text{upper limit} = 43 + 2.29$$
$$= 45.29$$

It can be claimed, with 95 percent confidence, that the interval between 40.71 and 45.29 includes the true mean gas mileage for all cars in the population.

The interpretation of this confidence interval is the same as that based on z. In the long run, 95 percent of all confidence intervals, similar to the one above, will include the unknown population mean. Although we never really know whether this particular confidence interval is true or false, we can be reasonably confident that the true mean for the entire population of cars is neither less than 40.71 miles per gallon nor more than 45.29 miles per gallon.

When constructing a confidence interval for the unknown population mean, use t *rather than* z *if, as is typically the case, the population standard deviation is unknown.* Strictly speaking, the use of t presupposes that the underlying population is normally distributed. Even when this normality assumption is violated, however, the resulting confidence intervals retain much of their accuracy as long as sample size is fairly large (greater than about 10).

ESTIMATING THE DIFFERENCE BETWEEN POPULATION MEANS

16.10 Blood-doping Experiment, Revisited

In Chapters 14 and 15, an experimenter attempted to determine whether the population mean endurance score for athletes who are blood doped exceeds that for athletes who are not blood doped. Essentially, the issue is whether or not blood doping increases endurance scores, and as has been seen, a test of the null hypothesis resolves this issue. If the null hypothesis is rejected, there's evidence that blood doping increases endurance; otherwise, if the null hypothesis is retained, there's a lack of evidence that blood doping increases endurance.

In this chapter, let's assume that the experimenter is not interested in determining whether or not blood doping increases endurance. Instead, he wishes to estimate, with the aid of point estimates and confidence intervals, the size of the effect of blood doping on endurance—whatever that effect might be. Under these circumstances, estimates must be obtained for the difference between population means, $\mu_1 - \mu_2$. In this experiment, $\mu_1 - \mu_2$ represents the difference between the population mean endurance score for athletes who are blood doped and the population mean endurance score for athletes who are not blood doped; the larger the estimated positive difference, the larger the estimated facilitative effect of blood doping on endurance scores.

16.11 Point Estimates for $\mu_1 - \mu_2$

A **point estimate for $\mu_1 - \mu_2$** specifies a single value that represents the unknown difference between population means. The best single point estimate for the difference between population means is simply the observed difference between sample means, $\bar{X}_1 - \bar{X}_2$ (for two independent samples), or the observed sample mean for difference scores, $\bar{D}$ (for two dependent samples). For example, if the blood-doping experiment reveals a two-minute difference between sample means—as, in fact, it did—then two minutes is the point estimate for the unknown difference between population means.

As in the one-sample case, point estimates ignore sampling variability and, therefore, tend to be inaccurate. Accordingly, confidence intervals are preferred to point estimates.

16.12 Confidence Intervals for $\mu_1 - \mu_2$

A **confidence interval for $\mu_1 - \mu_2$** specifies a range of values that includes the unknown population mean difference a certain percent of the time. For instance, using techniques described below, the investigator of blood doping might claim, with 95 percent confidence, that the interval between −0.54 and 4.54 minutes includes the true difference between population means (due to blood doping).

Under appropriate circumstances, a confidence interval for $\mu_1 - \mu_2$ can be constructed with the aid of z. The use of z, however, is hardly ever appropriate, since it requires that the population standard deviations be known. For this reason, subsequent sections ignore z in favor of t.

16.13 Confidence Intervals for $\mu_1 - \mu_2$ Based on t (Two Independent Samples)

Given that two samples are independent, as in the blood-doping experiment of Chapter 14, a confidence interval for $\mu_1 - \mu_2$ can be constructed from the following expression:

$$\bar{X}_1 - \bar{X}_2 \pm (t_{\text{conf}})(s_{\bar{X}_1 - \bar{X}_2}) \tag{16.3}$$

where $\bar{X}_1 - \bar{X}_2$ represents the difference between sample means, t_{conf} represents a number (distributed with $n_1 + n_2 - 2$ degrees of freedom) from the t tables, which satisfies the confidence specifications for the confidence interval, and $s_{\bar{X}_1 - \bar{X}_2}$ represents the estimated standard error defined in Formula 14.4.

To find the appropriate value of t_{conf} in Formula 16.3, refer to Table B and follow the same procedure as described previously for one sample. For example, if a 95 percent confidence interval is desired for the blood-doping experiment, first locate the row corresponding to 8 degrees of freedom (from $df = n_1 + n_2 - 2 = 5 + 5 - 2 = 8$), and then locate the column for the 95 percent level of confidence, that is, the column heading identified with a single asterisk. The intersected cell specifies that a value of 2.306 be entered in Formula 16.3.

Given this value for t_{conf}, as well as (from Table 14.1) values of 2 for $\bar{X}_1 - \bar{X}_2$, the difference between sample means, and 1.10 for $s_{\bar{X}_1 - \bar{X}_2}$, the estimated standard error, Formula 16.3 becomes

$$2 \pm (2.306)(1.10)$$

which, in turn, splits into

$$\text{lower limit} = 2 - 2.54$$
$$= -0.54$$

and

$$\text{upper limit} = 2 + 2.54$$
$$= 4.54$$

Now it can be claimed, with 95 percent confidence, that the interval between −0.54 and 4.54 includes the true difference between population means.

Notice that the numbers in this confidence interval refer to *differences* between population means. Otherwise, *the interpretation of a confidence interval for $\mu_1 - \mu_2$ is the same as that for μ.* In the long run, 95 percent of all confidence intervals, similar to the one above, will include the unknown difference between population means. Although we never really know whether this particular confidence interval is true or false, we can be reasonably confident that the true difference between population means (due to blood doping) is neither less than −0.54 minute nor more than 4.54 minutes.

No single interpretation adequately describes all of the possibilities included in this confidence interval. The appearance of negative differences, such as −0.54, indicates the possibility that blood doping might actually hinder endurance, while the appearance of positive differences, such as 4.54, indicates the possibility that blood doping might facilitate endurance. (The inclusion of a zero difference indicates the possibility that blood doping has no effect whatsoever on endurance.)

In spite of these contradictory possibilities, the relatively large upper limit of 4.54 minutes might stimulate additional research on blood doping. According to this possibility, blood doping might increase the population mean endurance score by 4.54 minutes, which, when compared with the sample mean for the control athletes (6.17 minutes), represents a huge 75 percent gain in endurance. Given this perspective, most investigators would probably choose to conduct another blood-doping experiment, using larger sample sizes, in order to produce a narrower, more precise confidence interval.

16.14 Confidence Intervals for $\mu_1 - \mu_2$ Based on t (Two Dependent Samples)

Given that two samples are dependent, as when athletes were matched for body weight in the blood-doping experiment of Chapter 15, a confidence interval for $\mu_1 - \mu_2$ can be constructed from the following expression:

$$\bar{D} \pm (t_{\text{conf}})(s_{\bar{D}}) \qquad (16.4)$$

where $\bar{D}$ represents the sample mean of the difference scores, t_{conf} represents a number (distributed with $n - 1$ degrees of freedom) from the t tables, which satisfies the confidence specifications for the confidence interval, and $s_{\bar{D}}$ represents the estimated standard error defined in Formula 15.4.

To find the appropriate value of t_{conf} in Formula 16.4, refer to Table B and follow the same procedure as described previously for two independent samples. If a 95 percent confidence interval is desired for the blood-doping experiment with matched athletes, first locate the row corresponding to 5 degrees of freedom (since there are 6 pairs of athletes matched for body weight, $df = n - 1 = 6 - 1 = 5$), and then locate the column for the 95 percent level of confidence, that is, the column heading identified with a single asterisk. The intersected cell specifies a value of 2.571 to be entered in Formula 16.4.

Given this value for t_{conf}, as well as (from Table 15.2) values of 2 for $\bar{D}$, the sample mean of the difference scores, and 0.68 for $s_{\bar{D}}$, the estimated standard error, Formula 16.4 becomes

$$2 \pm (2.571)(0.68)$$

which, in turn, splits into

$$\text{lower limit} = 2 - 1.75$$
$$= 0.25$$

and

$$\text{upper limit} = 2 + 1.75$$
$$= 3.75$$

It can be claimed, with 95 percent confidence, that the interval between 0.25 and 3.75 includes the true difference between population means.

In this case, a single interpretation does describe all of the possibilities included in the confidence interval. The appearance of only positive differences indicates that, when athletes are matched for body weight, blood doping facilitates endurance. Furthermore, we can be reasonably confident that, on the average, the true facilitative effect is neither less than 0.25 minute nor more than 3.75 minutes. It's instructive to compare the previous confidence interval (from -0.54 to 4.54) for two independent samples. The narrower, more precise interval for two dependent samples reflects a reduction in the estimated standard error caused by matching for body weight. In general, there's no guarantee that matching always will produce narrower, more precise confidence intervals. As suggested in Section 15.8, therefore, match only when a "worthy" variable has been identified.

GENERAL COMMENTS ABOUT CONFIDENCE INTERVALS

16.15 Hypothesis Tests or Confidence Intervals?

Ordinarily, data are used either to test a hypothesis or to construct a confidence interval, but not both. Traditionally, hypothesis tests have been preferred to confidence intervals in the behavioral sciences, and that emphasis is reflected in this book. As a matter of fact, however, *confidence intervals tend to be more informative than hypothesis tests. Hypothesis tests merely indicate whether or not an effect is present.* In the most recent blood-doping experiment, for instance, it was concluded, at the .05 level of significance, that blood doping has an effect on endurance. On the other hand, *confidence intervals indicate the possible size of the effect*. In the same blood-doping experiment, it was claimed, with 95 percent confidence, that the interval between 0.25 and 3.75 minutes includes the true effect of blood doping on endurance.

If the primary concern is whether or not an effect is present—as is often the case in relatively new research areas—use a hypothesis test. Otherwise, consider the use of a confidence interval. Also consider the use of a confidence interval whenever the null hypothesis is rejected. After it's been established that an effect is present, it makes sense to use the same data to estimate the possible size of that effect. For instance, after it's been established (by rejecting the null hypothesis) that blood doping has an effect on endurance scores, it makes sense to estimate, with a 95 percent confidence interval, that the interval between 0.25 and 3.75 minutes describes the possible size of that effect.

16.16 Other Types of Confidence Intervals

Confidence intervals and point estimates can be constructed not only for population means and differences between population means, but also for other characteristics of populations, including variances, correlation coefficients, and proportions. However, no additional discussion of confidence intervals appears in this book.

16.17 Summary

Rather than test a hypothesis about a single population mean or the difference between population means, you might choose to estimate these population characteristics, using point estimates and confidence intervals.

In point estimation, a single sample characteristic, such as a sample mean or the difference between two sample means, is used to estimate the corresponding population characteristic. Point estimates ignore sampling variability and, therefore, tend to be inaccurate.

Confidence intervals specify ranges of values that, in the long run, include the unknown population characteristic a certain percent of the time. For instance, given a 95 percent confidence interval, then, in the long run, 95 percent of all of these confidence intervals are true because they include the unknown population characteristic. Confidence intervals work because they are based on some well-established properties of sampling distributions.

Confidence intervals can be constructed for a single population mean and for the difference between two population means, using either z or t. In practice, z is rarely used since it requires knowledge of the population standard deviation(s).

Any level of confidence can be assigned to a confidence interval, but the 95 percent and 99 percent levels are most prevalent. Given one of these levels of confidence, then even though we can never know whether a particular confidence interval is true or false, we can be "reasonably confident" that the interval actually includes the unknown population characteristic.

Narrower, more precise confidence intervals are produced by lower levels of confidence (for example, 95 percent rather than 99 percent) and by larger sample sizes.

Confidence intervals tend to be more informative than hypothesis tests. Hypothesis tests merely indicate whether or not an effect is present, while confidence intervals indicate the possible size of the effect. Whenever appropriate—including whenever the null hypothesis has been rejected—consider the use of confidence intervals.

Important Terms & Symbols

Point estimate for μ
Confidence interval for μ
Level of confidence
Point estimate for $\mu_1 - \mu_2$
Confidence interval for $\mu_1 - \mu_2$

16.18 Exercises

1. Reading achievement scores are obtained for a group of fourth graders. A score of 4.0 indicates a level of achievement appropriate for fourth grade, while a score below 4.0 indicates underachievement and a score above 4.0 indicates overachievement. A random sample of 64 fourth graders reveals a mean achievement score of 3.82 with a sample standard deviation of 0.4.

   (a) What is the best single estimate of the unknown mean reading score for the entire population of fourth graders?
   (b) Construct a 95 percent confidence interval for the unknown population mean. (Remember to convert the sample standard deviation to a standard error.)
   (c) Interpret this confidence interval. (Do you find any consistent pattern among the possible values of the population mean?)

2. In the same school district, a total of 50 eighth graders are randomly assigned, in equal numbers, to either an experimental group, which receives special training in reading, or a control group, which receives the usual training in reading. After two months, the mean reading score for the experimental students $(\bar{X}_1)$ equals 8.46, and that for the control students $(\bar{X}_2)$ equals 8.20. Furthermore, the estimated standard error, $s_{\bar{X}_1 - \bar{X}_2}$, equals .09.

   (a) Estimate, with a single number, the difference between population means.
   (b) Construct a 95 percent confidence interval for the difference between population means.
   (c) Interpret the significance of this confidence interval for the special reading program.

3. After being matched for body weight, twenty-five pairs of overweight adults are randomly assigned, one pair at a time, to either an experimental group, which participates in a special weight control program modeled after Alcoholics Anonymous, or a control program consisting of diet information, calorie counting, and so on. After six months, it's found that, on the average, experimental subjects are 15.4 pounds lighter than control subjects. In other words, given that X_1 and X_2 designate the weights of experimental and control subjects, respectively, the sample mean of the difference scores, $\bar{D}$, equals -15.4 pounds. The sample standard deviation for the difference scores, s_D, equals 12 pounds.

   (a) Estimate, with a single number, the population mean of the difference scores.
   (b) Construct a 95 percent confidence interval for the population mean of the difference scores.
   (c) Interpret this confidence interval.

4. In Chapter 13, Exercise 1, the null hypothesis was rejected, and it was concluded that, for a large food chain, packages of ground beef weigh less than the advertised weight of one pound. Given a sample size of 10, a sample mean of 14.7 ounces, and an estimated standard error of 0.26 ounce, construct a 95 percent confidence interval for the true weight of all "one pound" packages of ground beef.

5. In Chapter 13, Exercise 3, the null hypothesis was rejected, and it was concluded that a mild electrical shock, administered to lab rats just prior to

each trial in a water maze, increases the number of trials required to learn that maze. Given a sample of 7 rats, a sample mean of 34.86 trials, and an estimated standard error of 1.15 trials, construct a 95 percent confidence interval for the true number of trials required to learn the water maze.

6. In Chapter 14, Exercise 3, an investigator tested a hypothesis to determine whether daily doses of vitamin C increase the IQ scores of high school students. Instead of a hypothesis test, a confidence interval could have been constructed—if, for instance, the investigator was primarily concerned about the size of the possible effects of vitamin C on IQ scores.

    (a) Given two groups, each consisting of 35 high school students; a mean IQ ($\bar{X}_1$) of 110 for the experimental group and a mean IQ ($\bar{X}_2$) of 108 for the control group; and an estimated standard error of 1.80; construct a 99 percent confidence interval for the true effect of daily doses of vitamin C on IQ.
    (b) Interpret this confidence interval.

7. In Chapter 14, Exercise 4, the null hypothesis was rejected, and it was concluded that, on the average, higher achievement scores are earned by students who receive letter grades in introductory biology than by students who receive a pass/fail. Given two groups, each consisting of 40 students, a mean achievement score ($\bar{X}_1$) of 86.2 for letter grade students, a mean achievement score ($\bar{X}_2$) of 81.6 for pass/fail students, and an estimated standard error ($s_{\bar{X}_1-\bar{X}_2}$) of 1.50, construct a 95 percent confidence interval for the true difference between the population mean achievement scores.

8. In Chapter 15, Exercise 3, the null hypothesis was rejected, and it was concluded that the anti-smoking film produces a decline in cigarette smoking among heavy smokers. For eight pairs of observations (each pair being taken for the same smoker before and after the film) it was found that, on the average, daily smoking decreases by 2 cigarettes. In other words, the sample mean of the difference scores, $\bar{D}$, equals 2. Furthermore, the estimated standard error, $s_{\bar{D}}$, equals 0.63 cigarette.

    (a) Construct a 95 percent confidence interval for the true effect of the anti-smoking film.
    (b) Interpret this confidence interval.

9. Imagine that one of the following 95 percent confidence intervals is based on the blood-doping experiment with two independent samples of subjects. (Remember, each of these numbers reflects a possible *difference,* in minutes, between the two population means.)

95% Confidence Interval	Lower Limit	Upper Limit
1	0.21	1.53
2	−2.53	1.78
3	1.89	2.21
4	−3.45	4.25
5	−1.54	−0.32

    (a) Which confidence interval is most precise?
    (b) Which most strongly supports the conclusion that blood doping *facilitates* endurance?
    (c) Which is least precise?
    (d) Which implies the largest sample sizes?
    (e) Which most strongly supports the conclusion that blood doping *hinders* endurance?
    (f) Which would most likely stimulate the investigator to conduct an additional experiment, using larger sample sizes?

17.1 –Introduction
17.2 –Two Sources of
Variability
17.3 –*F* Ratio
17.4 –*F* Test
17.5 –Variance Estimates
17.6 –Sum of Squares (*SS*)
17.7 –Degrees of Freedom (*df*)
17.8 –Mean Squares (*MS*) and
the *F* Ratio
17.9 –*F* Tables
17.10–Notes on Usage
17.11–Multiple Comparisons
17.12–Scheffé's Test
17.13–*F* Test Is Nondirectional
17.14–Assumptions
17.15–Summary
17.16–Exercises

17

Analysis of Variance (One Factor)

Surveys of recent psychological literature indicate that analysis of variance is used more often than any other type of statistical analysis.

Paradoxical though it might seem, variance (or variability) is analyzed to determine whether or not population means could be equal.

A single test, known as the F test, determines the fate of a null hypothesis for any number of population means—two, three, four, or more.

17.1 Introduction

Do crowds affect our willingness, either positively or negatively, to assume responsibility for the welfare of ourselves and others? For instance, does the presence of other people either facilitate or inhibit our reaction to potentially dangerous smoke seeping from a wall vent? Hoping to answer this question, a social psychologist measures any delay in a subject's alarm reaction as smoke gradually fills a waiting room occupied only by the subject, plus "crowds" of either zero, two, or four experimental confederates (who act as regular subjects but, in fact, ignore the smoke).

As usual, the null hypothesis is tested with experimental findings. The present null hypothesis states that, on the average, equal delays characterize the three populations of subjects who are exposed to smoke in rooms with zero, two, or four confederates. Expressed symbolically, the null hypothesis reads:

$$H_0: \mu_0 = \mu_2 = \mu_4$$

where μ_0 represents the mean delay for the population of subjects who are exposed to smoke in rooms with zero confederates, and so forth. Rejection of the null hypothesis implies, most generally, that crowd size affects our willingness to assume responsibility.

Resist any urge to test this null hypothesis with *t*. The *t* test can't handle null hypotheses for more than two population means. *An overall test of the null hypothesis for three (or more) population means requires a new statistical procedure known as* **analysis of variance,** which is often abbreviated as **ANOVA.** Later sections treat the computational procedures for ANOVA; the next few sections emphasize the intuitive basis for ANOVA within the context of the social psychologist's experiment.

For simplicity, let's assume that the social psychologist randomly assigns only three subjects to each of the three groups, to be tested (one subject at a time) in rooms with zero, two, or four confederates, and she measures the delay of each subject's alarm reaction to the nearest minute. Using just your intuition, make a preliminary decision about whether the null hypothesis should be retained or rejected for each of the two possible experimental outcomes presented in Table 17.1. Do this before reading further.

Your intuition is correct if you decided that the null hypothesis probably should not be rejected for outcome A, but it probably should be rejected for outcome B.

Your decisions for outcomes A and B most likely were based on the differences between group means for each outcome. Observed mean differences have been a major ingredient in previous tests of the null hypothesis involving t, and these differences are just as important in tests of the null hypothesis involving ANOVA. It's easy to lose sight of this fact because observed mean differences appear, somewhat disguised, as one type of variability in ANOVA. It takes extra effort to view ANOVA—with its emphasis on the analysis of several sources of variability—as related to tests in previous chapters. Reminders of this fact appear throughout the present chapter.

17.2 Two Sources of Variability

First, let's look more closely at one source of variability—the differences between group means in outcome A and the differences between group means in outcome B. Relatively small differences occur between group means in outcome A (13, 11, and 12). As noted in previous chapters, relatively small differences between group means often can be attributed to chance. Even though the null hypothesis is true (because crowd size doesn't affect the subjects' alarm reactions), group means tend to differ merely as a result of chance sampling variability. In the case of outcome A, the null hypothesis probably

TABLE 17.1
TWO POSSIBLE EXPERIMENTAL OUTCOMES: ALARM REACTIONS IN MINUTES

OUTCOME A

Groups (Number of Confederates)

	(Zero)	(Two)	(Four)	
	16	9	10	
	12	10	12	
	11	14	14	
Group mean:	13	11	12	Overall mean = 12

OUTCOME B

Groups (Number of Confederates)

	(Zero)	(Two)	(Four)	
	8	11	16	
	4	12	18	
	3	16	20	
Group mean:	5	13	18	Overall mean = 12

should not be rejected; there is a lack of evidence that the experimental treatments affect the subjects' alarm reactions.

Relatively large differences occur between the group means for outcome B (5, 13, and 18). These large differences cannot readily be attributed to chance. Instead they probably indicate that the null hypothesis is false (because crowd size affects the subjects' alarm reactions). In the case of outcome B, the null hypothesis probably should be rejected. Now there is evidence of a **treatment effect,** that is, at *least one difference between the population means for the various experimental conditions* (crowd sizes).

A more definitive decision about the null hypothesis requires that differences between group means be viewed as one source of variability that, when adjusted appropriately, can be compared with a second source of variability. In particular, an estimate of **variability between groups,** that is, the *variation among scores of subjects treated differently,* must be compared with another, completely independent estimate of **variability within groups,** that is, the *variation among scores of subjects treated alike.* As will be seen, the more variability between groups exceeds variability within groups, the more suspect is the null hypothesis.

Let's focus on the second source of variability—the variability within groups for subjects treated alike. Beginning with the first set of three scores (16, 12, 11) in Table 17.1, inspect the differences among the scores of the three subjects treated alike in this group. Continue this procedure, one group at a time, to obtain an overall impression of variability within groups for all three groups in outcome A and for all three groups in outcome B. Notice the relative stability of the differences among the three scores within each of the various groups, regardless of whether the group happens to be in outcome A or outcome B. For instance, one crude measure of variability, the range, equals either 4 or 5 for each group shown in Table 17.1. The key point is that variability within each group depends entirely on the scores of subjects treated alike (exposed to the same crowd size), and it never involves the scores of subjects treated differently (exposed to different crowd sizes). In contrast to variability between groups, variability within groups never reflects the presence of a treatment effect. Regardless of whether the null hypothesis is true or false, variability within groups reflects only **random error,** that is, *the combined effects* (on the scores of individual subjects) *of all uncontrolled factors,* such as "individual differences" among subjects, slight variations in experimental conditions, errors in measurement, and so forth. In ANOVA, the within-group estimate often is referred to simply as the "error term."

17.3 *F* Ratio

Heretofore, the null hypothesis has been tested with a *t* (or *z*) ratio. In the two-sample case, *t* reflects the ratio between the observed sample mean difference in the numerator and the estimated standard error in the denominator. In the case of three or more samples, the null hypothesis is tested with a new ratio, the *F* ratio. Essentially, *F* reflects the ratio of the observed sample mean differences (measured as variability between groups) in the numerator and the estimated error term (measured as variability within groups) in the denominator term, that is

$$F = \frac{\text{variability between groups}}{\text{variability within groups}} \qquad (17.1)$$

Like *t*, *F* has its own family of sampling distributions that can be consulted, as described in Section 17.9, to test the null hypothesis. The resulting test is known as an "*F* test."

An **F** *test* of the null hypothesis is based on the notion that, if the null hypothesis is true, both the numerator and denominator of the F ratio tend to be about the same, but if the null hypothesis is false, the numerator tends to be larger than the denominator.

If the null hypothesis is true (because there is no treatment effect due to different crowd sizes) both estimates of variability between groups and variability within groups reflect only random error. In this case,

$$F = \frac{\text{random error}}{\text{random error}}$$

Except for chance, estimates in both the numerator and denominator are similar, and generally speaking, *F* varies about a value of one.

If the null hypothesis is false (because there is a treatment effect due to different crowd sizes) both estimates still reflect random error, but the estimate of variability between groups also reflects the treatment effect, that is, the differences between population means. In this case,

$$F = \frac{\text{random error} + \text{treatment effect}}{\text{random error}}$$

When the null hypothesis is false, the presence of a treatment effect tends to cause a chain reaction: observed differences between group means are large, as also is the variability between groups. Accordingly, the numerator term exceeds the denominator term, producing an *F* whose value is larger than one. When the null hypothesis is *seriously* false, the sizable treatment effect tends to cause an even more pronounced chain reaction, beginning with very large observed differences between group means and ending with an *F* whose value is *considerably* larger than one.

In practice, of course, we never really know whether the null hypothesis is true or false. Following the usual procedure, we assume the null hypothesis to be true and view the observed *F* within the context of its hypothesized sampling distribution, as shown in Figure 17.1. If the observed *F* appears to emerge from the dense concentration of possible *F* ratios smaller than the critical *F*, the experimental outcome is not too improbable (that is, the observed differences between group means are not too large), on the assumption that the null hypothesis is true. Therefore, the null hypothesis is retained. On the other hand, if the observed *F* appears to emerge from the sparse concentration of possible *F* ratios equal to or greater than the critical *F*, the experimental outcome is too improbable, and the null hypothesis is rejected. In the latter case, the value of the observed *F* is presumed to be inflated by a treatment effect.

Full-fledged hypothesis tests for outcomes A and B agree with the earlier intuitive decisions. Given the .05 level of significance, the null hypothesis should be retained for outcome A since the observed *F* of 0.50 is smaller than the critical *F* of 5.14. But the null hypothesis should be rejected for outcome B since the observed *F* of 21.50 exceeds the critical *F*. The hypothesis test for outcome B, as summarized below, will be described in more detail in later sections of this chapter.

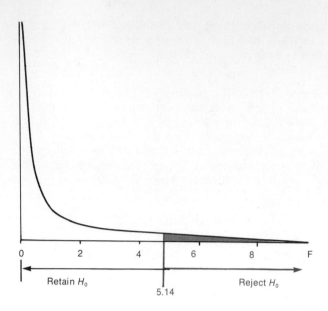

FIGURE 17.1

Hypothesized sampling distribution of F.

Hypothesis Test Summary:
F Test For One Factor
(Smoke Alarm Experiment, Outcome B)

Problem: On the average, are subjects' alarm reactions to potentially dangerous smoke affected by crowds of zero, two, or four confederates?

Statistical Hypotheses:

$$H_0: \mu_0 = \mu_2 = \mu_4$$
$$H_1: H_0 \text{ is not true.}$$

Statistical Test:

$$F = \frac{MS_{\text{between}}}{MS_{\text{within}}}$$

Decision Rule:
    Reject H_0 at .05 level of significance if $F \geq 5.14$ (from Table C, given $df_{\text{between}} = 2$ and $df_{\text{within}} = 6$).

Calculations:
    $F = 21.5$ (See Tables 17.3 and 17.4 for additional details.)

Decision:
    Reject H_0

Interpretation:

    Crowd size affects the subjects' mean alarm reactions.

17.5 Variance Estimates

As its name implies, analysis of variance uses variance estimates to measure variability between groups and within groups. Described briefly in Section 5.6, the term *variance* refers to the square of the standard deviation, and it can be obtained from any formula for the standard deviation merely by eliminating the square root sign. A *vari-*

ance estimate signifies that information from a sample is used to determine the unknown variance of the population.

The most common example of a variance estimate is the square of the sample standard deviation defined in Formula 13.1:

$$s^2 = \frac{\Sigma(X - \bar{X})^2}{n - 1}$$

This estimate is designed for use with a single sample and, therefore, can't be used in analysis of variance. It can be used, however, to identify two general features of variance estimates in ANOVA. First, in the above expression, the numerator term represents the sum of the squared deviations about the sample mean, $\bar{X}$. In ANOVA, the numerator term of a variance estimate always consists of the **sum of squares,** that is, *the sum of squared deviations for some set of scores about their mean.* Second, in the above expression, the denominator represents the number of degrees of freedom for these deviations. (Remember, as discussed in Section 13.8, only $n - 1$ of these deviations are free to vary. One degree of freedom is lost because the sum of n deviations about their own mean always must equal zero.) In ANOVA, the denominator term of a variance estimate always consists of the number of **degrees of freedom,** that is, *the number of deviations in the numerator that are free to vary.*

Essentially, *a variance estimate in ANOVA consists of some sum of squares divided by its degrees of freedom.* This operation always produces a number equal to the mean of the squared deviations—hence the designation **mean square,** abbreviated as *MS.* In ANOVA, the latter term is most common, and it will be used in subsequent discussions. A most general expression for any variance estimate reads:

$$MS = \frac{SS}{df} \tag{17.2}$$

where MS represents the variance estimate, SS denotes the sum of squared deviations about their mean, and df equals the corresponding number of degrees of freedom. Formula 17.2 should be read as "the mean square equals the sum of squares divided by its degrees of freedom."

Eventually, an F test of the null hypothesis will be based on a ratio involving two variance estimates: the mean square for variability between groups and the mean square for variability within groups. Prior to using these mean squares, we must calculate their sums of squares, as described in Section 17.6, and their degrees of freedom, as described in Section 17.7.

17.6 Sum of Squares (*SS*)

Most computational effort in ANOVA is directed toward the various sum of squares terms: the sum of squares for variability between groups, $SS_{between}$; the sum of squares for variability within groups, SS_{within}; and the sum of squares for the total of these two, SS_{total}. Remember, any sum of squares always equals the sum of the squared deviations of some set of scores about their mean. $SS_{between}$ equals the sum of the squared deviations of group means about their mean, the overall mean. SS_{within} equals the sum of the squared deviations of all scores about their respective group means. SS_{total} equals the sum of the squared deviations of all scores about the overall mean.

When calculating the various SS terms, we do not deal directly with these sets of squared deviations. Instead, it's more convenient to

TABLE 17.2
COMPUTATIONAL
FORMULAS FOR
SS TERMS

SS_{between} = sum of squared deviations of group means about the overall mean

$$= \left[\frac{(\text{1st group total})^2}{\text{1st sample size}} + \cdots + \frac{(\text{last group total})^2}{\text{last sample size}}\right] - \frac{(\text{overall total})^2}{\text{overall sample size}}$$

SS_{within} = sum of squared deviations of scores about their respective group means

$$= \text{sum of all squared scores} - \left[\frac{(\text{1st group total})^2}{\text{1st sample size}} + \cdots + \frac{(\text{last group total})^2}{\text{last sample size}}\right]$$

SS_{total} = sum of squared deviations of scores about the overall mean

$$= \text{sum of all squared scores} - \frac{(\text{overall total})^2}{\text{overall sample size}}$$

**FORMULAS FOR
df TERMS**

df_{between} = number of groups − 1

df_{within} = number of scores − number of groups

df_{total} = number of scores − 1

use the equivalent expressions listed in the top half of Table 17.2. Study these expressions; they comply with a highly predictable computational pattern that, once learned, is easily remembered:

(i) Each SS term consists of two main components: first a positive component, then a negative component.

(ii) Totals replace group means and the overall mean.

(iii) Each score—whether a total score or an original score—is squared and, in the case of a total score, divided by its sample size.

Table 17.3 indicates how to use these computational formulas for the data in outcome B. To minimize computational errors, calculate from scratch each of the three SS terms, even though this entails some duplication of effort (not shown in Table 17.3 to save space). Then, as an almost foolproof check of your computations, as shown at the bottom of Table 17.3, verify that SS_{total} equals the sum of the various SS terms, that is,

$$SS_{\text{total}} = SS_{\text{between}} + SS_{\text{within}} \tag{17.3}$$

17.7 Degrees of Freedom (*df*)

The number of degrees of freedom differs for each SS term in ANOVA, and for convenience the various *df* formulas are listed in the bottom half of Table 17.2. To determine the *df* for any SS, simply substitute the appropriate numbers and subtract. For outcome B, which consists of three groups and a total of nine scores:

$$df_{\text{between}} = 3 - 1 = 2$$
$$df_{\text{within}} = 9 - 3 = 6$$
$$df_{\text{total}} = 9 - 1 = 8$$

Remember, degrees of freedom reflect the number of deviations that are free to vary in the corresponding SS term. The value of 2 for df_{between} reflects the loss of one degree of freedom because the three group means are expressed as deviations about the one overall mean. The value of 6 for df_{within} reflects the loss of three degrees of freedom because all nine scores are expressed as deviations about their three respective group means. Finally, the value of 8 for df_{total} reflects the loss of one degree of freedom because all nine scores are expressed as deviations about the one overall mean.

TABLE 17.3
CALCULATION OF SS TERMS

A. COMPUTATIONAL SEQUENCE

Find each group total and also the overall total for all groups ①.
Substitute numbers into computational formula ② and solve for SS_{between}.
Substitute numbers into computational formula ③ and solve for SS_{within}.
Substitute numbers into computational formula ④ and solve for SS_{total}.
Do computational check ⑤.

B. DATA AND COMPUTATIONS

Outcome B

(Zero)	(Two)	(Four)
8	11	16
4	12	18
3	16	20

① Group Totals: 15   39   54   Overall Total = 108

② $SS_{\text{between}} = \dfrac{(\text{1st group total})^2}{\text{1st sample size}} + \dfrac{(\text{2nd group total})^2}{\text{2nd sample size}} + \dfrac{(\text{3rd group total})^2}{\text{3rd sample size}} - \dfrac{(\text{overall total})^2}{\text{overall sample size}}$

$= \dfrac{(15)^2}{3} + \dfrac{(39)^2}{3} + \dfrac{(54)^2}{3} - \dfrac{(108)^2}{9}$

$= \dfrac{225}{3} + \dfrac{1521}{3} + \dfrac{2916}{3} - \dfrac{11664}{9}$

$= 75 + 507 + 972 - 1296$

$= 1554 - 1296 = 258$

③ $SS_{\text{within}} = \text{sum of all squared scores} - \left[\dfrac{(\text{1st group total})^2}{\text{1st sample size}} + \dfrac{(\text{2nd group total})^2}{\text{2nd sample size}} + \dfrac{(\text{3rd group total})^2}{\text{3rd sample size}}\right]$

$= (8)^2 + (4)^2 + (3)^2 + (11)^2 + (12)^2 + (16)^2 + (16)^2 + (18)^2 + (20)^2 - \left[\dfrac{(15)^2}{3} + \dfrac{(39)^2}{3} + \dfrac{(54)^2}{3}\right]$

$= 1590 - 1554 = 36$

④ $SS_{\text{total}} = \text{sum of all squared scores} - \dfrac{(\text{overall total})^2}{\text{overall sample size}}$

$= (8)^2 + (4)^2 + (3)^2 + (11)^2 + (12)^2 + (16)^2 + (16)^2 + (18)^2 + (20)^2 - \dfrac{(108)^2}{9}$

$= 1590 - 1296 = 294$

⑤ $SS_{\text{total}} = SS_{\text{between}} + SS_{\text{within}}$

$294 = 258 + 36$

$294 = 294$

In ANOVA, the degrees of freedom for SS_{total} always equals the combined degrees of freedom for the remaining SS terms, that is,

$$df_{\text{total}} = df_{\text{between}} + df_{\text{within}} \qquad (17.4)$$

This formula can be used to verify that the correct number of degrees of freedom has been assigned to each of the SS terms in outcome B.

Having found the values of the various SS terms and their degrees of freedom, we can determine the values of the mean squares

17.8 Mean Squares (MS) and the F Ratio

197

for variability between groups and variability within groups, and then calculate the value of F, as suggested in Figure 17.2.

The value of the mean square for variability between groups, $MS_{between}$, is given by the following expression:

$$MS_{between} = \frac{SS_{between}}{df_{between}} \qquad (17.5)$$

$MS_{between}$ reflects the variability between means for groups of subjects who are treated differently. Relatively large values of $MS_{between}$ suggest the presence of a treatment effect.

For outcome B,

$$MS_{between} = \frac{258}{2} = 129$$

The value of the mean square for variability within groups, MS_{within}, is given by the following expression:

$$MS_{within} = \frac{SS_{within}}{df_{within}} \qquad (17.6)$$

MS_{within} reflects the variability between scores for subjects who are treated alike within each group, pooled across all groups. Regardless of whether a treatment effect is present, MS_{within} measures only random error.

For outcome B,

$$MS_{within} = \frac{36}{6} = 6$$

Finally, Formula 7.1 for F can be rewritten as:

$$F = \frac{MS_{between}}{MS_{within}} \qquad (17.7)$$

As mentioned previously, if the null hypothesis is true (because alarm reactions are not affected by crowd size), the value of F varies about a value of approximately one, but if the null hypothesis is false, the value of F tends to be larger than one.

For outcome B, the null hypothesis is suspect since

$$F = \frac{129}{6} = 21.5$$

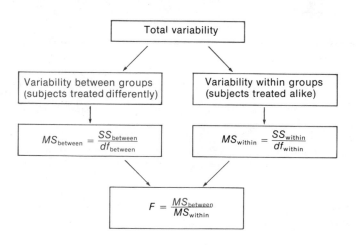

FIGURE 17.2
Sources of variability in *ANOVA* and the *F* ratio.

A decision about the null hypothesis requires that the observed F be compared with a critical F. As has been noted, there's not one but a family of F sampling distributions. Any particular F distribution is uniquely specified by the pair of degrees of freedom associated with the mean squares in its numerator and denominator.

Critical F values for hypothesis tests at the .05 level (light numbers) and the .01 level (dark numbers) are listed in Table 17.4 (for a few F distributions) and in Table C of the Appendix (for a wide variety of F distributions). To read either table, simply find the cell intersected by the column with degrees of freedom equal to those in the numerator of F, $df_{between}$, and by the row with degrees of freedom equal to those in the denominator of F, df_{within}. Table 17.4 illustrates this procedure when, as in outcome B, 2 and 6 degrees of freedom are associated with the numerator and denominator of F, respectively. In this case, the column with 2 degrees of freedom and the row with 6 degrees of freedom intersect a cell (shaded in red) that lists a critical F value of 5.14 for a hypothesis test at the .05 level of significance. As has been anticipated in previous sections, since the observed F of 21.5 for outcome B exceeds this critical F, the overall null hypothesis can be rejected. There is evidence that crowd size affects the subjects' alarm reactions.

17.9 F Tables

ANOVA results are usually reported as in Table 17.5. "Source" refers to the source of variability: between groups, within groups, and total. Notice the arrangement of column headings from SS and df to MS and F. Also notice that the bottom row for total variability contains entries only for SS and df. Ordinarily, the red numbers in parentheses don't appear in ANOVA tables, but in Table 17.5 they show the origin of each MS and of F. The asterisk in Table 17.5 spotlights the fact that the observed F of 21.50 exceeds the critical F of 5.14 and, therefore, causes the null hypothesis to be rejected at the .05 level of significance.

Sometimes ANOVA tables appear with labels other than those shown in Table 17.5. For instance, "Between" might be replaced with "Treatment" since the variability between groups reflects any treatment

17.10 Notes on Usage

DEGREES OF FREEDOM IN DENOMINATOR	DEGREES OF FREEDOM IN NUMERATOR					
	1	2	3	4	•	•
1	161	200	216			
	4052	4999	5403			
2	18.51	19.00	19.16			
	98.49	99.01	99.17			
3	10.13	9.55	9.28			
	34.12	30.81	29.46			
4	7.17	6.94	6.59			
	21.20	18.00	16.09			
5	6.61	5.79	5.41			
	16.26	13.27	12.06			
6	5.99	5.14	4.76			
	13.74	10.92	9.78			
7	5.59	4.74	4.35			
	12.25	9.55	3.45			
8						
•						
•						
•						

TABLE 17.4
SPECIMEN TABLE FROM TABLE C OF APPENDIX
CRITICAL VALUES OF F
.05 LEVEL OF
   SIGNIFICANCE
   (LIGHT NUMBERS)
.01 LEVEL OF
   SIGNIFICANCE
   (DARK NUMBERS)

TABLE 17.5
ANOVA TABLE
(OUTCOME B)

SOURCE	SS	df	MS	F
Between	258	2	$\left(\frac{258}{2}=\right)129$	$\left(\frac{129}{6}=\right)21.5^*$
Within	36	6	$\left(\frac{36}{6}=\right)6$	
Total	294	8		

*Significant at .05 level.

effect. Likewise, "Within" might be replaced with "Error" since variability within groups reflects the presence of random error.

In addition to an ANOVA table, published reports of a hypothesis test might be limited to an interpretative comment, plus a parenthetical statement that summarizes the statistical analysis. For example, a published report of the hypothesis test for outcome B might read as follows:

"There is evidence that, on the average, crowd size affects the alarm reactions of subjects to potentially dangerous smoke [F (2,6) = 21.50, $p < .05$]."

The parenthetical statement indicates that an F based on 2 and 6 degrees of freedom was found to equal 21.50. Furthermore, p, the probability of the observed F—on the assumption that H_0 is true—is less than .05. Since p is less than .05, the observed F qualifies as an improbable outcome, and the null hypothesis is rejected, as implied in the interpretative comment.

17.11 Multiple Comparisons

Rejection of the overall null hypothesis indicates only that not all population means are equal. In the case of outcome B, rejection of H_0 signals the presence of one or more inequalities between the mean delays for populations of subjects exposed to crowds of zero, two, and four confederates, that is, between μ_0, μ_2 and μ_4. To pinpoint the one or more differences between pairs of population means that contribute to the rejection of the overall H_0, you must use a test of multiple comparisons. A test of multiple comparisons is designed to evaluate not just one but a series of differences between population means. In the case of outcome B, each of the three possible differences between pairs of population means, $\mu_0 - \mu_2$, $\mu_0 - \mu_4$, $\mu_2 - \mu_4$, should be evaluated.

These differences can't be evaluated with a series of t tests. Essentially, the t test is designed to evaluate a single comparison using a pair of randomly selected samples, not multiple comparisons using all possible pairs of samples. Among other complications, the use of multiple t tests increases the probability of a type I error (rejecting a true null hypothesis) beyond that value specified by the level of significance.

A coin tossing example might clarify this problem. When a fair coin is tossed only once, the probability of heads equals .50—just as when a single t test is to be conducted at the .05 level of significance, the probability of a type I error equals .05. When a fair coin is tossed three times, however, heads can appear not only on the first toss but on the second or third toss as well, and hence the probability of heads on *at least one* of the three tosses exceeds .50. By the same token, when a series of three t tests are conducted at the .05 level of significance, a type

I error can be committed not only on the first test but on the second or third test as well, and hence the probability of committing a type I error on *at least one* of the three tests exceeds .05. In fact, the cumulative probability of at least one type I error could be as large as .15 for this series of three *t* tests.

The shortcoming just described does not apply to a number of specially designed multiple comparison tests, including the test by Scheffé (pronounced Shef-fay) emphasized in this book. Once the overall null hypothesis has been rejected in ANOVA, Scheffé's test can be used for all possible comparisons, and yet the cumulative probability of at least one type I error never exceeds the specified level of significance.

One version of Scheffé's test supplies a critical value for evaluating the observed mean difference between any pair of groups, for instance, the groups with zero and two confederates. If the observed mean difference, $\bar{X}_0 - \bar{X}_2$, is either more positive or more negative than Scheffé's critical value, $(\bar{X}_0 - \bar{X}_2)_{crit}$, the null hypothesis for that comparison can be rejected. To determine the critical value for groups with zero and two confederates, use the following expression, along with numerical data from the ANOVA for outcome B:

$$(\bar{X}_0 - \bar{X}_2)_{crit} = \pm \sqrt{(df_{between})(F_{crit})(MS_{within})\left(\frac{1}{n_0} + \frac{1}{n_2}\right)}$$

$$= \pm \sqrt{(2)(5.14)(6)\left(\frac{1}{3} + \frac{1}{3}\right)} = \pm \sqrt{(2)(5.14)(6)\left(\frac{2}{3}\right)}$$

$$= \pm \sqrt{41.12} = \pm 6.41$$

Since the observed mean difference $\bar{X}_0 - \bar{X}_2$ equals -8 (from $5 - 13$, as shown in Table 17.1), and this is more negative than the critical mean difference -6.41, the null hypothesis for $\mu_0 - \mu_2$ can be rejected at the .05 level. In other words, there is evidence that the mean alarm reactions differ between populations of subjects exposed to crowds of zero and two confederates.

When sample sizes are unequal, Scheffé's critical value must be calculated for each comparison. When all sample sizes are equal, as in outcome B, the critical mean difference need be calculated for only one comparison, then used to evaluate all remaining comparisons. Table 17.6 summarizes the results of Scheffé's test for each of the three differences between pairs of means in outcome B, using the critical value of ± 6.41.

Table 17.6 reveals that two of the three comparisons reach the .05 level of significance. It can be concluded that differences between population means for groups of zero and two confederates, and for zero

POPULATION COMPARISON	OBSERVED MEAN DIFFERENCE	SCHEFFÉ'S CRITICAL VALUE
$\mu_0 - \mu_2$	-8*	± 6.41
$\mu_0 - \mu_4$	-13*	± 6.41
$\mu_2 - \mu_4$	-5	± 6.41

TABLE 17.6
SUMMARY TABLE FOR SCHEFFÉ'S TEST

*Significant at .05 level.

and four confederates, contribute to the original rejection of the overall null hypothesis. Thus, there is evidence that the alarm reactions of subjects in crowds of zero confederates differ—that is, they are shorter than—the reactions of subjects in crowds of either two or four confederates. There is no evidence, however, that the reactions of subjects in crowds of two confederates differ from those of subjects in crowds of four confederates.

For any comparison of differences between pairs of group means, $\bar{X}_i - \bar{X}_j$, Scheffé's critical value is given by the following expression:

$$(\bar{X}_i - \bar{X}_j)_{\text{crit}} = \pm \sqrt{(df_{\text{between}})(F_{\text{crit}})(MS_{\text{within}})\left(\frac{1}{n_i} + \frac{1}{n_j}\right)} \quad (17.8)$$

where the subscripts i and j refer to *any* two groups, and all terms under the square root refer back to the original ANOVA.

Scheffé's test should be used only if the overall null hypothesis has been rejected. It is designed to evaluate all possible comparisons of interest, not just differences for pairs of means. For example, when modified appropriately, Scheffé's test can evaluate other types of comparisons, such as the difference between the mean for those subjects in crowds of zero confederates and the combined mean for those subjects in crowds of two and four confederates.*

17.13 F Test Is Nondirectional

It might seem strange that, even though the entire rejection region for the null hypothesis appears only in the upper tail of the F sampling distribution, as in Figure 17.1, *the F test in ANOVA is the equivalent of a nondirectional test.* Recall that all variations in ANOVA are squared. When squared, all values become positive, regardless of whether the original differences between groups (or group means) are positive or negative. All squared differences between groups have a cumulative positive effect on the observed F and thereby insure that F is a nondirectional test, even though only the upper tail of its sampling distribution contains the rejection region.

A similar effect could be produced by squaring the t test. When squared, all values of t become positive, regardless of whether the original value for the observed t was positive or negative. Hence, the t^2 test also qualifies as a nondirectional test, even though the entire rejection region appears only in the upper tail of the t^2 sampling distribution. As a matter of fact, the values of t^2 and F are identical when both tests are applied to the same data for two independent groups. When only two independent groups are involved, the t^2 test can be viewed as a special case of the more general F test in ANOVA for two or more independent groups.

17.14 Assumptions

The assumptions for F tests in ANOVA are the same as those for t tests of mean differences for two independent samples. All underlying populations are assumed to be normally distributed with equal variances. You can proceed without the slightest fear as long as the sample sizes are equal and each is fairly large—say greater than about 10.

---

*For a discussion of how to modify Scheffé's test for these types of comparisons, see Minium, E.W., *Statistical Reasoning in Psychology and Education,* 2nd Ed., Wiley, New York, 1978, Chap. 22. For more information about tests of multiple comparisons, including test of special "planned" comparisons that can replace the overall F test in ANOVA, see Keppel, G., *Design and Analysis: a Researcher's Handbook,* Prentice-Hall, Englewood Cliffs, N.J., 1973, Chaps. 6 and 8.

Otherwise, watch out for any conspicuous departures from either normality or equal variances among sets of observations. In those cases where either of these assumptions seems to be seriously violated, either increase the sample sizes before attempting a hypothesis test, or use a less sensitive test, such as the Kruskal-Wallis *H* test described in Chapter 20.

One final caution: the ANOVA techniques described in this book presume that all scores are independent. In other words, subjects are not matched across groups, and each subject contributes just one score to the overall analysis. Special ANOVA techniques must be used when scores lack independence.

17.15 Summary

Analysis of variance, abbreviated as ANOVA, tests the null hypothesis for two, three, or more population means by classifying total variability into two independent components: variability between groups and variability within groups. Both components reflect only random error if the null hypothesis is true, and the resulting *F* ratio (variability between groups divided by variability within groups) tends toward a value of approximately one. If, however, the null hypothesis is false, variability between groups reflects both random error and a treatment effect, while variability within groups still reflects only random error, and the resulting *F* ratio tends toward a value greater than one.

Each variance estimate or mean square (*MS*) is found by dividing the appropriate sum of squares (*SS*) term by its degrees of freedom (*df*) (see Tables 17.3, 17.4, and 17.5). In practice, once a value of *F* has been obtained, it's compared with a critical *F* from Table C in the Appendix. If the observed *F* equals or exceeds the critical *F*, the observed sample mean differences are too large (given that the null hypothesis is true), and therefore, the null hypothesis is rejected. Otherwise, for all smaller observed values of *F*, the null hypothesis is retained.

ANOVA results are often summarized in tabular form (see Table 17.5).

To pinpoint differences between specific pairs of population means that contribute to the rejection of the overall H_0, use Scheffé's test for multiple comparisons as described in Formula 17.8. Only use Scheffé's test if the overall H_0 has been rejected.

Because all variations in ANOVA are squared, all variance estimates are positive. Thus, the *F* test is a nondirectional test.

F tests in ANOVA assume that all underlying populations are normally distributed with equal variances. Only serious violations of these assumptions greatly compromise the accuracy of the *F* test.

Important Terms & Symbols

> Analysis of variance (ANOVA)
> Treatment effect
> Random error
> Variability between groups
> Variability within groups
> Sum of squares (*SS*)
> Degrees of freedom (*df*)
> Mean square (*MS*)
> *F* test
> Scheffé's test

Note: To simplify computations, most of the following problems employ unrealistically small sample sizes. When testing hypotheses, follow the customary step-by-step procedure.

17.16 Exercises

1. Imagine a simple experiment with three groups, each containing four observations. For each of the following outcomes, indicate whether or not variability between groups is present, and also whether or not variability within groups is present.

(a)

GROUP 1	GROUP 2	GROUP 3
8	8	8
8	8	8
8	8	8
8	8	8

(b)

GROUP 1	GROUP 2	GROUP 3
8	4	12
8	4	12
8	4	12
8	4	12

(c)

GROUP 1	GROUP 2	GROUP 3
4	6	5
6	6	7
8	10	9
14	10	11

(d)

GROUP 1	GROUP 2	GROUP 3
6	11	20
8	12	18
8	14	23
10	15	25

2. (a) Use the data for outcome A in Table 17.1 to test the null hypothesis at the .01 level of significance.
   (b) Summarize results with an ANOVA table—as in Table 17.5.

3. Another social psychologist conducts the smoke alarm experiment described in this chapter. For reasons beyond his control, unequal numbers of "real" subjects occupy the different groups.

   (a) Given the results below, test the null hypothesis at the .05 level of significance.
   (b) Summarize results with an ANOVA table.
   (c) If appropriate, use Scheffé's test and interpret the results.

ALARM REACTIONS IN MINUTES

ZERO	TWO	TWELVE
1	4	7
3	7	12
6	5	10
2		9
1		

4. A third social psychologist modifies the above design to include five different crowd sizes.

   (a) Given the results below, test the null hypothesis at the .05 level of significance.
   (b) Summarize results with an ANOVA table.
   (c) If appropriate, use Scheffé's test and interpret the results.

ALARM REACTIONS IN MINUTES

ZERO	TWO	FOUR	EIGHT	TWELVE
1	4	6	15	20
1	3	1	6	25
3	1	2	9	10
6	7	10	17	10

5. For some experiment, imagine four possible outcomes, as described below in the ANOVA tables.

A.

SOURCE	SS	df	MS	F
Between	900	3	300	6
Within	800	16	50	
Total	1700	19		

B.

SOURCE	SS	df	MS	F
Between	600	3	200	4
Within	800	16	50	
Total	1400	19		

C.

SOURCE	SS	df	MS	F
Between	300	3	100	1
Within	8000	80	100	
Total	8300	83		

D.

SOURCE	SS	df	MS	F
Between	300	3	100	1
Within	400	4	100	
Total	700	7		

   (a) How many groups are in outcome D?
   (b) Assuming groups of equal size, what's the size of each group in outcome C?
   (c) Which outcome(s) would cause the null hypothesis to be rejected at the .05 level of significance?
   (d) Which outcome is based on the largest total number of subjects?
   (e) Which outcome would you most prefer if you were the experimenter?
   (f) Which outcome provides the least information about a possible treatment effect?
   (g) Which outcome would least likely stimulate additional research?

6. Twenty-three overweight male volunteers are randomly assigned to three different treatment programs, designed to produce a weight loss by focusing

on diet, exercise, or the modification of eating behavior. Weight changes were recorded, to the nearest pound, for all participants who completed the two-month experiment. Positive scores signify a weight drop; negative scores, a weight gain.

(a) Using the results listed below, test the null hypothesis at the .05 level of significance. Note: Negative signs should be honored when calculating group and overall totals.
(b) Summarize results with an ANOVA table.
(c) How might these results appear in a published report?
(d) If appropriate, use Scheffé's test and interpret the results.

WEIGHT CHANGES

DIET	EXERCISE	BEHAVIOR MODIFICATION
3	−1	7
4	8	1
0	4	10
−3	2	0
5	2	18
10	−3	12
3		4
0		6
		5

7. A common assumption is that sleep deprivation influences aggression. To test this assumption, volunteer subjects are randomly assigned to sleep deprivation periods of 0, 24, 48, and 72 hours, and subsequently tested for aggressive behavior in a controlled social situation. Aggressive scores signify the total number of different aggressive behaviors, such as "put downs," arguments, or verbal interruptions, demonstrated by subjects during the test period.

(a) Test the null hypothesis at the .05 level of significance.
(b) Summarize results with an ANOVA table.
(c) How might these results appear in a published report?
(d) If appropriate, use Scheffé's test and interpret the results.

AGGRESSIVE SCORES

0	24	48	72
0	1	5	7
1	3	4	1
0	2	7	6
3	2	8	9
1	4	6	10
2	6	3	12
4	3	2	8
2	4	5	7

18

Analysis of Variance (Two Factors)

18.1 –Data Interpretation with Graphs
18.2 –Three F Ratios
18.3 –Variance Estimates
18.4 –Sum of Squares (SS)
18.5 –Degrees of Freedom (df)
18.6 –Mean Squares (MS) and F Ratios
18.7 –F Tables
18.8 –Interaction
18.9 –Multiple Comparisons
18.10–Assumptions and a Final Caution
18.11–Other Types of ANOVA
18.12–Summary
18.13–Exercises

In psychological jargon, a "significant interaction" can refer to an important exchange between people. Although it means something entirely different in statistical terminology, interaction emerges as the most striking feature of two-factor ANOVA.

Two-factor ANOVA represents a statistical "best buy." It allows three different null hypotheses to be tested with the same experimental data.

The previous chapter described an attempt to determine the effect of crowd size on the alarm reactions of subjects—presumably both males and females—to potentially dangerous smoke. If gender were viewed as important, possibly because of the noticeably different reactions of males and females during a pilot study, the social psychologist might design an experiment with two factors: crowd size (zero, two, four confederates), and gender (female, male). Using this design, she still can test the original null hypothesis about the effect of crowd size on subjects' alarm reactions. As a bonus, she also can test two new null hypotheses: one about the effect of gender and the other about the combined effect of crowd size and gender on subjects' alarm reactions.

MODERATE DRINKING CAN BE FUN... AS CAN A DRIVE IN THE COUNTRY... BUT NOT THE INTERACTION

For simplicity, let's assume that the social psychologist randomly assigns two female subjects each to be tested (one subject at a time) with zero, two, or four confederates, and then repeats the random assignment for an equal number of male subjects. The resulting six groups, each consisting of two subjects, represent all possible combinations of the two factors.

18.1 Data Interpretation with Graphs

Table 18.1 shows one possible set of outcomes for the two-factor experiment. Numbers in red represent the various means. Rather than plunge directly into a statistical analysis, let's attempt to decipher—with the aid of graphs—the messages embodied in Table 18.1.

The slanted line in Figure 18.1 depicts the large differences between column means, that is, between mean alarm reactions of subjects in crowds of zero, two, and four confederates. This finding suggests that the null hypothesis for crowd size probably should be rejected. The steeper the slant, the larger the observed mean differences and the more suspect the null hypothesis. A fairly level line in Figure 18.1 would have reflected the relative absence of any differences between column means.

The slanted line in Figure 18.2 depicts the large difference between row means, that is, between mean alarm reactions of male and female subjects. This finding suggests that the null hypothesis for gender probably should be rejected.

These preliminary conclusions must be qualified, however, because of a complication due to the combined effect or interaction of crowd size and gender. **Interaction** *occurs whenever the effects of one factor are not consistent for all values of the second factor.* For instance, the effect of crowd size is not consistent for male and female subjects, as revealed by a comparison of the first row of three group means for female subjects and the second row of three group means for male subjects, originally listed in Table 18.1 and plotted in Figure 18.3. While the line for female subjects in Figure 18.3 remains fairly level, that for male subjects is slanted, suggesting that the alarm reactions of male subjects, but not those of female subjects, are influenced by crowd size. Given this lack of a consistent effect—portrayed in Figure 18.3 by the pronounced lack of parallelism between the lines for male and female subjects—the null hypothesis (that there is no interaction between the two factors) probably should be rejected. Parallelism between the lines for male and female subjects in Figure 18.3 would have reflected the absence of any interaction in the two-factor experiment. The more extreme the nonparallelism, the more inconsistent the effects

**TABLE 18.1
OUTCOME OF TWO-FACTOR EXPERIMENT (ALARM REACTIONS IN MINUTES)**

GENDER	CROWD SIZE			Row Mean
	Zero	Two	Four	
Female	8 8 8	8 6 7	10 8 9	8
Male	9 11 10	15 19 17	24 18 21	16
Column Mean	9	12	15	Overall mean = 12

Note: Red numbers are means.

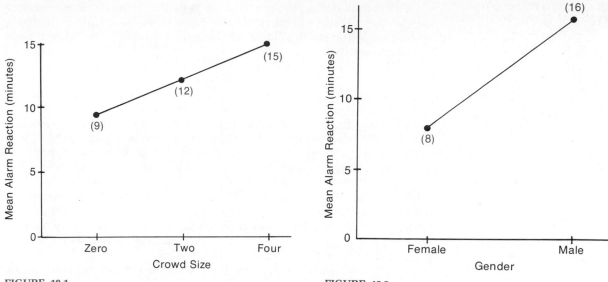

FIGURE 18.1

Effect of crowd size.

FIGURE 18.2

Effect of gender.

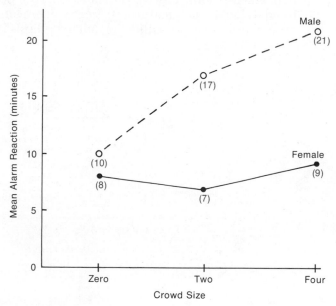

FIGURE 18.3

Combined effect (interaction) of crowd size and gender.

of one factor for different values of the second factor and the more suspect the null hypothesis for interaction. (Section 18.8 contains additional comments about interaction.)

To summarize, graphs of data for the two-factor experiment suggest a number of preliminary conclusions. Each of the three null hypotheses about the effects of crowd size, gender, and the interaction of these factors probably should be rejected. Because of the suspected interaction, however, generalizations about the effects of one factor must be associated with specific values of the second factor. Pending the outcome of the statistical analysis, you could state that crowd size appears to influence the alarm reactions of male subjects but not those of female subjects.

18.2 Three F Ratios

As suggested in Figure 18.4, F ratios in both one- and two-factor ANOVA always consist of a numerator, shown in red, that measures some aspect of variability between groups and a denominator, shown in black, that measures variability within groups. In one-factor ANOVA, as you'll recall, a single null hypothesis is tested with one F ratio. *In two-factor ANOVA, three different null hypotheses are tested, one at a time, with three F ratios:* F_{column}, F_{row}, *and* $F_{interaction}$. The numerator of each of these three F ratios reflects a different aspect of variability between groups: variability between columns (crowd size), variability between rows (gender), and interaction—any remaining variability between groups not attributable to either variability between columns (crowd size) or rows (gender).

The red numerator terms for the three F ratios in the bottom panel of Figure 18.4 estimate random error and, if present, a treatment effect (for subjects treated differently by the experiment, in the case of crowd size, and by "nature," in the case of gender). But the black denominator term always estimates only random error (for subjects treated alike in the same group). In practice, a large F value is viewed as most improbable, given that the null hypothesis is true, and therefore, it

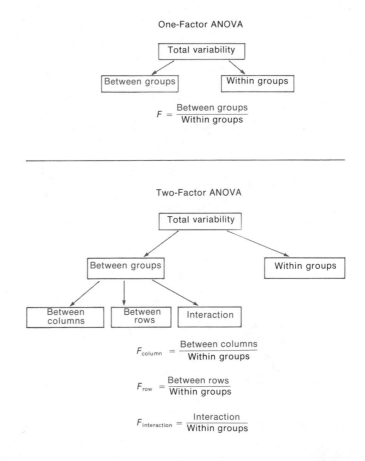

FIGURE 18.4

Sources of variability and F ratios on one- and two-factor ANOVA.

leads to the rejection of that particular null hypothesis. Otherwise, the null hypothesis is retained.

Hypothesis tests for the results of the smoke alarm experiment agree with the previous analysis based on graphs. As summarized below, each of the three null hypotheses in the experiment can be rejected at the .05 level of significance. Rejection of the null hypotheses for column and row effects indicate that both crowd size and gender influence the alarm reactions of subjects to smoke. Rejection of the null hypothesis for interaction, however, spotlights the inconsistent effect of crowd size on the alarm reactions of male and female subjects.

Hypothesis Test Summary:
F Test for Two Factors
(Smoke Alarm Experiment)

Problem:
Do crowd size and gender, as well as the interaction of these two factors, influence the mean alarm reactions of subjects to potentially dangerous smoke?

Statistical Hypotheses:
H_0: no treatment effect due to columns or crowd size
(or $\mu_0 = \mu_2 = \mu_4$).
H_0: no treatment effect due to rows or gender
(or $\mu_{female} = \mu_{male}$).
H_0: no interaction.
H_1: H_0 is not true.
(Same H_1 accommodates each H_0.)

Statistical Tests:

$$F_{column} = \frac{MS_{column}}{MS_{within}}$$

$$F_{row} = \frac{MS_{row}}{MS_{within}}$$

$$F_{interaction} = \frac{MS_{interaction}}{MS_{within}}$$

Decision Rule:
Reject H_0 at the .05 level of significance if F_{column} or $F_{interaction} \geq 5.14$ (from Table C, given 2 and 6 degrees of freedom), and if $F_{row} \geq 5.99$ (given 1 and 6 degrees of freedom).

Calculations:

$$F_{column} = 6.75$$
$$F_{row} = 36.02$$
$$F_{interaction} = 5.25$$

(See Table 18.3 for calculations.)

Decision:
Reject all three null hypotheses at the .05 level of significance.

Interpretation:
Both crowd size and gender influence the mean alarm reactions of subjects to smoke. The interaction indicates that the influence of crowd size differs for male and female subjects.

18.3 Variance Estimates

Considerable similarity exists between measures of variability in one- and two-factor ANOVA. In both cases, every measure of variability consists of a variance estimate or mean square obtained by dividing a particular sum of squares (SS) by its degrees of freedom (df). Let's concentrate first on the various SS, then on their associated df.

18.4 Sum of Squares (SS)

Once again, most computational effort is directed toward the various SS terms: the column sum of squares, SS_{column}; the row sum of squares, SS_{row}; the interaction sum of squares, $SS_{interaction}$; the within-group sum of squares, SS_{within}; and the total sum of squares, SS_{total}. Although all of these SS terms can be calculated directly, it's more convenient to calculate $SS_{interaction}$ indirectly, as shown below, because SS_{total} always equals the sum of the various SS terms, that is,

$$SS_{total} = SS_{column} + SS_{row} + SS_{interaction} + SS_{within} \qquad (18.1)$$

The top part of Table 18.2 lists the various computational formulas for each SS term, including the formula for calculating $SS_{interaction}$ indirectly. Table 18.3 illustrates the application of these formulas to the data for the two-factor experiment. Notice the highly predictable computational pattern first described in Section 17.6. Each score is always squared, and each total, whether for a group, column, or row total, or for the overall total, is always squared and then divided by its sample size.

To minimize computational errors, repeat all computations from scratch a second time and proceed only if all results agree. This is particularly important since the indirect determination of $SS_{interaction}$ eliminates the value of Formula 18.1 as a computational check.

TABLE 18.2 COMPUTATIONAL FORMULAS FOR SS TERMS

SS_{column} = sum of squared deviations of column means about the overall mean

$$= \frac{(\text{1st column total})^2}{\text{column sample size}} + \cdots + \frac{(\text{last column total})^2}{\text{column sample size}} - \frac{(\text{overall total})^2}{\text{overall sample size}}$$

SS_{row} = sum of squared deviations of row means about the overall mean

$$= \frac{(\text{1st row total})^2}{\text{row sample size}} + \cdots + \frac{(\text{last row total})^2}{\text{row sample size}} - \frac{(\text{overall total})^2}{\text{overall sample size}}$$

SS_{within} = sum of squared deviations of scores about their respective group means

$$= \text{sum of all squared scores} - \left[\frac{(\text{1st group total})^2}{\text{group sample size}} + \cdots + \frac{(\text{last group total})^2}{\text{group sample size}} \right]$$

SS_{total} = sum of squared deviations of scores about the overall mean

$$= \text{sum of all squared scores} - \frac{(\text{overall total})^2}{\text{overall sample size}}$$

$SS_{interaction}$ = sum of squared deviations of group means about the overall mean (not attributable to variations in column and row means)

$$= SS_{total} - [SS_{column} + SS_{row} + SS_{within}]$$

FORMULAS for df TERMS

df_{column} = number of columns − 1

df_{row} = number of rows − 1

$df_{interaction}$ = (number of columns − 1)(number of rows − 1)

df_{within} = number of scores − number of groups

df_{total} = number of scores − 1

TABLE 18.3
CALCULATION OF SS TERMS

A. COMPUTATIONAL SEQUENCE

Find (and circle) each group total ①.
Find each column and row total and also the overall total ②.
Substitute numbers into computational formula ③ and solve for SS_{column}.
Substitute numbers into computational formula ④ and solve for SS_{row}.
Substitute numbers into computational formula ⑤ and solve for SS_{within}.
Substitute numbers into computational formula ⑥ and solve for SS_{total}.
Substitute numbers into formula ⑦ and solve for $SS_{interaction}$.

B. DATA AND COMPUTATIONS

Gender	Crowd Size ①			Row Totals ②
	Zero	Two	Four	
Female	8 / 8 ⑯	8 / 6 ⑭	10 / 8 ⑱	48
Male	9 / 11 ⑳	15 / 19 ㉞	24 / 18 ㊷	96
② Column Totals	36	48	60	② Overall Total = 144

③ $SS_{column} = \dfrac{(1st\ column\ total)^2}{column\ sample\ size} + \dfrac{(2nd\ column\ total)^2}{column\ sample\ size} + \dfrac{(3rd\ column\ total)^2}{column\ sample\ size} - \dfrac{(overall\ total)^2}{overall\ sample\ size}$

$= \dfrac{(36)^2}{4} + \dfrac{(48)^2}{4} + \dfrac{(60)^2}{4} - \dfrac{(144)^2}{12}$

$= 1800 - 1728 = 72$

④ $SS_{row} = \dfrac{(1st\ row\ total)^2}{row\ sample\ size} + \dfrac{(2nd\ row\ total)^2}{row\ sample\ size} - \dfrac{(overall\ total)^2}{overall\ sample\ size}$

$= \dfrac{(48)^2}{6} + \dfrac{(96)^2}{6} - \dfrac{(144)^2}{12}$

$= 1920 - 1728 = 192$

⑤ $SS_{within} = $ sum of all squared scores $- \left[\dfrac{(1st\ group\ total)^2}{group\ sample\ size} + \dfrac{(2nd\ group\ total)^2}{group\ sample\ size} + \cdots + \dfrac{(6th\ group\ total)^2}{group\ sample\ size} \right]$

$= (8)^2 + (8)^2 + (9)^2 + (11)^2 + (8)^2 + (6)^2 + (15)^2 + (19)^2 + (10)^2 + (8)^2 + (24)^2 + (18)^2$

$- \left[\dfrac{(16)^2}{2} + \dfrac{(20)^2}{2} + \dfrac{(14)^2}{2} + \dfrac{(34)^2}{2} + \dfrac{(18)^2}{2} + \dfrac{(42)^2}{2} \right]$

$= 2080 - 2048 = 32$

⑥ $SS_{total} = $ sum of all squared scores $- \dfrac{(overall\ total)^2}{overall\ sample\ size}$

$= (8)^2 + (8)^2 + (9)^2 + (11)^2 + (8)^2 + (6)^2 + (15)^2 + (19)^2 + (10)^2 + (8)^2 + (24)^2 + (18)^2 - \dfrac{(144)^2}{12}$

$= 2080 - 1728$

$= 352$

⑦ $SS_{interaction} = SS_{total} - [SS_{column} + SS_{row} + SS_{within}]$

$= 352 - [72 + 192 + 32]$

$= 352 - 296 = 56$

18.5 Degrees of Freedom (df)

The number of degrees of freedom differs for each SS term in two-factor ANOVA, and for convenience, the various df formulas are listed in the bottom half of Table 18.3. Notice that $df_{\text{interaction}}$ represents the product of df_{column} and df_{row}.

The df for the present experiment are as follows:

$$df_{\text{column}} = 3 - 1 = 2$$
$$df_{\text{row}} = 2 - 1 = 1$$
$$df_{\text{interaction}} = (3 - 1)(2 - 1) = 2$$
$$df_{\text{within}} = 12 - 6 = 6$$
$$df_{\text{total}} = 12 - 1 = 11$$

Recall the general rule that the degrees of freedom for SS_{total} equals the combined degrees of freedom for all remaining SS terms, that is,

$$df_{\text{total}} = df_{\text{column}} + df_{\text{row}} + df_{\text{interaction}} + df_{\text{within}} \tag{18.2}$$

This formula can be used to verify that the correct number of degrees of freedom has been assigned to each SS term.

18.6 Mean Squares (MS) and F Ratios

Having found values for the various SS terms and their df, we can determine values for the corresponding MS terms and then calculate the three F ratios using formulas in Table 18.4. Notice that MS_{within}—the estimate of random error—appears in the denominator of each of these three F ratios.

ANOVA results for the two-factor experiment are summarized in Table 18.5. The origin of each MS term and of each F ratio is suggested by the red numbers (which ordinarily don't appear in ANOVA tables).

Other labels also might have appeared in Table 18.5. For instance, "Column" and "Row" might have been replaced by descriptions of the treatment variables, in this case, "Crowd Size" and "Gender." By the same token, "Interaction" might have been replaced by "Crowd Size × Gender" or by some abbreviation, such as "C × G" (for Crowd Size × Gender), and "Within" might have been replaced by "Error."

18.7 F Tables

Each of the three F ratios in Table 18.5 exceeds its respective critical F ratio. To obtain critical F ratios, refer to Table C. Follow the same procedure described in Section 17.9 to verify that, when 2 and 6 degrees of freedom are associated with the numerator and denominator of F, as for F_{column} and $F_{\text{interaction}}$, the critical F equals 5.14. Also verify

TABLE 18.4
FORMULAS FOR MEAN SQUARES (MS) AND F RATIOS

SOURCE	MS	F
Column	$MS_{\text{column}} = \dfrac{SS_{\text{column}}}{df_{\text{column}}}$	$F_{\text{column}} = \dfrac{MS_{\text{column}}}{MS_{\text{within}}}$
Row	$MS_{\text{row}} = \dfrac{SS_{\text{row}}}{df_{\text{row}}}$	$F_{\text{row}} = \dfrac{MS_{\text{row}}}{MS_{\text{within}}}$
Interaction	$MS_{\text{interaction}} = \dfrac{SS_{\text{interaction}}}{df_{\text{interaction}}}$	$F_{\text{interaction}} = \dfrac{MS_{\text{interaction}}}{MS_{\text{within}}}$
Within	$MS_{\text{within}} = \dfrac{SS_{\text{within}}}{df_{\text{within}}}$	

SOURCE	SS	df	MS	F	
Column	72	2	$\left(\frac{72}{2}=\right)$ 36	$\left(\frac{36}{5.33}=\right)$ 6.75*	**TABLE 18.5** **ANOVA TABLE (TWO-FACTOR EXPERIMENT)**
Row	192	1	$\left(\frac{192}{1}=\right)$192	$\left(\frac{192}{5.33}=\right)$ 36.02*	
Interaction	56	2	$\left(\frac{56}{2}=\right)$ 28	$\left(\frac{28}{5.33}=\right)$ 5.25*	
Within	32	6	$\left(\frac{32}{6}=\right)$ 5.33		
Total	352	11			

*Significant at .05 level.

that, when 1 and 6 degrees of freedom are associated with F, as for F_{row}, the critical F equals 5.99.

18.8 Interaction

Interaction emerges as the most striking feature of two-factor ANOVA. As has been noted previously, two factors interact if the effects of one factor are not consistent for all values (or levels) of a second factor. Rather than being a complication to be avoided if possible, an interaction often highlights pertinent issues for future research. For example, the interaction between crowd size and gender suggests that subsequent experiments might attempt to identify those factors that cause the alarm reactions of male subjects, but not female subjects, to be relatively sensitive to crowd size.

In the two-factor experiment, the combined effect of crowd size and gender could have differed from that described in Figure 18.3. Examples of some other possible effects are shown in Figure 18.5. The two top panels in Figure 18.5 describe outcomes that, because of their

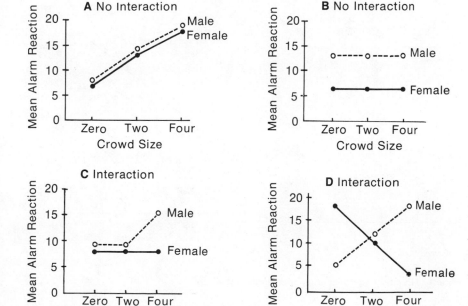

FIGURE 18.5
Some possible outcomes (two-factor experiment).

215

consistency, would cause the retention of the null hypothesis for interaction. No interaction appears in Figure 18.5A, where the parallel lines reflect the consistent effect of crowd size on the alarm reactions of both female and male subjects. Likewise, no interaction appears in Figure 18.5B, where the parallel lines reflect the consistent tendency for alarm reactions of female subjects to be shorter than those for male subjects.

The two bottom panels in Figure 18.5 describe outcomes that, because of their inconsistency, probably would cause the rejection of the null hypothesis for interaction. A relatively weak interaction appears in Figure 18.5C, where the nonparallel lines reflect the inconsistent effect of the largest crowd size on the reactions of female and male subjects. A strong interaction appears in Figure 18.5D, where the non-parallel lines reflect the totally inconsistent (opposite) effect of crowd size on the alarm reactions of female and male subjects.

The original interaction between crowd size and gender could have been described in two different ways. First, we could have pointed out the inconsistent effect of crowd size on the alarm reactions of female and male subjects, as depicted in Figure 18.6A (and originally shown in Figure 18.3). Alternately, the interaction could have been described as the inconsistent effect of gender on the alarm reactions of subjects in crowd sizes of zero, two, and four confederates, as depicted in Figure 18.6B. Although different, the configurations in both Figure 18.6A and B show the lack of parallelism that typifies an interaction between two factors. Furthermore, the interpretation is the same for both configurations—crowd size influences the alarm reactions of male subjects, but not those of female subjects. You needn't worry, therefore, about misinterpreting an interaction because of an arbitrary decision about how to look at it.

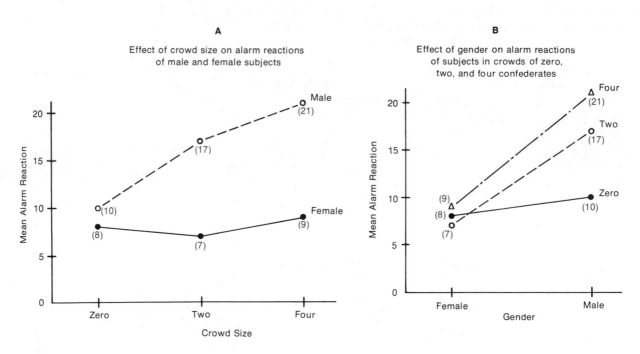

Note: Red numbers represent group means from Table 18.1

FIGURE 18.6

Two versions of the same interaction.

A modification of Scheffé's test for multiple comparisons, described in the previous chapter, may be used to pinpoint important differences between pairs of column, row, or group means whenever the corresponding overall null hypothesis has been rejected. Ordinarily, when all three null hypotheses have been rejected, as in the two-factor experiment for the present chapter, Scheffé's test would have been used to evaluate differences between pairs of group means such as, for instance, the difference between mean alarm reactions for *only* male subjects with zero and two confederates (that is, 10 − 17 in Table 18.1). In this case, the objective is to identify those specific differences that contribute to the rejection of the overall null hypothesis for interaction. The use of Scheffé's test in two-factor experiments is not described in this book.

18.9 Multiple Comparisons

The assumptions for *F* tests in two-factor ANOVA are the same as those for one-factor ANOVA; all underlying populations are assumed to be normally distributed with equal variances. To protect yourelf against violations of these assumptions, use equal group sizes, each greater than 10.

It cannot be overemphasized that *all groups in two-factor experiments should have equal sample sizes*. Even small departures from equal sample sizes cause extensive computational and interpretation problems in ANOVA. If you are forced to analyze data based on unequal sample sizes—possibly because of a missing subject, equipment breakdown, or recording error—consult a more advanced statistics book.*

18.10 Assumptions and a Final Caution

One- and two-factor experiments don't exhaust the possibilities for ANOVA. For instance, you could use ANOVA to analyze the results of an experiment with three or more factors, where each subject contributes not one, but many observations. Although the basic concepts described in this book transfer almost intact to a wide assortment of more intricate experimental designs, computational procedures grow more complex, and the interpretation of results often is more difficult. Intricate experimental designs, requiring the use of complex types of ANOVA, provide the skilled researcher with powerful tools for evaluating complicated situations. Under no circumstances, however, should an experiment be valued simply because of the complexity of its design and statistical analysis. Strive to use the least complex design and analysis that will answer your research question.

18.11 Other Types of ANOVA

Prior to any statistical analysis, and particularly prior to complex analyses, such as two-factor ANOVA, it's often instructive to construct graphs of the various possible effects; slanted lines forecast that the null hypothesis for a single factor might be rejected, and nonparallel lines serve advance notice that the null hypothesis for interaction might be rejected.

In two-factor analysis of variance, three null hypotheses are tested with three different *F* ratios. The numerator of each *F* ratio measures a different aspect of variability between groups: variability

18.12 Summary

---

*See Keppel, G., *Design and Analysis: a Researcher's Handbook*, Prentice-Hall, Englewood Cliffs, N.J., 1973.

between columns; variability between rows; and any remaining variability between groups due to interaction. The denominator of each F ratio measures the variability within groups.

Each measure of variability or mean square (MS) is calculated by dividing the appropriate sum of squares (SS) by its degrees of freedom (df). (See Table 18.2.) Once a calculated value of F has been obtained, it's compared with a critical F. As usual, if the calculated F exceeds the critical F, the corresponding null hypothesis is rejected.

ANOVA results are often summarized in tabular form. (See Table 18.5.)

Two factors interact if the effects of one factor are not consistent for all values of a second factor. Interaction emerges as the most striking feature of two-factor ANOVA.

The assumptions for F tests in two-factor ANOVA are the same as those for one-factor ANOVA. For a variety of reasons, all groups in two-factor experiments should have equal sample sizes.

Important Terms

Two-factor ANOVA
Interaction

Note: To simplify computations, most of the following problems employ unrealistically small sample sizes.

18.13   Exercises

1. A home economist wishes to determine whether college students prefer a commercially produced pizza with a particular topping (plain, sausage, salami, or everything) and one type of crust (thick or thin). One hundred sixty volunteers are randomly assigned to one of the eight groups defined by this two-factor experiment. After eating their assigned pizza, the twenty subjects in each group rate their preference on a scale ranging from 0 (inedible) to 10 (the best). Results, in the form of means for groups, rows, and columns, are as follows:

MEAN PREFERENCE SCORES FOR PIZZA AS A FUNCTION OF TOPPING AND CRUST

CRUST	Plain	Sausage	Salami	Everything	Row
Thick	7.2	5.7	4.8	6.1	6.0
Thin	8.9	4.8	8.4	1.3	5.9
Column	8.1	5.3	6.6	3.7	

Construct graphs for each of the three possible effects. Use this information to interpret the outcome of the experiment. (Ordinarily, of course, you would verify these speculations by performing ANOVA—a task that cannot be performed for these data, since only means are supplied.)

2. A psychologist randomly assigns eight rats to each group in a two-factor experiment designed to determine the effect of food deprivation (0, 24, 48, and 72 hours) and reward magnitude (1, 2, and 3 food pellets) on rate of bar pressing.

   (a) Specify the various sources of variability and their degrees of freedom.
   (b) With rough sketches, indicate the types of graphs that could be obtained if rate of bar pressing increases with food deprivation,

remains relatively stable for reward magnitude, and shows little or no interaction.

18.13 EXERCISES

3. Each of the following (incomplete) ANOVA tables represents some experiment. Determine how many values (or levels) of each factor were used; the total number of groups; and, on the assumption that all groups have equal numbers of subjects, the number of subjects in each group. Then, using the .05 level of significance for all hypothesis tests, complete the ANOVA table.

(a)

SOURCE	SS	df	MS	F
Column	790	1		
Row	326	2		
Interaction	1887			
Within	14702	60		
Total				

(b)

SOURCE	SS	df	MS	F
Treatment A	142	2		
Treatment B	480	2		
A × B	209			
Error	5030	81		
Total				

4. A health educator suspects that the "days of discomfort" due to common colds can be reduced with ingestion of large doses of vitamin C and daily visits to a sauna. Using a two-factor design, subjects with new colds are randomly assigned to one of four different daily dosages of vitamin C (0, 500, 1000, 1500 milligrams) and to one of three different daily exposures to a sauna (0, $\frac{1}{2}$, and 1 hour).

NUMBER OF DAYS OF DISCOMFORT DUE TO COLDS

SAUNA EXPOSURE (HOURS)	VITAMIN C DOSAGE (MILLIGRAMS)			
	0	500	1000	1500
0	6	5	4	2
	4	3	2	3
	5	3	3	2
$\frac{1}{2}$	5	4	3	2
	4	3	2	1
	5	2	3	2
1	4	4	3	1
	3	2	2	2
	4	3	2	1

(a) Using appropriate sets of means, graph the various possible effects and tentatively interpret the experimental outcomes.
(b) Test the various null hypotheses at the .05 level of significance, using the customary step-by-step procedure.
(c) Summarize results with an ANOVA table.

5. A psychologist employs a two-factor experiment to study the combined effect of sleep deprivation and alcoholic consumption on the performance

of automobile drivers. Prior to the driving test, subjects go without sleep for various time periods and then drink a glass of orange juice laced with controlled amounts of vodka. Performance is measured by the number of errors made on a driving simulator. Two subjects are randomly assigned to each possible combination of sleep deprivation (0, 24, 48, 72 hours) and alcoholic consumption (0, 1, 2, 3 ounces), yielding the following results:

NUMBER OF DRIVING ERRORS

ALCOHOLIC CONSUMPTION (OUNCES)	SLEEP DEPRIVATION (HOURS)			
	0	24	48	72
0	0	2	5	5
	3	4	4	6
1	1	3	6	5
	3	3	7	8
2	3	2	8	7
	5	5	11	12
3	4	4	10	9
	6	7	13	15

(a) Construct graphs for the various effects and attempt to anticipate the results of the statistical analysis.
(b) Test the various null hypotheses at the .05 level of significance.
(c) Summarize results with an ANOVA table.

6. Does the type of instruction in a college sociology class (lecture or self-paced) and its grading policy (letter or pass/fail) influence the performance of students, as measured by the number of quizzes successfully completed during the semester? Six students are randomly assigned to each of the four possible combinations, yielding the following results:

NUMBER OF QUIZZES SUCCESSFULLY COMPLETED

GRADING POLICY	TYPE OF INSTRUCTION	
	Lecture	Self-Paced
Letter grades	4	7
	3	8
	5	6
	2	4
	6	10
	5	12
Pass/Fail	8	4
	9	1
	4	3
	5	2
	8	6
	10	8

(a) Graph and interpret the three effects.
(b) Test the various null hypotheses at the .01 level of significance.
(c) Summarize results with an ANOVA table.

19

Chi-Square (χ²) Test for Qualitative Data

ONE VARIABLE

19.1 –Observed and Expected Frequencies
19.2 –Calculation of χ^2
19.3 –χ^2 Tables and Degrees of Freedom
19.4 –χ^2 Test
19.5 –χ^2 Test Is Nondirectional
19.6 –Special Case: $df = 1$

TWO VARIABLES

19.7 –Observed and Expected Proportions
19.8 –Expected Frequencies
19.9 –χ^2 Tables and Degrees of Freedom
19.10–χ^2 Test
19.11–Special Case: $df = 1$
19.12–Some Precautions
19.13–Summary
19.14–Exercises

Want to test a hypothesis for some qualitative data? More often than not, chi-square (χ^2) is the appropriate test.

In everyday life, if large discrepancies appear between our expectations and reality, one possible solution is to relax or even drop our expectations. In statistics, if sufficiently large discrepancies appear between expected frequencies and observed frequencies, the strategy is always the same—reject the null hypothesis (from which the expected frequencies are derived).

When observations are merely classified into various categories—yes and no; male and female; Republican, Democrat, and Independent; Afro-American, Asian-American, Euro-American, and so forth—measurement is nominal and the data are qualitative, as discussed in Sections 1.8 and 1.9. Hypothesis tests for qualitative data, expressed as frequencies, require the use of a new test known as **chi-square** (symbolized as χ^2 and pronounced ki square). Our discussion of chi-square differentiates between the **one-variable case,** where observations are classified in terms of a single qualitative variable, and the **two-variable case,** where observations are classified in terms of two qualitative variables.

ONE VARIABLE

19.1 Observed and Expected Frequencies

Your blood belongs to one of four genetically determined types: O, A, B, or AB. A recent bulletin issued by a large blood bank claims that these four blood types are distributed in the U.S. population according to the following proportions: .44 are type O; .42 are type A; .10 are type B; and .04 are type AB. Let's treat this claim as a null hypothesis to be tested with a random sample of 100 students from a large university in California. If the null hypothesis is true, then, except for chance, the hypothesized proportions should be reflected in the sample. In the sample of 100 students, 44 students should have type O (from the product of .44 and 100); 42 should have type A; 10 should have type B; and only 4 should have type AB. In Table 19.1, each of these numbers is referred to as an **expected frequency** (f_e), that is, *the theoretical frequency for each category of the qualitative variable if, in fact, the null hypothesis is true.* To obtain the expected frequency for any category,

| | | BLOOD TYPE | | | |
FREQUENCY	O	A	B	AB	Total
Observed (f_o)	38	38	20	4	100
Expected (f_e)	44	42	10	4	100

find the product of the hypothesized or expected proportion for that category and the total sample size, namely,

$$f_e = (\text{expected proportion})(\text{total sample size}) \qquad (19.1)$$

where f_e represents the expected frequency.

It's most unlikely that a random sample exactly reflects the characteristics of its population. Even though the null hypothesis is true, discrepancies will appear between observed and expected frequencies, as in Table 19.1. *The crucial question is whether the discrepancies between observed and expected frequencies are small enough to qualify as a probable outcome,* given that the null hypothesis is true. If so, the null hypothesis is retained. Otherwise, if the discrepancies are too large, the null hypothesis is rejected.

19.2 Calculation of χ^2

To determine whether discrepancies between observed and expected frequencies qualify as a probable or an improbable outcome, a value is calculated for χ^2 and compared with its hypothesized sampling distribution. To calculate the value of χ^2, use the following expression:

$$\chi^2 = \Sigma \frac{(f_o - f_e)^2}{f_e} \qquad (19.2)$$

where f_o denotes the observed frequency and f_e denotes the expected frequency for each category of the qualitative variable.

Table 19.2 illustrates how to use Formula 19.2 to calculate χ^2 for the present example. Notice several features of Formula 19.2. The larger the discrepancies between observed and expected frequencies, $f_o - f_e$, the larger the value of χ^2 and, therefore, as will be seen, the more suspect the null hypothesis. Because of the squaring of each discrepancy, negative discrepancies become positive, and the value of χ^2 never can be negative.

19.3 χ^2 Tables and Degrees of Freedom

As with t and F, χ^2 has not one, but a family of sampling distributions. Table D of the Appendix supplies critical values from various χ^2 sampling distributions for hypothesis tests at the .10, .05, .01, and .001 levels of significance.

To locate the appropriate row in Table D, first identify the correct number of degrees of freedom. In the one-variable case, degrees of freedom for χ^2 can be obtained from the following expression:

$$df = C - 1 \qquad (19.3)$$

where C refers to the total number of categories of the qualitative variable.

To understand Formula 19.3, focus on the set of observed frequencies for 100 students in Table 19.1. In practice, of course, the observed frequencies of the four (C) categories have equal status, and

TABLE 19.2
CALCULATION OF χ^2
(ONE VARIABLE)

A. COMPUTATIONAL SEQUENCE

Find an expected frequency for each expected proportion ①.
List observed and expected frequencies ②.
Substitute numbers in formula ③ and solve for χ^2.

B. DATA AND COMPUTATIONS

① f_e = (expected proportion) (sample size)
$f_e(O)$ = (.44)(100) = 44
$f_e(A)$ = (.42)(100) = 42
$f_e(B)$ = (.10)(100) = 10
$f_e(AB)$ = (.04)(100) = 4

②

Frequency	O	A	B	AB	Total
f_o	38	38	20	4	100
f_e	44	42	10	4	100

③
$$\chi^2 = \sum \frac{(f_o - f_e)^2}{f_e}$$

$$= \frac{(38-44)^2}{44} + \frac{(38-42)^2}{42} + \frac{(20-10)^2}{10} + \frac{(4-4)^2}{4}$$

$$= \frac{(-6)^2}{44} + \frac{(-4)^2}{42} + \frac{(10)^2}{10} + \frac{(0)^2}{4}$$

$$= \frac{36}{44} + \frac{16}{42} + \frac{100}{10} + 0$$

$$= .82 + .38 + 10.00 + 0$$

$$= 11.20$$

any combination of four frequencies that sums to 100 is possible. From the more abstract perspective of degrees of freedom, however, only three $(C - 1)$ of these frequencies are free to vary—because of the mathematical restriction that, when calculating χ^2 for the present data, all observed (or expected) frequencies must sum to 100. Although observed frequencies of any three of the four categories are free to vary, the frequency of the fourth category must be some number that, when combined with the other three frequencies, yields a sum of 100. By the same token, if there had been five categories, the frequencies of any four categories would have been free to vary, but not that of the fifth category. In the one-variable case for χ^2, the number of degrees of freedom always equals one less than the total number of categories (C), as indicated in Formula 19.3.

In the present example, where the categories consist of the four blood types,

$$df = 4 - 1 = 3$$

To find the critical χ^2 for a hypothesis test at the .05 level of significance, locate the cell in Table D intersected by the row for 3 degrees of freedom and the column for the .05 level of significance. This cell lists a value of 7.81 for the critical χ^2.

19.4 χ^2 Test

Following the usual procedure, assume the null hypothesis to be true and view the observed χ^2 within the context of the hypothesized sampling distribution shown in Figure 19.1. If the observed χ^2 appears to emerge from the dense concentration of possible χ^2 values smaller than the critical χ^2 of 7.81, the observed outcome is not too improbable

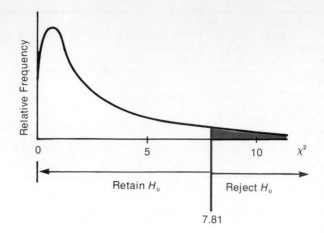

FIGURE 19.1

Hypothesized sampling distribution of χ^2 (3 degrees of freedom).

(that is, the discrepancies between observed and expected frequencies are not too large), given the assumption that the null hypothesis is true. Therefore, the null hypothesis is retained. On the other hand, if the observed χ^2 appears to emerge from the sparse concentration of possible χ^2 values equal to or greater than the critical χ^2 of 7.81, the observed outcome is too improbable, and the null hypothesis is rejected.

In fact, since the observed χ^2 of 11.20 exceeds the critical χ^2 of 7.81, the null hypothesis should be rejected; there is evidence that the distribution of blood types in the student population differs from that claimed for the U.S. population.

Hypothesis Test Summary:
χ^2 Test for One Variable
(Survey of blood types)

Problem:

Does the distribution of blood types in a population of college students comply with that described (for the U.S. population) in a blood bank bulletin?

Statistical Hypotheses:

$$H_0: P_O = .44; \; P_A = .42; \; P_B = .10; \; P_{AB} = .04$$

(where P_O represents the proportion of type O blood in the population of students, and so forth)

$$H_1: H_0 \text{ is not true.}$$

Statistical Test:

$$\chi^2 = \Sigma \frac{(f_o - f_e)^2}{f_e}$$

Decision Rule:

Reject H_0 at the .05 level of significance if $\chi^2 \geq 7.81$ (from Table D, given $df = C - 1 = 4 - 1 = 3$).

Calculations:
$\chi^2 = 11.20$ (See Table 19.2.)

Decision:
  Reject H_0.

Interpretation:
  The distribution of blood types in a student population differs from that claimed for the U.S. population.

Since all discrepancies contribute to the rejection of the null hypothesis, there are no statistical grounds for identifying the largest discrepancy as *the* discrepancy that causes the null hypothesis to be rejected. However, it is sometimes helpful to speculate about one or more of the most obvious discrepancies between observed and expected frequencies. As can be seen in Table 19.1, the present survey contains an unexpectedly large number of students with type B blood. Subsequent investigation revealed that the sample included a large number of Asian-American students, a group that has an established high incidence of type B blood. This might explain why the hypothesized distribution of blood types fails to describe that for the population of students from which the random sample was taken. Certainly a random sample should be taken from a much broader population before questioning the blood bank's claim about the distribution of blood types in the U.S. population.

19.5 χ^2 Test Is Nondirectional

As has been noted previously, the F test is nondirectional since all measures of variability are squared in ANOVA. In the same way, *the χ^2 test is nondirectional* since all discrepancies between observed and expected frequencies are squared. All squared discrepancies have a cumulative positive effect on the value of the observed χ^2 and thereby insure that χ^2 is a nondirectional test even though, as illustrated in Figure 19.1, only the upper tail of its sampling distribution contains the rejection region.

19.6 Special Case: $df = 1$

Hypothesis tests with χ^2 often involve observations that are classified into *only two* categories, such as yes and no, male and female, defective and nondefective. Under these circumstances,

$$df = C - 1 = 2 - 1 = 1$$

Whenever $df = 1$, Formula 19.2 should be modified slightly as described in the following example.

Under the U.S. criminal code, defendents have the right to a court trial before a jury of peers. Strictly speaking, this implies that prospective jurors should be selected randomly from among all eligible adults within the judicial district. In practice, prospective jurors often are selected randomly from some convenient roster of names, such as the list of registered voters. An investigator is concerned that this procedure discriminates against eligible black adults and, therefore, violates the intent of the law.

The most recent U.S. census reveals that blacks compose 30 percent or .30 of all eligible adults within a judicial district. If the selection process doesn't discriminate against blacks, they also should compose .30 of all members of the population from which prospective jurors are selected. Accordingly, a value of .30 appears in the null hypothesis for the population of registered voters (used to identify

TABLE 19.3
OBSERVED AND
EXPECTED FREQUENCIES

| FREQUENCY | RACE OF PROSPECTIVE JURORS | | Total |
	Black	Nonblack	
Observed (f_o)	51	149	200
Expected (f_e)	60	140	200

prospective jurors in that judicial district). Retention of the null hypothesis indicates a lack of evidence of any discrimination. Rejection of the null hypothesis indicates discrimination.

A random sample of 200 names is selected from the list of registered voters, and it's subsequently verified, as indicated in Table 19.3, that only 51 members of the sample are black.

To test the null hypothesis with χ^2, observed frequencies must be compared with expected frequencies. Shown in Table 19.3, expected frequencies are calculated in the usual manner from Formula 19.1: merely find the product of the expected proportion of blacks (.30) and the sample size (200), and also the product of the expected proportion of nonblacks $(1 - .30 = .70)$ and the sample size (200). To determine whether the discrepancies between observed and expected frequencies qualify as a probable or improbable outcome, given that the null hypothesis is true, calculate a value for χ^2 and compare it with the value of the critical χ^2.

Since there are only two categories (black and nonblack) for the present qualitative variable, $df = 1$. Whenever $df = 1$, an important technical adjustment officially known as Yates' correction requires that, before being squared, each discrepancy between observed and expected frequencies be reduced by 0.5. The adjusted formula for χ^2 reads:

$$\chi^2 = \Sigma\frac{(|f_o - f_e| - 0.5)^2}{f_e} \tag{19.4}$$

where the vertical strokes indicate that, whether positive or negative, each discrepancy should be treated as positive and reduced by 0.5 before being squared. Otherwise, all other terms are defined as before.

In the one-variable case, use Formula 19.4 whenever $df = 1$. Essentially, it corrects for an exaggeration produced by the presence of only two categories and yields a more accurate value of χ^2. Table 19.4 illustrates how to use Formula 19.4 to calculate χ^2 for the present data.

Hypothesis Test Summary:
χ^2 Test for One Variable When $df = 1$
(Discrimination against blacks)

Problem:
    Does racial discrimination occur when prospective jurors are selected from the list of registered voters?

Statistical Hypotheses:

H_0: $P_b = .30$
(where P_b represents the proportion of blacks
in the population of registered voters)

H_1: H_0 is not true.

Statistical Test:

$$\chi^2 = \Sigma \frac{(|f_o - f_e| - 0.5)^2}{f_e}$$

Decision Rule:
Reject H_0 at the .05 level of significance if $\chi^2 \geq 3.84$ (from Table D, given $df = C - 1 = 2 - 1 = 1$).

Calculations:
$\chi^2 = 1.72$ (See Table 19.4.)

Decision:
Retain H_0.

Interpretation:
There is no evidence that racial discrimination occurs when prospective jurors are selected from the list of registered voters.

One measure of "social responsibility" focuses on the return of self-addressed, stamped envelopes intentionally "lost" in the vicinity of mail boxes. A returned letter is assumed to indicate the presence of social responsibility, while a nonreturned letter indicates the absence of social responsibility. An investigator suspects that the degree of social responsibility, as measured by the lost letter technique, differs in three types of environment: downtown, suburbia, and campus. A total of 200 letters are "lost"—100 in downtown and 50 each in suburbia and on campus—according to procedures that control for possible contaminating factors, such as mailbox accessibility. After a 10-day wait, each letter is classified on the basis of where it was lost and whether or not it

TWO VARIABLES

TABLE 19.4
CALCULATION OF χ^2
($df = 1$)

A. COMPUTATIONAL SEQUENCE
Find an expected frequency for each expected proportion ①.
List observed and expected frequencies ②.
Substitute numbers in formula ③ and solve for χ^2.

B. DATA AND COMPUTATIONS
① f_e = (expected proportion)(sample size)
f_e (black) = (.30)(200) = 60
f_e (nonblack) = (.70)(200) = 140

② Frequency	Black	Nonblack	Total
f_o	51	149	200
f_e	60	140	200

③ $\chi^2 = \sum \dfrac{(|f_o - f_e| - 0.5)^2}{f_e}$

$= \dfrac{(|51 - 60| - 0.5)^2}{60} + \dfrac{(|149 - 140| - 0.5)^2}{140}$

$= \dfrac{(|-9| - 0.5)^2}{60} + \dfrac{(|9| - 0.5)^2}{140} = \dfrac{(9 - 0.5)^2}{60} + \dfrac{(9 - 0.5)^2}{140}$

$= \dfrac{(8.5)^2}{60} + \dfrac{(8.5)^2}{140} = \dfrac{(72.25)^2}{60} + \dfrac{(72.25)^2}{140} = 1.20 + 0.52$

$= 1.72$

TABLE 19.5
OBSERVED FREQUENCIES

SOCIAL RESPONSIBILITY	TYPE OF ENVIRONMENT			Total
	Downtown	Suburbia	Campus	
Present (Letter returned)	48	32	40	120
Absent (Letter not returned)	52	18	10	80
Total	100	50	50	200

was returned, as shown in Table 19.5. For instance, 48 letters were returned from downtown, 32 letters were returned from suburbia, and so forth.

19.7 Observed and Expected Proportions

According to the null hypothesis, degree of social responsibility and type of environment are not related. In other words, the same degree of social responsibility exists in all three types of environment. If the null hypothesis is true, then, except for chance, the proportion of returned cards for each of the three environments should equal the overall proportion of returned cards for all three environments. To gain a preliminary impression about the fate of the null hypothesis, it's helpful to convert the frequencies in Table 19.5 into the proportions shown in Table 19.6. Notice that these proportions are based on the various column totals and that the proportions within each column always sum to 1.00.

The top row of Table 19.6 contains three observed proportions and one overall proportion. Each of the three observed proportions describes the proportion of returned cards from among those lost in that environment. The overall proportion describes the proportion of returned cards from among those lost in *all* three environments. Its value is obtained by dividing the total number of returned cards from all three environments (120) by the total number of lost cards for all three environments (200).

If the null hypothesis is true, then, except for chance, each of the three observed proportions (.48, .64, and .80) should equal the overall proportion (.60), which often is referred to as the *expected* proportion. In fact, relatively large discrepancies appear between the observed and expected proportions for downtown and for campus, suggesting that the null hypothesis probably should be rejected. There is preliminary support for the investigator's hunch that different degrees of social responsibility exist in the three environments.

TABLE 19.6
OBSERVED
PROPORTIONS AND
OVERALL (EXPECTED)
PROPORTIONS

SOCIAL RESPONSIBILITY	TYPE OF ENVIRONMENT Observed Proportions			Overall (Expected) Proportion Total
	Downtown	Suburbia	Campus	
Present (Letter returned)	$\frac{48}{100} = .48$	$\frac{32}{50} = .64$	$\frac{40}{50} = .80$	$\frac{120}{200} = .60$
Absent (Letter not returned)	$\frac{52}{100} = .52$	$\frac{18}{50} = .36$	$\frac{10}{50} = .20$	$\frac{80}{200} = .40$
Total	1.00	1.00	1.00	1.00

When the row variable consists of only two categories, as in Table 19.6, there are only two rows of observed and expected proportions. In this case, you need to inspect the discrepancies in just one of the two rows—since discrepancies in the remaining row merely complement those in the first row and no new information is gained. When the row variable consists of more than two categories, as in Exercise 7 at the end of this chapter, follow the same procedure illustrated in Table 19.6 to obtain a set of observed and expected proportions for each row. In this case, however, you should inspect the discrepancies for all rows in order to gain a preliminary impression about whether the null hypothesis should be rejected.

19.8 Expected Frequencies

Although a χ^2 test could be based on the discrepancies between observed and expected *proportions*, it's more efficient to use the discrepancies between observed and expected *frequencies*, as described originally in Formula 19.2. Observed frequencies for the present study have been listed in Table 19.5, and only the expected frequencies need to be determined.

As has been noted, if the three environments are considered together, 120/200 or .60 of all letters were returned (while 80/200 or .40 of all letters were not returned). If the null hypothesis is true, these same proportions should be reflected in the number of letters returned (or not returned) for *each* of the three environments. More specifically, given that 100 letters were lost downtown, .60 of this total, that is, (.60)(100) or 60 letters, should be returned, and 60 is the expected frequency of returned letters for downtown. Given that 50 letters were lost in suburbia, .60 of this total, or 30, is the expected frequency of returned letters for suburbia. Finally, given that 50 letters were lost on campus, .60 of this total, or 30, is the expected frequency of returned letters for campus.

Calculations for the entire set of expected frequencies are summarized in Table 19.7. Notice the consistent manner in which each expected frequency is obtained: simply find the product of expected proportion for any cell and the total frequency for the column occupied by that cell.

Expected frequencies have been derived from expected proportions in order to spotlight the reasoning behind the χ^2 test. In the long run, it's more efficient to calculate expected frequencies directly from the various marginal totals according to the following formula:

$$f_e = \frac{(\text{column total})(\text{row total})}{\text{overall total}} \qquad (19.4)$$

TABLE 19.7
OBSERVED AND EXPECTED FREQUENCIES

SOCIAL RESPONSIBILITY	TYPE OF ENVIRONMENT			Total
	Downtown	Suburbia	Campus	
Present (Letter returned)	$f_o = 48$ $f_e = (.60)(100) = 60$	$f_o = 32$ $f_e = (.60)(50) = 30$	$f_o = 40$ $f_e = (.60)(50) = 30$	120
Absent (Letter not returned)	$f_o = 52$ $f_e = (.40)(100) = 40$	$f_o = 18$ $f_e = (.40)(50) = 20$	$f_o = 10$ $f_e = (.40)(50) = 20$	80
Total	100	50	50	200

where f_e refers to the expected frequency for any cell, column total refers to the total frequency for the column occupied by that cell, row total refers to the total frequency for the row occupied by that cell, and overall total refers to the grand total for all columns (or all rows).

Using the marginal totals in Table 19.7, we may verify that Formula 19.4 yields the expected frequencies shown in that table. For example, the expected frequency of returned letters for downtown is

$$f_e = \frac{(100)(120)}{200} = \frac{12000}{200} = 60$$

And the expected frequency of returned letters for suburbia is

$$f_e = \frac{(50)(120)}{200} = \frac{6000}{200} = 30$$

Before reading on, use Formula 19.4 to obtain the expected frequencies for the four remaining cells in Table 19.7.

Having determined the set of expected frequencies, you may use Formula 19.2 to calculate the value of χ^2, as described in Table 19.8.

TABLE 19.8
CALCULATION OF χ^2
(TWO VARIABLES)

A. COMPUTATIONAL SEQUENCE

Use formula ① to obtain all expected frequencies from table of observed frequencies.

Construct a table of observed and expected frequencies ②.

Substitute numbers in formula ③ and solve for χ^2.

B. DATA AND COMPUTATIONS

① $f_e = \dfrac{\text{(column total)(row total)}}{\text{overall total}}$

$f_e(\text{letters returned, downtown}) = \dfrac{(100)(120)}{200} = 60$

$f_e(\text{letters returned, suburbia}) = \dfrac{(50)(120)}{200} = 30$

$f_e(\text{letters returned, campus}) = \dfrac{(50)(120)}{200} = 30$

$f_e(\text{letters not returned, downtown}) = \dfrac{(100)(80)}{200} = 40$

$f_e(\text{letters not returned, suburbia}) = \dfrac{(50)(80)}{200} = 20$

$f_e(\text{letters not returned, campus}) = \dfrac{(50)(80)}{200} = 20$

②

	Downtown	Suburbia	Campus	Total
Present (Letter returned)	$f_o = 48$ $f_e = 60$	$f_o = 32$ $f_e = 30$	$f_o = 40$ $f_e = 30$	120
Absent (Letter not returned)	$f_o = 52$ $f_e = 40$	$f_o = 18$ $f_e = 20$	$f_o = 10$ $f_e = 20$	80
Total	100	50	50	200

③ $\chi^2 = \sum \dfrac{(f_o - f_e)^2}{f_e}$

$= \dfrac{(48-60)^2}{60} + \dfrac{(32-30)^2}{30} + \dfrac{(40-30)^2}{30} + \dfrac{(52-40)^2}{40} + \dfrac{(18-20)^2}{20} + \dfrac{(10-20)^2}{20}$

$= 2.40 + 0.13 + 3.33 + 3.60 + 0.20 + 5.00$

$= 14.66$

Location of a critical χ^2 value in Table D requires that you know the correct number of degrees of freedom. In the two-variable case, degrees of freedom for χ^2 can be obtained from the following expression:

$$df = (C - 1)(R - 1) \qquad (19.5)$$

where C equals the number of categories for the column variable and R equals the number of categories for the row variable. In the present example, where there are three columns (downtown, suburbia, and campus) and two rows (present and absent),

$$df = (3 - 1)(2 - 1) = (2)(1) = 2$$

To find the critical χ^2 for a test at the .05 level of significance, locate the cell in Table D intersected by the row for 2 degrees of freedom and the column for the .05 level. In this case, the value of the critical χ^2 equals 5.99.

To understand Formula 19.5, focus on the set of observed frequencies in Table 19.5. In practice, of course, the observed frequencies of the six cells (inside the black lines) have equal status, and any combination of six frequencies that sum to the various marginal totals is possible. From the more abstract perspective of degrees of freedom, however, only two of these frequencies are free to vary—because of the mathematical restriction that, within each column, all observed (or expected) cell frequencies must sum to the column total and, within each row, all observed (or expected) cell frequencies must sum to the row total. Although observed frequencies of any two of the six cells are free to vary, the frequencies of the four remaining cells are determined by the various marginal totals. As indicated in Table 19.9A, any frequencies could be assigned to the first two cells in the top row, but the frequency of the third cell in the top row would be determined by the row total, while the frequencies of the three cells in the second row would be determined by their respective column totals.

Imagine a study involving three columns and three rows, yielding a total of nine cells with observed frequencies. According to Formula 19.5,

$$df = (3 - 1)(3 - 1) = (2)(2) = 4$$

This implies that after frequencies have been assigned to four of the cells, the frequencies of the five remaining cells are determined by the various marginal totals. As indicated in Table 19.9B, any frequencies could be assigned to the first two cells in the top row and to the first two cells in the middle row. The frequency of the remaining cell within each of these rows would be determined by its row total, while the frequencies of the three cells in the bottom row would be determined by their respective column totals.

Since the calculated χ^2 of 14.66 exceeds the critical χ^2 of 5.99, the null hypothesis should be rejected; there is evidence that degree of social responsibility and type of environment are related. In other words, knowledge about the type of environment supplies extra information about the degree of social responsibility—it appears to be highest on campus and lowest downtown.

19.9 χ^2 Tables and Degrees of Freedom

TABLE 19.9
DEGREES OF FREEDOM (TWO VARIABLES)

A. THREE COLUMNS AND TWO ROWS:
$df = (3 - 1)(2 - 1) = 2$

✔	✔	✗	120
✗	✗	✗	80
100	50	50	200

B. THREE COLUMNS AND THREE ROWS:
$df = (3 - 1)(3 - 1) = 4$

✔	✔	✗	20
✔	✔	✗	100
✗	✗	✗	80
100	50	50	200

✔ Cell frequency is free to vary.

✗ Cell frequency is not free to vary but fixed by marginal total.

19.10 χ^2 Test

Hypothesis Test Summary:
χ^2 Test for Two Variables
(Lost letter study)

Problem:
Do different degrees of social responsibility, as measured by the lost letter technique, exist in three types of environment: downtown, suburbia, and campus?

Statistical Hypotheses:
H_0: Degree of social responsibility and type of environment are not related.

H_1: H_0 is false.

Statistical Test:

$$\chi^2 = \Sigma \frac{(f_o - f_e)^2}{f_e}$$

Decision Rule:
Reject H_0 at .05 level of significance if $\chi^2 \geq 5.99$ (from Table D, given $df = (C - 1)(R - 1) = (3 - 1)(2 - 1) = 2$).

Calculations:
$\chi^2 = 14.66$ (See Table 19.8.)

Decision:
Reject H_0.

Interpretation:
Degree of social responsibility and type of environment are related.

A published report of this hypothesis test might be limited to an interpretative comment, plus a parenthetical statement that summarizes the statistical analysis, as follows:

"There is evidence that degree of social responsibility is related to the three types of environment [$\chi^2(2) = 14.66$, $p < .05$]."

The parenthetical statement indicates that a χ^2 based on 2 degrees of freedom was found to equal 14.66. Furthermore, p, the probability of the observed χ^2—on the assumption that H_0 is true—is less than .05. Since p is less than .05, the observed χ^2 qualifies as an improbable outcome, and the null hypothesis is rejected, as implied in the interpretative comment.

19.11 Special Case: $df = 1$

It's fairly common for data to be classified for two qualitative variables, each of which contains only two categories. In a survey, for example, people might be classified on the basis of gender (male, female) and affiliation with a major political party (Democrat, Republican). Only one degree of freedom exists in this type of situation since

$$df = (2 - 1)(2 - 1) = 1$$

Under these circumstances, it has been standard practice to use the adjusted formula (19.4) for χ^2. Recent computer studies suggest that, in the two-variable case, this practice is less accurate than one based on the original unadjusted formula (19.2) for χ^2* *Always use Formula 19.2 in the two-variable case—even when df = 1.*

19.12   Some Precautions

The valid use of χ^2 requires that *observations should be independent.* One outcome should have no influence on another; for instance, when tossing a pair of dice, the appearance of a six spot on one die has no influence on the number of spots displayed by the other die. A violation of independence would have occurred in the blood type survey, for example, if any student's blood type had been counted more than once in the observed frequencies of Table 19.1.

The valid use of χ^2 also requires that expected frequencies should not be too small. A conservative rule specifies that *all expected frequencies should be 5 or more when df = 1, and at least four fifths of all expected frequencies should be 5 or more when df > 1.* Excessively small expected frequencies need not necessarily lead to a statistical dead end; sometimes it's possible to create a larger expected frequency from the combination of smaller expected frequencies. (See Exercise 8.)

19.13   Summary

χ^2 is designed to test the null hypothesis for qualitative data, expressed as frequencies. In the one-variable case, the null hypothesis claims that the population distribution complies with a set of hypothesized proportions. In the two-variable case, the null hypothesis claims, most generally, that there is no relationship between the two qualitative variables. In either case, the null hypothesis is used to generate sets of expected frequencies.

Essentially, as defined in Formula 19.2, χ^2 reflects the size of the discrepancies between observed and expected frequencies, and the larger the value of χ^2, the more suspect the null hypothesis.

In the one-variable case, when df = 1, use an adjusted formula (19.4) for χ^2.

To obtain critical values for χ^2, Table D must be consulted with the appropriate number of degrees of freedom, given by Formulas 19.3 and 19.5 for the one- and two-variable cases, respectively. Because all discrepancies are squared, the χ^2 test is nondirectional, even though only the upper tail of the χ^2 sampling distribution contains the rejection region.

Use of the χ^2 test requires that observations be independent and that expected frequencies be sufficiently large.

Important Terms & Symbols

Chi-square (χ^2)	Observed frequency (f_o)
One-variable case	Expected frequency (f_e)
Two-variable case	

*For example, Camilli, G., and Hopkins, K.D., Applicability of chi-square to 2×2 contingency tables with small expected cell frequencies, *Psychol. Bull.*, 1978, *85*, 163–167.

19.14 Exercises

1. In a random sample, 300 college students indicate whether they *most* desire love, wealth, power, health, fame, or family happiness. Using the .05 level of significance and the results below, test the null hypothesis that, in the underlying population, the various desires are equally popular.

DESIRES OF COLLEGE STUDENTS

FREQUENCY	LOVE	WEALTH	POWER	HEALTH	FAME	FAMILY HAP.	TOTAL
Observed (f_o)	65	55	45	60	40	35	300

2. Randomly selected records of 140 criminals reveal that the crimes (for which they were convicted) were committed on the following days of the week:

DAYS WHEN CRIMES WERE COMMITTED

FREQUENCY	MON.	TUE.	WED.	THU.	FRI.	SAT.	SUN.	TOTAL
Observed (f_o)	17	21	22	18	23	24	15	140

Using the .01 level of significance, test the null hypothesis that, in the underlying population, crimes were equally likely to be committed on any day of the week.

3. While playing a coin-tossing game in which you're to guess whether heads or tails will appear, you observe 30 heads in a string of 50 coin tosses. Test the null hypothesis that this coin is unbiased, that is, heads and tails are equally likely to appear in the long run.

4. In Chapter 1, Table 1.1 lists the weights of 53 female statistics students. Although students were asked to report their weights to the nearest pound, inspection of Table 1.1 reveals that a disproportionately large number (27) reported weights ending in either a zero or a five. This suggests that many students probably report their weights rounded to the nearest five or ten pounds rather than to the nearest pound. Using the .05 level of significance, test the null hypothesis that, in the underlying population, weights are rounded to the nearest pound. Hint: if the null hypothesis is true, two tenths of all weights should end in either a zero or a five, while eight tenths of all weights should end in either a one, two, three, four, six, seven, eight, or nine. Therefore, the situation involves a single variable with only two categories, and $df = 1$.

5. A researcher suspects that there might be a relationship, possibly based on genetic factors, between hair color and susceptibility to poison oak. Three hundred volunteer subjects are exposed to a small amount of poison oak and then classified according to their susceptibility (rash or no rash) and their hair color (red, blond, brown, or black), yielding the frequencies shown below.

HAIR COLOR AND SUSCEPTIBILITY TO POISON OAK

SUSCEPTIBILITY	HAIR COLOR				Total
	Red	Blond	Brown	Black	
Rash	10	30	60	80	180
No Rash	20	30	30	40	120
Total	30	60	90	120	300

(a) Convert the frequencies in the top row of the table into a set of observed proportions and a single expected proportion. Inspect the discrepancies between observed and expected proportions in the top row and indicate, as a preliminary impression, whether the null hypothesis (that hair color and susceptibility to poison oak are not related) should be rejected.

(b) Test the null hypothesis at the .01 level of significance.

6. Students are classified according to religious preference (Buddhist, Jewish, Protestant, Roman Catholic, and Other) and political affiliation (Democrat, Republican, Independent, and Other).

RELIGIOUS PREFERENCE AND POLITICAL AFFILIATION

POLITICAL AFFILIATION	Buddhist	Jewish	Protestant	Rom. Cath.	Other	Total
Democrat	30	30	40	60	40	200
Republican	10	10	40	20	20	100
Independent	10	10	20	20	40	100
Other	0	0	0	0	100	100
Total	50	50	100	100	200	500

(a) Convert the frequencies in each row of the table into a set of observed proportions and a single expected proportion. Inspect the discrepancies, row by row, and indicate, as a preliminary impression, whether the null hypothesis should be rejected.

(b) The above table consists of five columns and four rows. Identify those cell frequencies that are free to vary and those that are fixed, as in Table 19.9, and determine the number of degrees of freedom for χ^2 in this situation.

(c) Using the .05 level of significance, test the null hypothesis that these two variables are independent in the underlying population.

7. Do two different generations of college students respond similarly to the question, "Have you ever smoked marijuana?" The table below shows answers to this question for samples of students who attended statistics classes during the early 1970s and late 1970s. Using the .05 level of significance, test the null hypothesis that there is no relationship between replies and the two generations of college students.

MARIJUANA SMOKING RESPONSES OF TWO GENERATIONS

RESPONSE	Early 1970s	Late 1970s	Total
Yes	62	55	117
No	38	25	63
Total	100	80	180

8. Test the null hypothesis at the .01 level of significance that the distribution of blood types for college students complies with the proportions described in the blood bank bulletin, namely, .44 for O; .42 for A; .10 for B; and .04 for AB. Now, however, assume that results are available for a random sample of only 60 students. The results are: 27 for O; 24 for A; 5 for B; and 4 for AB. (Note: Because of the small sample size, the expected frequency for AB is too small. Create a sufficiently large expected frequency by combining B and AB blood types.)

20.1 –Three Tests for Ranked
Data

**MANN-WHITNEY *U* TEST
(TWO INDEPENDENT
SAMPLES)**

20.2 –Why Not a *t* Test?
20.3 –Calculation of *U*
20.4 –*U* Tables
20.5 –Decision Rule
20.6 –Null Hypothesis
20.7 –Directional Tests
20.8 –Large Sample
   Approximation for *U*

**WILCOXON *T* TEST
(TWO DEPENDENT
SAMPLES)**

20.9 –Why Not a *t* Test?
20.10–Calculation of *T*
20.11–*T* Tables
20.12–Decision Rule
20.13–Null Hypothesis
20.14–Directional Tests
20.15–Large Sample
   Approximation for *T*

**KRUSKAL-WALLIS *H* TEST
(THREE OR MORE
INDEPENDENT SAMPLES)**

20.16–Why Not an *F* Test?
20.17–Calculation of *H*
20.18–χ^2 Tables
20.19–Decision Rule
20.20–Null Hypothesis
20.21–*H* Test Is Nondirectional

GENERAL COMMENTS

20.22–Ties
20.23–A Note on Terminology
20.24–Final Caution
20.25–Summary
20.26–Exercises

20
Tests for Ranked Data

Impress your friends with the less familiar U, T, and H tests designed for use with ranked data—and as replacements for the better known t and F tests when certain assumptions appear to be violated.

In the topsy-turvy world of some tests for ranked data, smaller is better, that is, the smaller the measure, the more suspect the null hypothesis.

JIM FREAKS OUT ON ALPHABET SOUP EVER SINCE HIS STATISTICS CLASS

20.1   Three Tests for Ranked Data

When observations are numbers that indicate relative standings within a set of observations, such as the top ten batters in the National League, measurement is ordinal and the data are ranked, as discussed in Sections 1.6 and 1.7. This chapter describes tests of ranked data for two independent samples, two dependent samples, and more than two independent samples.

Being relatively free of assumptions, these tests often serve as substitutes for the traditional *t* and *F* tests whenever populations can't be assumed to be normally distributed with equal variances. Under these circumstances, tests are conducted after quantitative data have been converted to ranks.

A common theme describes the tests for ranked data. If the null hypothesis is true—that is, if there is no difference between the underlying populations—then, once a single set of ranks has been assigned to all observations in an experiment or survey, the collective ranks for each of the various groups should tend to be about the same. On the other hand, if the null hypothesis is false, the collective ranks for the various groups should tend to differ.

In practice, the greater the difference in ranks between groups, the more suspect the null hypothesis. Each test has its own special measure for the difference in ranks, as well as tables of critical values to be consulted prior to the decision about the null hypothesis. If the difference in ranks is too improbable, given that the null hypothesis is true, the null hypothesis is rejected. Otherwise, the null hypothesis is retained.

MANN-WHITNEY *U* TEST (TWO INDEPENDENT SAMPLES)

When asked to estimate the number of hours spent watching TV each week, are the anonymous replies of high school students influenced by whether TV viewing is depicted favorably or unfavorably? More specifically, one half of the members of a social studies class are selected at random to receive questionnaires that depict TV viewing favorably (as the preferred activity of better students), while the other one half of the class receive questionnaires that depict TV viewing unfavorably (as the preferred activity of poorer students). After discarding the replies of several students, who responded not with numbers but with words, such as "a lot" and "hardly at all," the results were listed in Table 20.1.

20.2   Why Not a *t* Test?

When taken at face value, it might appear that the estimates in Table 20.1 could be tested with the customary *t* test for two independent samples. Closer inspection reveals a complication. Each group of estimates includes one or two very large values, suggesting that the underlying populations are positively skewed rather than normal. When sample sizes are small, as in the present experiment, violations of the normality assumption could seriously impair the accuracy of the *t* test and cause the probability of a type I error to differ considerably from that specified in the level of significance.

A standard remedy is to convert all estimates in Table 20.1 into ranks and to analyze the new ranked data with the Mann-Whitney *U* test for two independent samples. As is true of all tests for ranked data, the *U* test is immune to violations of assumptions about normality and equal variances.

20.3   Calculation of *U*

Table 20.2 indicates how to convert the estimates in Table 20.1 into ranks. Before assigning numerical ranks to the two groups, coded as groups 1 and 2, list all observations from smallest to largest for the combined groups. Beginning with the smallest estimate, assign the consecutive numerical ranks 1, 2, 3, and so forth, until all estimates have been converted to ranks. When two or more estimates are the same, assign the mean of the numerical ranks that would have been assigned if the estimates had been different. For example, each of the two estimates of 0 hours receives a rank of 1.5, the mean of the ranks 1 and 2, while each of the three estimates of 7 hours receives a rank of 8, the mean of the ranks 7, 8, and 9.

Although not mentioned in Table 20.2, it's wise to pause at this point and to form a preliminary impression about any differences in ranks between groups 1 and 2. The more one group tends to outrank the other, the larger the difference between the median ranks for the two groups and the more suspect the null hypothesis. To locate the median rank for either group, identify the value of the middle rank within the

TABLE 20.1
ESTIMATES OF WEEKLY TV VIEWING TIME (HOURS)

TV FAVORABLE	TV UNFAVORABLE
12	43
4	14
7	42
20	1
7	2
7	0
10	0
49	

237

TABLE 20.2
CALCULATION OF U

A. COMPUTATIONAL SEQUENCE

Identify the sample sizes of group 1, n_1, group 2, n_2, and the combined groups, n ①.
List observations from smallest to largest for the combined groups ②.
Assign numerical ranks to the ordered observations for the combined groups ③.
Sum the ranks for group 1 ④ and for group 2 ⑤.
Substitute numbers in formula ⑥ and verify that ranks have been assigned and added correctly.
Substitute numbers in formula ⑦ and solve for U_1.
Substitute numbers in formula ⑧ and solve for U_2.
Set U equal to whichever is smaller—U_1 or U_2 ⑨.

B. DATA AND COMPUTATIONS

① $n_1 = 9$; $n_2 = 7$; $n = 9 + 7 = 16$

② Observations		③ Ranks	
(1) TV Favorable	(2) TV Unfavorable	(1) TV Favorable	(2) TV Unfavorable
	0		1.5
	0		1.5
	1		3
	2		4
4		5	
5		6	
7		8	
7		8	
7		8	
10		10	
12		11	
	14		12
20		13	
	42		14
	43		15
49		16	
		④ $R_1 = \overline{85}$	⑤ $R_2 = \overline{51}$

⑥ Computational check:

$$R_1 + R_2 = \frac{n(n + 1)}{2}$$

$$96 + 40 = \frac{16(16 + 1)}{2}$$

$$136 = 136$$

⑦ $U_1 = n_1 n_2 + \dfrac{n_1(n_1 + 1)}{2} - R_1$

$\quad = (9)(7) + \dfrac{9(9 + 1)}{2} - 85$

$\quad = 63 + 45 - 85$

$\quad = 23$

⑧ $U_2 = n_1 n_2 + \dfrac{n_2(n_2 + 1)}{2} - R_2$

$\quad = (9)(7) + \dfrac{7(7 + 1)}{2} - 51$

$\quad = 63 + 28 - 51$

$\quad = 40$

⑨ U = whichever is smaller—U_1 or U_2
$\quad = 23$

list of existing ranks for that group. In Table 20.2, the median rank equals 8 for group 1 and 4 for group 2. (If you had difficulty determining these median ranks, refer to Section 4.2.) There is a tendency for group 1 to outrank group 2. In other words, estimates in the TV-favorable group tend to be larger than those in the TV-unfavorable group. It remains to be seen whether this result will cause the null hypothesis to be rejected.

Once all observations have been ranked, find the sum of ranks for group 1, R_1, and the sum of the ranks for group 2, R_2. To verify that ranks have been assigned and added correctly, perform the computational check shown in Table 20.2. Finally, calculate values for both U_1 and U_2, and set the smaller of these two values equal to U, that is,

$$U_1 = n_1 n_2 + \frac{n_1(n_1 + 1)}{2} - R_1$$

$$U_2 = n_1 n_2 + \frac{n_2(n_2 + 1)}{2} - R_2 \qquad (20.1)$$

$$U = \text{the smaller of } U_1 \text{ or } U_2$$

where n_1 and n_2 represent the sample sizes of groups 1 and 2, and R_1 and R_2 represent the sum of ranks for groups 1 and 2. The value of U equals 23 for the present study.

Critical values of U are supplied in Table E of the Appendix. Notice that there are two sets of tables—one for nondirectional tests and one for directional tests. Both tables supply critical values of U for hypothesis tests at the .05 level (light numbers) and the .01 level (dark numbers).

20.4 U Tables

To find the correct critical U, identify the entry in the cell intersected by n_1 and n_2, the sample sizes of groups 1 and 2. For the present study, given a nondirectional test at the .05 level of significance with an n_1 of 9 and an n_2 of 7, the value of the critical U equals 12.

An unusual feature of hypothesis tests involving U (and also T, described later in the chapter) is that *the null hypothesis is rejected only if the observed U is less than or equal to the critical U.* Otherwise, if the observed U exceeds the critical U, the null hypothesis is retained.

20.5 Decision Rule

To appreciate the topsy-turvy decision rule for the U test, let's look more closely at U. Although not apparent in Formula 20.1, U *represents the number of times that individual ranks in the lower ranking group exceed individual ranks in the higher ranking group.* When a maximum difference separates two groups—because no rank in the lower ranking group exceeds any rank in the higher ranking group—U equals 0. At the other extreme, when a minimum difference separates two groups—because, as often as not, individual ranks in the lower ranking group exceed individual ranks in the higher ranking group—U equals a large number given by the expression

$$\frac{n_1 n_2}{2}$$

which is 31.5 for the present study.

Ordinarily, the difference in ranks between groups is neither maximum nor minimum, and U equals some intermediate value that, to be interpreted, must be compared with the appropriate critical *U* value. In the present study, since the observed *U* of 23 *exceeds* the critical *U* of 12, only a moderate difference separates the two groups, and the null hypothesis is retained.

Hypothesis Test Summary:
U Test for Two Independent Samples
(Estimates of TV viewing)

Problem:
　　Are high school students' estimates of their weekly TV viewing time influenced by depicting TV viewing as a favorable (1) or an unfavorable (2) activity?

Statistical Hypotheses:
　　H_0: Population distribution 1 = Population distribution 2
　　H_1: Population distribution 1 $\neq$ Population distribution 2

Statistical Test:
　　Mann-Whitney *U* test (Formula 20.1)

Decision Rule:
　　Reject H_0 at the .05 level of significance if $U \leq 12$ (from Table E, given $n_1 = 9$ and $n_2 = 7$).

Calculations:
　　$U = 23$ (See Table 20.2 for computations.)

Decision:
　　Retain H_0.

Interpretation:
　　There is no evidence that high school students' estimates of their weekly TV viewing times are influenced by depicting TV viewing as a favorable or an unfavorable activity.

20.6　Null Hypothesis

　　In the above summary, the null hypothesis equates two *entire* population distributions. Any type of inequality between population distributions, whether caused by differences in central tendency, variability, or shape, could contribute to the rejection of H_0. Strictly speaking, *the rejection of H_0 signifies only that the two populations differ because of some unspecified inequality, or combination of inequalities,* between the original population distributions.

　　More precise conclusions are possible if it can be assumed that both population distributions have roughly similar shapes, that is, for instance, if both population distributions are symmetrical or if both are similarly skewed. Under these circumstances, rejection of H_0 signifies that the two population distributions occupy different locations and, therefore, possess different central tendencies (which usually is interpreted as a difference between population medians since the test is based on ranked data).

20.7　Directional Tests

　　The assumption that population distributions have similar shapes also is required whenever, because of an exclusive concern about population differences in a particular direction, a directional test is desired. In the absence of this assumption, a false H_0 could reflect a complex pattern of inequalities—rather than a simple difference in location—between population distributions, and therefore, only a less precise, nondirectional test would be appropriate.

Judging from the estimates for groups 1 and 2 in Table 20.1, both population distributions could have roughly similar (positively skewed) shapes. Accordingly, if there had been an exclusive concern that population distribution 1 exceeds population distribution 2, a directional test would have been possible in this study.

Prior to conducting a directional test, always verify that the observed differences are in the direction of concern. For instance, if the above directional test had been used in the previous study, you should have verified that the median rank for group 1 exceeds that for group 2 (as it actually does in this study). Otherwise, if the observed differences between median ranks had been in the direction of no concern, the hypothesis test should have been halted and H_0 retained.

Table E can't be used when either n_1 or n_2 exceeds 20. Under these circumstances, U should be approximated with a z ratio that, as usual, must be compared with a critical z from the standard normal distribution (Table A). In particular,

20.8 Large Sample Approximation for U

$$z = \frac{U - \dfrac{n_1 n_2}{2}}{\sqrt{\dfrac{n_1 n_2(n_1 + n_2 + 1)}{12}}} \qquad (20.2)$$

where U, n_1, and n_2 have been defined previously. The ratio described in Formula 20.2 qualifies as a z ratio since U is expressed as a deviation from its mean value of $n_1 n_2/2$ (given that the null hypothesis is true) and then is divided by its standard error, the expression in the denominator.

Formula 20.2 *always* yields an observed z whose value is negative or equal to zero. Given the .05 level of significance, H_0 should be rejected if the observed z is more negative than or equal to the critical z of -1.96 for a nondirectional test (or -1.65 for a directional test). Otherwise, retain H_0.

The previous experiment failed to support the investigator's hunch that estimates of TV viewing time can be influenced by depicting it as a favorable or an unfavorable activity. Noting the large differences among the estimates of students *within* the same group, the investigator might attempt to reduce this variability—and improve the precision of the analysis—by matching students with the aid of some relevant variable. (See Section 15.8.) For instance, some of the variability among estimates might be due to differences in home environment. The investigator could match for home environment by using pairs of students who are siblings. One member of each pair is assigned randomly to one group, and the other sibling is assigned automatically to the second group. As in the previous experiment, questionnaries depict TV viewing as either a favorable or an unfavorable activity. Results for the eight pairs of students are listed in the middle portion of Table 20.3.

WILCOXON T TEST (TWO DEPENDENT SAMPLES)

TABLE 20.3
CALCULATION OF T

A. COMPUTATIONAL SEQUENCE

For each pair of observations, subtract the second observation from the first observation to obtain a difference score ①.

Ignore difference scores of zero, and without regard to sign, list the remaining difference scores from smallest to largest ②.

Assign numerical ranks to the ordered difference scores (still without regard to sign) ③.

List the ranks for positive difference scores in the plus ranks column ④, and list the ranks for negative difference scores in the minus ranks column ⑤.

Sum the ranks for positive differences, R_+ ⑥, and sum the ranks for negative differences, R_- ⑦.

Determine n, the number of nonzero difference scores ⑧.

Substitute numbers in formula ⑨ to verify that ranks have been assigned and added correctly.

Set T equal to whichever is smaller—R_+ or R_- ⑩.

B. DATA AND COMPUTATIONS

Pairs of Students	Observations (1) TV Favorable	Observations (2) TV Unfavorable	① Difference Scores	② Ordered Scores	③ Ranks	④ Plus Ranks	⑤ Minus Ranks
A	2	0	2	2	1.5	1.5	
B	11	5	6	−2	1.5		1.5
C	10	12	−2	3	3	3	
D	6	6	0	5	4	4	
E	7	2	5	6	5	5	
F	43	33	10	10	6	6	
G	30	5	25	25	7	7	
H	5	2	3			⑥ $R_+ = 26.5$	⑦ $R_- = 1.5$

⑧ $n = 7$

⑨ Computational check:

$$R_+ + R_- = \frac{n(n + 1)}{2}$$

$$26.5 + 1.5 = \frac{7(7 + 1)}{2}$$

$$28 = 28$$

⑩ T = whichever is smaller—R_+ or R_-
$T = 1.5$

20.9 Why Not a t Test?

It might appear that the eight difference scores in Table 20.3 could be tested with the t test for two dependent samples. Once again, there's a complication. The set of difference scores contains one very large score, suggesting that the population of difference scores might be positively skewed rather than normal. When sample sizes are small, as in the present experiment, violations of the normality assumption could seriously impair the accuracy of the t test for two dependent samples. A standard remedy is to rank all difference scores and to analyze the resulting ranked data with the Wilcoxon T test.

20.10 Calculation of T

Table 20.3 shows how to calculate T. When ordering difference scores from smallest to largest, ignore all difference scores of zero, and *temporarily treat all negative difference scores as though they were positive*. Beginning with the smallest difference score, assign the consecutive numerical ranks, 1, 2, 3, and so forth, until all nonzero difference scores have been ranked. When two or more difference scores are the same (regardless of sign), assign them the mean of the numerical ranks that would have been assigned if the scores had been different. For example, each of the two difference scores 2 and -2 receives a rank of 1.5, the mean of ranks 1 and 2.

Once numerical ranks have been assigned, those ranks associated with positive difference scores are listed in the plus ranks column, while those associated with negative difference scores are listed in the minus ranks column. Next, find the sum of all ranks for positive difference scores, R_+, and the sum of all ranks for negative difference scores, R_-. (Notice that the more one group of difference scores outranks the other, the larger the discrepancy between the two sums of ranks, R_+ and R_-, and the more suspect the null hypothesis.) To verify that ranks have been assigned and added correctly, perform the computational check in Table 20.3. Finally, the value of T equals the smaller value, either R_+ or R_-, that is,

$$T = \text{the smaller of } R_+ \text{ or } R_- \qquad (20.3)$$

where R_+ and R_- represent the sum of the ranks for positive and negative difference scores. The value of T equals 1.5 for the present study.

20.11 T Tables

Critical values of T are supplied in Table F of the Appendix. There are two sets of tables—one for nondirectional tests and one for directional tests. Both tables supply critical values of T for hypothesis tests at the .05 and .01 levels of significance.

To find the correct critical T value, locate the cell intersected by n, the number of nonzero difference scores, and the desired level of significance, given either a nondirectional or directional test. In the present example, where n equals 7, the critical T equals 2 for a nondirectional test at the .05 level of significance.

20.12 Decision Rule

As with U, *the null hypothesis is rejected only if the observed T is less than or equal to the critical T.* Otherwise, if the observed T exceeds the critical T, the null hypothesis is retained. The properties of T are similar to those of U. The greater the discrepancy in ranks between positive and negative difference scores, the smaller the value of T. In effect, T represents the sum of the ranks for the lower ranking set of difference scores. For example, when the lower ranking set of difference scores fails to appear in the rankings—because all difference scores have the same sign—the value of T equals zero, and the null hypothesis is suspect. In the present study, since the calculated T of 1.5 is less than the critical T of 2, the null hypothesis is rejected.

Hypothesis Test Summary:
T Test for Two Dependent Samples
(Estimates of TV viewing)

Problem:
If high school students are matched for home environment, are their estimates of weekly TV viewing time influenced by depicting TV viewing as a favorable (1) or an unfavorable (2) activity?

Statistical Hypotheses:

H_0: Population distribution 1 = Population distribution 2
H_1: Population distribution 1 $\neq$ Population distribution 2

Statistical Test:
Wilcoxon *T* test (Formula 20.3)

Decision Rule:
Reject H_0 at the .05 level if $T \leq 2$ (from Table F, given $n = 7$).

Calculations:
$T = 1.5$ (See Table 20.3 for computations.)

Decision:
Reject H_0.

Interpretation:
If high school students are matched for home environment, their estimates of TV viewing time are influenced by depicting TV viewing as a favorable or unfavorable activity. Estimates tend to be larger when TV viewing is depicted favorably rather than unfavorably.

20.13 Null Hypothesis

As with the *U* test, the *T* test equates two entire population distributions. Strictly speaking, the rejection of H_0 signifies only that the two populations differ because of some unspecified inequality, or combination of inequalities, between the original population distributions. More precise conclusions about central tendencies are possible only if it can be assumed that both population distributions have roughly similar shapes.

Judging from the estimates for groups 1 and 2 in Table 20.3, both population distributions could have similar (positively skewed) shapes. Therefore, the rejection of H_0 signifies that, on the average, estimates tend to be larger when TV viewing is depicted favorably rather than unfavorably, given that students have been matched for home environment.

20.14 Directional Tests

In the previous study, if there had been an exclusive concern that population distribution 1 exceeds population distribution 2, a directional test would have been possible—on the assumption that both population distributions are positively skewed. Prior to conducting this directional test, you would have to verify that R_+, the sum of the positive ranks in favor of group 1, exceeds R_-, the sum of the minus ranks in favor of group 2 (as it actually does in the present study, where R_+ equals 26.5 and R_- equals only 1.5). Otherwise, if the observed difference between sums of ranks had been in the direction of no concern, the hypothesis test should have been halted and H_0 retained.

20.15 Large Sample Approximation for *T*

Table F can't be used when *n*, the number of nonzero difference scores, exceeds 50. Under these circumstances, *T* should be approximated with a z ratio that has the standard normal distribution (Table A):

$$z = \frac{T - \dfrac{n(n + 1)}{4}}{\sqrt{\dfrac{n(n + 1)(2n + 1)}{24}}} \tag{20.4}$$

where T and n have been defined previously. The ratio described in Formula 20.4 qualifies as a z ratio since T is expressed as a deviation from its mean value of $n(n + 1)/4$ (given that the null hypothesis is true) and then divided by its standard error, the expression in the denominator.

Formula 20.4 *always* yields an observed z whose value is negative or equal to zero. Given the .05 level of significance, H_o should be rejected if the observed z is more negative than or equal to the critical z of -1.96 for a nondirectional test (or -1.65 for a directional test).

20.15 LARGE SAMPLE APPROXIMATION FOR T

Some parents are concerned about the amount of violence in TV cartoons for children. During five consecutive Saturday mornings, 20-minute cartoon sequences are randomly selected and videotaped from the offerings of each of the major TV networks, coded as A, B, and C. A child psychologist, who can not identify the network source of each cartoon, ranks the 15 videotapes from least violent (1) to most violent (15). On the basis of these ranks, as shown in Table 20.4, can it

KRUSKAL-WALLIS H TEST (THREE OR MORE INDEPENDENT SAMPLES)

A. COMPUTATIONAL SEQUENCE

Find the sum of ranks for each group ①.

Identify the sizes of group 1, n_1, group 2, n_2, group 3, n_3, and the combined groups, n ②.

Substitute numbers in formula ③ and verify that ranks have been added correctly.
Substitute numbers in formula ④ and solve for H.

TABLE 20.4 CALCULATION OF H

B. DATA AND COMPUTATIONS

	Ranks	
(1)	(2)	(3)
A	B	C
8	4.5	10
4.5	14	15
2	12	6
13	7	1
10	3	10

① $R_1 = 37.5$ $\quad$ $R_2 = 40.5$ $\quad$ $R_3 = 42$

② $n_1 = 5$ $\quad$ $n_2 = 5$ $\quad$ $n_3 = 5$ $\quad$ $n = 5 + 5 + 5 = 15$

③ Computational check:

$$R_1 + R_2 + R_3 = \frac{n(n + 1)}{2}$$

$$37.5 + 40.5 + 42 = \frac{15(15 + 1)}{2}$$

$$120 = 120$$

④ $H = \dfrac{12}{n(n + 1)}\left[\dfrac{R_1^2}{n_1} + \dfrac{R_2^2}{n_2} + \dfrac{R_3^2}{n_3}\right] - 3(n + 1)$

$\quad = \dfrac{12}{15(15 + 1)}\left[\dfrac{(37.5)^2}{5} + \dfrac{(40.5)^2}{5} + \dfrac{(42)^2}{5}\right] - 3(15 + 1)$

$\quad = \dfrac{12}{240}\left[\dfrac{4810.5}{5}\right] - 48$

$\quad = .05[962.1] - 48$

$\quad = 48.11 - 48 = 0.11$

be concluded that the underlying populations of cartoons for the three networks rank differently in terms of violence?

20.16 Why Not an *F* Test?

An inspection of the numerical ranks in Table 20.4 might suggest an *F* test for three independent samples within the context of a one-factor ANOVA. However, when original observations are numerical ranks, as in the present example, there is no basis for speculating about whether the underlying populations are normally distributed with equal variances, as assumed in ANOVA. It is advisable to use a test, such as the Kruskel-Wallis *H*, that retains its accuracy even though these assumptions might be violated.

20.17 Calculation of *H*

Table 20.4 shows how to calculate *H*. If the original data had been quantitative rather than ranked, then the first step would have been to assign numerical ranks—beginning with 1 for the smallest and so forth—for the *combined* three groups. In other words, the same ranking procedure is followed for *H* as for *U*. When ties occur between ranks, assign a mean rank. In Table 20.4, two cartoons are assigned ranks of 4.5, the mean of ranks 4 and 5.

Find the sums of ranks for groups 1, 2, and 3, that is, R_1, R_2, and R_3. (Notice that, when sample sizes are equal, the larger the differences between these three sums, the more the three groups differ from each other, and the more suspect the null hypothesis. Otherwise, to gain a preliminary impression when sample sizes are unequal, compare the median ranks of the various groups, as suggested in Section 20.3.) Use the computational check in Table 20.4 to verify that ranks have been added correctly. Finally, the value of *H* can be determined from the following formula:

$$H = \frac{12}{n(n+1)}\left[\Sigma \frac{R_i^2}{n_i}\right] - 3(n+1) \tag{20.5}$$

where *n* equals the combined sample size of all groups, R_i represents the sum of ranks of the *i*th group and n_i represents the sample size of the *i*th group. Each sum of ranks, R_i, is squared and divided by its sample size—like the predecessor of the *H* test, the *F* test in ANOVA. The value of *H* equals 0.105 for the present study.

20.18 χ^2 Tables

When sample sizes are very small, critical values of *H* must be obtained from special tables. When each sample size consists of at least four observations—as is ordinarily the case—relatively accurate critical values can be obtained from the χ^2 distribution (Table D in the Appendix). As usual, the value of the critical χ^2 appears in the cell intersected by the desired level of significance and the number of degrees of freedom. The number of degrees of freedom, *df*, can be determined from

$$df = \text{number of groups} - 1 \tag{20.6}$$

Since there are three groups in the present study, *df* equals 3 − 1 or 2, and the critical χ^2 equals 5.991 for a test at the .05 level of significance.

20.19 Decision Rule

In contrast to the decision rules for *U* and *T*, *the null hypothesis is rejected only if the observed H is larger than or equal to the critical χ^2.*

The larger the differences in ranks among groups, the larger the value of H and the more suspect the null hypothesis. In the present study, since the observed H of 0.105 is less than the critical χ^2 of 5.991, the null hypothesis is retained.

Hypothesis Test Summary:
H Test for Three Independent Samples
(Violence in TV cartoons)

Problem:
Does violence in cartoon programming, as judged by a child psychologist, differ for the three major TV networks, coded as A, B, and C?

Statistical Hypotheses:
H_0: Population A = Population B = Population C
H_1: H_0 is not true.

Statistical Test:
Kruskel-Wallis H test (Formula 20.5)

Decision Rule:
Reject H_0 at the .05 level if $H \geq 5.991$ (from Table D, given $df = 2$).

Calculations:
$H = 0.105$ (See Table 20.6 for calculations.)

Decision:
Retain H_0.

Interpretation:
There is no evidence that violence in cartoon programming differs for TV networks A, B, and C.

20.20 Null Hypothesis

The above null hypothesis equates three entire population distributions. Unless population distributions can be assumed to have roughly similar shapes, rejection of H_0 signifies only that two or more populations differ in some unspecified manner—because of differences in central tendency, variability, shape, or some combination of these factors. When original observations consist of numerical ranks, as in the present example, there's no obvious basis for speculating that the population distributions have similar shapes. Therefore, if H_0 had been rejected, it would have been impossible to pinpoint the precise nature of the differences among populations.

20.21 H Test Is Nondirectional

Since the sum of ranks for the ith group, R_i, is squared in Formula 20.5, the H test—like the F test in ANOVA—is nondirectional.

GENERAL COMMENTS

20.22 Ties

In addition to the customary assumption about random sampling, all tests in this chapter assume that no two observations are exactly the same. In other words, there shouldn't be any ties in ranks. Refer to more advanced statistics books for a possible adjustment if (1) the observed value of U, T, or H is in the vicinity of its critical value *and* (2) there's at least one *critical tie.** Critical ties occur between samples (rather than within samples) for the U and H tests, and they occur between plus and minus ranks for the T test. Neither of the two ties in Table 20.2 for the U test is critical; the one tie in Table 20.3 for the T test is critical, as also is the one tie in Table 20.4 for the H test.

20.23 A Note on Terminology

The U, T, and H tests of this chapter, as well as the χ^2 test of the previous chapter and other tests not described in this book, are often referred to as *nonparametric tests*. Parameter refers to any descriptive measure of a population such as the population mean. Nonparametric tests—U, T, and H—evaluate hypotheses for *entire* population distributions, while parametric tests—t and F—evaluate hypotheses for a particular parameter, usually the population mean.

Nonparametric tests also are referred to as *distribution-free* tests. This name highlights the fact that these tests require no assumptions about the precise form of the population distribution. As has been noted, U, T, and H tests can be conducted in the absence of assumptions about the underlying population distributions. In contrast, other types of tests, such as t and F, require populations to be normally distributed with equal variances.

Although widely used, these labels can be misleading. For example, the T test for TV viewing estimates, described earlier in this chapter, qualifies neither as nonparametric nor as distribution-free. In effect, when the two population distributions were assumed to have similar (positively skewed) shapes, we sacrificed the distribution-free status of this test to gain a more precise, parametric test of any difference in central tendency.

20.24 Final Caution

Use the U, T, and H tests only under appropriate circumstances, that is, (1) when data are ranked *or* (2) when data are quantitative but don't seem to originate from normally distributed populations with equal variances. Beware of nonnormality when sample sizes are small (less than about 10), and beware of unequal variances when sample sizes are small and unequal.

When data are quantitative and populations appear to be normally distributed with equal variances, use the t and F tests. Under these circumstances, the t and F tests are less likely to cause the retention of a false null hypothesis, and thus they minimize the probability of a type II error.

20.25 Summary

This chapter describes three different tests of the null hypothesis, using ranked data for two independent samples (Mann-Whitney U test), two dependent samples (Wilcoxon T test), and three or more independent samples (Kruskel-Wallis H test). Being relatively free of assumptions, these tests often replace the t and F tests whenever populations can not be assumed to be normally distributed with equal variances.

Once observations have been expressed as ranks, each test prescribes its own special measure (Formulas 20.1, 20.3, and 20.5) of the difference in ranks between groups, as well as tables of critical values for evaluating significance.

Strictly speaking, U, T, and H test the null hypothesis that entire

---

*Bradley, J., *Distribution-free Statistical Tests*, Prentice-Hall, Englewood Cliffs, N.J., 1968, Ch. 3.

population distributions are equal. Rejection of H_0 signifies merely that populations differ in some unspecified manner. If populations are assumed to have similar shapes, then rejection of H_0 signifies that populations differ in their central tendencies.

Given an exclusive concern about population differences in a particular direction, the U and T tests can be directional—if it can be assumed that populations have similar shapes.

The U, T, and H tests assume that there are no ties in ranks. Except in cases of borderline significance, however, the occurrence of ties can be ignored.

Important Terms & Symbols

Mann-Whitney test (U)
Wilcoxon test (T)
Kruskel-Wallis test (H)

1. Does it matter whether encounter group leaders adopt either an aggressive or a supportive role to facilitate growth among group members? One randomly selected set of six graduate trainees is taught to be aggressive, while the other set of six trainees is taught to be supportive. Subsequently, each trainee is randomly assigned to lead a small encounter group. Without being aware of the nature of the experiment, a panel of experienced group leaders ranks each encounter group from least (1) to most (12) growth promoting, on the basis of anonymous diaries submitted by all members of each group. The results are:

GROWTH-PROMOTING RANKS OF
ENCOUNTER GROUPS

AGGRESSIVE LEADER	SUPPORTIVE LEADER
1	9
2	6
4.5	12
11	10
3	7
4.5	8

Use U to test the null hypothesis at the .05 level.

2. A random sample of "high risk" automobile drivers (three moving violations in one year) are required, according to random assignment, either to attend a traffic school or to perform supervised volunteer work. During the subsequent five-year period, these same drivers were cited as follows:

NUMBER OF MOVING VIOLATIONS

TRAFFIC SCHOOL	VOLUNTEER WORK
0	26
0	7
15	4
9	1
7	1
0	14
2	6
23	10
7	
8	

(a) Why might the Mann-Whitney U test be preferred to the t test for these data?

(b) Use U to test the null hypothesis at the .05 level of significance.

3. Does an anti-smoking workshop cause a decline in cigarette smoking? The daily consumption of cigarettes is estimated for a random sample of nine smokers during each month before (1) and after (2) their attendance at an anti-smoking workshop, consisting of several hours of films, lectures, and testimonials. The results are:

DAILY CIGARETTE CONSUMPTION

SMOKER	BEFORE(1)	AFTER(2)
A	22	17
B	15	13
C	40	0
D	83	95
E	14	10
F	3	3
G	70	9
H	8	7
I	15	13

(a) Why might the Wilcoxon T test be preferred to the customary t test for these data?

(b) Use T to test the null hypothesis at the .05 level.

4. A social psychologist wishes to test the assertion that our attitude toward other people tends to reflect our perception of their attitude toward us. A randomly selected member of each of 12 couples, who live together, is told (in private) that their partner has rated them at the high end of a 0 to 100 scale of trustworthiness, while the other member is told (also in private) that their partner has rated them at the low end of the trustworthiness scale. Each person is then asked to estimate the trustworthiness of their partner, yielding the results below. (According to the original assertion, people in the trustworthy condition should tend to give higher ratings than their partners in the untrustworthy condition.) Use T to test the null hypothesis at the .01 level.

TRUSTWORTHINESS RATINGS

COUPLE	TRUSTWORTHY(1)	UNTRUSTWORTHY(2)
A	75	60
B	35	30
C	50	55
D	93	20
E	74	12
F	47	34
G	95	22
H	63	63
I	44	43
J	88	79
K	56	33
L	86	72

5. A consumers' group wishes to determine whether motion picture ratings are, in any sense, associated with the number of violent or sexually explicit scenes in films. Five films are randomly selected from among each of the four ratings (X, R, PG, and G), and a trained observer counts the number of incidents in each film to obtain the following results:

X–ADULTS ONLY	R–RESTRICTED	PG–PARENTAL GUIDANCE	G–GENERAL
15	8	7	3
20	16	11	0
9	14	6	3
13	10	4	0
25	6	9	2

(a) Why might the H test be preferred to the F test for these data?

(b) Use H to test the null hypothesis at the .05 level of significance.

6. Does background music influence the scores of college students on a reading comprehension test? Sets of ten randomly selected students take a reading comprehension test with rock, country-western, or classical music in the background. The results are as follows: (higher scores reflect better comprehension)

READING COMPREHENSION SCORES

ROCK (1)	COUNTRY-WESTERN (2)	CLASSICAL (3)
90	99	52
11	94	75
82	95	91
67	23	94
98	72	97
93	81	31
73	79	83
90	28	85
87	94	100
84	77	69

(a) Why might the H test be preferred to the F test for these data?

(b) Use H to test the null hypothesis at the .05 level.

7. Use U rather than t to test the results in Chapter 14, Exercise 1.

8. Use T rather than t to test the effects of "ABC" meditation described in Chapter 15, Exercise 2.

9. Use H rather than F to test the weight change data recorded in Chapter 17, Exercise 6.

10. Noting that the calculations for the H test tend to be much easier than those for the F test, one person always uses the H test. Any objection?

APPENDIX A
Math Review

A.1 PRETEST
A.2 COMMON SYMBOLS
A.3 ORDER OF OPERATIONS
A.4 POSITIVE AND NEGATIVE NUMBERS
A.5 FRACTIONS
A.6 SQUARE ROOT RADICALS ($\sqrt{}$)
A.7 ROUNDING NUMBERS
A.8 POSTTEST
A.9 ANSWERS (WITH RELEVANT REVIEW SECTIONS)

This appendix summarizes many of the basic math symbols and operations used in this book. Little, if any, of this material will be entirely new, but—possibly because of years of little or no use—much may seem only slightly familiar. In any event, it's important that you master this material.

First, take the pretest in Section A.1, comparing your answers with those in Section A.9. Whenever errors occur, study the review section indicated for that set of answers. Then, after browsing through all review sections, take the posttest in Section A.8, again checking your answers with those in Section A.9. If you're still making lots of errors, repeat the entire process spending even more time studying the various review sections. If errors persist, consult your instructor for additional guidance.

A.1 PRETEST

Questions 1–6: Are the following statements true or false?

1. $(5)(4) = 20$ **2.** $4 > 6$ **3.** $7 \leq 10$ **4.** $|-5| = 5$ **5.** $(8)^2 = 56$ **6.** $\sqrt{9} = 3$

Questions 7–30: Find the answers.

7. $\dfrac{5-3}{2-1} =$ **8.** $\sqrt{5+4+7} =$ **9.** $3(4+3) =$ **10.** $16 - \dfrac{10}{\sqrt{25}} =$

11. $(3)^2(10) - 4 =$ **12.** $[3^2 + 2^2]^2 =$ **13.** $\sqrt{\dfrac{2(3)-2^2}{5-3}} =$ **14.** $\sqrt{\dfrac{(8-6)^2 + (5-3)^2}{2}} =$

15. $2 + 4 + (-1) =$ **16.** $5 - (3) =$ **17.** $2 + 7 + (-8) + (-3) =$ **18.** $5 - (-1) =$

19. $(-4)(-3) =$ **20.** $(-5)(6) =$ **21.** $\dfrac{-10}{2} =$ **22.** $\dfrac{4}{5} - \dfrac{1}{5} =$ **23.** $\dfrac{1}{4} + \dfrac{2}{5} =$

24. $\dfrac{2^2}{4} + \dfrac{3^2}{3} - \dfrac{2^2}{8} =$ **25.** $\left(\dfrac{2}{3}\right)\left(\dfrac{6}{7}\right) =$ **26.** $\sqrt{16+9} =$ **27.** $\sqrt{(4)(9)} =$

28. $\sqrt{4}\,\sqrt{9} =$ **29.** $\dfrac{\sqrt{25}}{\sqrt{100}} =$ **30.** $\sqrt{\dfrac{25}{100}} =$

Questions 31–35: Round to the nearest hundredth.

31. 98.769 **32.** 3.274 **33.** 23.765 **34.** 5476.375003 **35.** 54.1499

A.2 COMMON SYMBOLS

SYMBOL	MEANING	EXAMPLE
$=$	equals	$4 = 4$
$\neq$	doesn't equal	$4 \neq 2$
$+$	plus (addition)	$2 + 3 = 5$
$-$	minus (subtraction)	$3 - 2 = 1$
$\pm$	plus and minus	$4 \pm 2 = 4 + 2$ and $4 - 2$
$(\ )(\ )$	times (multiplication)*	$(3)(2) = 3(2) = 6$
$/, \dfrac{(\ )}{(\ )}$	divided by (division)	$6/2 = 3, \dfrac{(8)}{(2)} = 4$
$>$	is greater than	$4 > 3$
$<$	is less than	$5 < 8$
$\geq$	equals or is greater than	$z \geq 2$
$\leq$	equals or is less than	$t \leq 4$
$\sqrt{}$	the square root of†	$\sqrt{9} = 3$
$(\ )^2$	the square of	$(4)^2 = (4)(4) = 16$
$\mid\ \mid$	the absolute (positive) value of	$\mid 4 \mid = 4, \ \mid -4 \mid = 4$
$\ldots$	continuing the pattern	$1, 2, 3, \ldots, 8$
		translates as: 1, 2, 3, 4, 5, 6, 7, 8

*When multiplication involves symbols, parentheses can be dropped. For instance,

$$(X)(Y) = X(Y) = XY$$

†The square root of a number is that number which, when multiplied by itself, yields the original number.

A.3 ORDER OF OPERATIONS

Expressions should be treated as single numbers when they appear in parentheses, square root signs, or in the top (or bottom) of fractions.

Examples:

$$2(4 - 1) = 2(3) = 6$$
$$\sqrt{12 - 8} = \sqrt{4} = 2$$
$$\frac{8 - 4}{2 + 2} = \frac{4}{4} = 1$$

If all expressions contain single numbers, the order for performing operations is as follows:
  (a) square or square root
  (b) multiplication or division
  (c) addition or subtraction.

Examples:

$$10 + \frac{6}{\sqrt{4}} = 10 + \frac{6}{2} = 10 + 3 = 13$$
$$(3)(2)^2 - 1 = (3)(4) - 1 = 12 - 1 = 11$$

When expressions are nested, one within the other, work outward from the inside.

Examples:

$$\sqrt{\frac{(6-3)^2 + (5-2)^2}{2}} = \sqrt{\frac{(3)^2 + (3)^2}{2}} = \sqrt{\frac{9+9}{2}} = \sqrt{\frac{18}{2}} = \sqrt{9} = 3$$

$$\sqrt{\frac{3(4) - (2)^2}{4-2}} = \sqrt{\frac{12-4}{2}} = \sqrt{\frac{8}{2}} = \sqrt{4} = 2$$

A.4 POSITIVE AND NEGATIVE NUMBERS

In the absence of any sign, a number is understood to be positive.

Example:

$$8 = \quad +8$$

To *add* numbers with unlike signs,
(a) find two separate sums, one for all positive numbers and the other for all negative numbers
(b) find the difference between these two sums
(c) attach the sign of the larger sum.

Example:

$$2 + 3 + (-4) + (-3) = 5 + (-7) = -2$$

To *subtract* one number from another,
(a) change the sign of the number to be subtracted
(b) proceed as in addition.

Examples:

$$4 - (3) = 4 + (-3) = 1$$
$$4 - (-3) = 4 + 3 = 7$$

To *multiply* (or *divide*) two signed numbers,
(a) obtain the numerical result
(b) attach a positive sign if the two original numbers have like signs or a negative sign if the two original numbers have unlike signs.

Examples:

$$(-4)(-2) = 8; \quad (4)(-2) = -8$$

$$\frac{4}{2} = 2; \quad \frac{-4}{2} = -2$$

A.5 FRACTIONS

A fraction consists of an upper part, the numerator, and a lower part, the denominator.

To *add* (or *subtract*) fractions, their denominators must be the same.

(a) If denominators are the same, merely add (or subtract) numbers in the numerators, and leave the number in the denominator unchanged.

Examples:

$$\frac{3}{5} + \frac{1}{5} = \frac{3+1}{5} = \frac{4}{5}$$

$$\frac{7}{10} - \frac{3}{10} = \frac{7+(-3)}{10} = \frac{4}{10}$$

(b) If denominators are different, first find a common denominator. To obtain a common denominator, multiply both parts of each fraction by the denominators of all remaining fractions. Then proceed as above.

Examples:

$$\frac{2}{3} + \frac{1}{4} = \frac{(4)2}{(4)3} + \frac{(3)1}{(3)4} = \frac{8}{12} + \frac{3}{12} = \frac{11}{12}$$

$$\frac{4}{6} + \frac{2}{5} = \frac{(5)4}{(5)6} + \frac{(6)2}{(6)5} = \frac{20}{30} + \frac{12}{30} = \frac{32}{30}$$

To *add* (or *subtract*) fractions, sometimes it's more efficient to follow a different procedure. First, express each fraction as a decimal number—by dividing denominator into numerator—and then, merely add (or subtract) the resulting decimal numbers.

Examples:

$$\frac{3}{4} - \frac{1}{4} = .75 - .25 = .50$$

$$\frac{3}{10} + \frac{2}{6} + \frac{1}{5} = .30 + .33 + .20 = .83$$

To multiply fractions, multiply all numerators to obtain the new numerator, and multiply all denominators to obtain the new denominator.

$$\left(\frac{2}{3}\right)\left(\frac{3}{5}\right) = \frac{6}{15}$$

$$\left(\frac{2}{4}\right)\left(\frac{3}{4}\right) = \frac{6}{16}$$

A.6 SQUARE ROOT RADICALS ($\sqrt{}$)

The square root of a sum *doesn't* equal the sum of the square roots.

Examples:

$$\sqrt{16+9} \neq \sqrt{16} + \sqrt{9}$$
$$5 \neq 4 + 3$$

The square root of a product equals the product of the square roots.

Example:

$$\sqrt{(4)(9)} = (\sqrt{4})(\sqrt{9}) = (2)(3) = 6$$

The square root of a fraction equals the square root of the numerator divided by the **MATH REVIEW**
square root of the denominator.

Example:

$$\sqrt{\frac{4}{16}} = \frac{\sqrt{4}}{\sqrt{16}} = \frac{2}{4}$$

A.7 ROUNDING NUMBERS

When the first term of the number to be dropped is 5 or more, increase the remaining
number by one unit. Otherwise, leave the remaining number unchanged.

Examples:

When rounding to the nearest whole number,
17.2 rounds to 17
4.7 rounds to 5
23.4999 rounds to 23
8.573 rounds to 9
When rounding to the nearest hundredth,
21.866 rounds to 21.87
37.364 rounds to 37.36
102.645332 rounds to 102.65
87.98497 rounds to 87.98
52.105000 rounds to 52.11

A.8 POSTTEST

Questions 101–112: Find the answers.

101. $\sqrt{36} =$ **102.** $|24| =$ **103.** $(7)^2 =$ **104.** $5 \pm 3 =$ **105.** $3\sqrt{8 - (2)^2} =$

106. $\dfrac{1^2 + 4^2 + 5^2}{4^2 - 3^2} =$ **107.** $18 - (-3) =$ **108.** $(-10)(-8) =$ **109.** $\dfrac{3}{5} + \dfrac{2}{8} =$

110. $\dfrac{(2 - 3)^2}{2} + \dfrac{(6 - 4)^2}{3} =$ **111.** $\sqrt{9 + 9 + 9 + 9} =$ **112.** $\sqrt{25}\sqrt{4} =$

Questions 113–114: Round to the nearest tenth.

113. 107.45   **114.** 3.2499

A.9 ANSWERS (WITH RELEVANT REVIEW SECTIONS)

Pretest

1. True
2. False
3. True   review Section A.2
4. True
5. False
6. True

7. 2
8. $\sqrt{16} = 4$
9. 21
10. 14
11. 86   review Section A.3
12. $(13)^2 = 169$
13. $\sqrt{1} = 1$
14. $\sqrt{4} = 2$

15. 5
16. 2
17. -2
18. 6   review Section A.4
19. 12
20. -30
21. -5

22. $\dfrac{3}{5}$ or .60
23. $\dfrac{13}{20}$ or .65
24. $\dfrac{84}{24}$ or 3.5   review Section A.5
25. $\dfrac{12}{21}$

26. 5 ⎫
27. 6 ⎪
28. 6 ⎬ review Section A.6
29. ½ ⎪
30. ½ ⎭

31. 98.77   ⎫
32. 3.27   ⎪
33. 23.77   ⎬ review Section A.7
34. 5476.38 ⎪
35. 54.15   ⎭

Posttest

101. 6   ⎫
102. 24   ⎪
103. 49   ⎬ review Section A.2
104. 8 and 2 ⎭

105. 6 ⎫
106. 6 ⎬ review Section A.3

107. 21 ⎫
108. 80 ⎬ review Section A.4

109. $\dfrac{34}{40}$ or .85 ⎫
110. $\dfrac{11}{6}$ or 1.83 ⎬ review Section A.5

111. 6 ⎫
112. 10 ⎬ review Section A.6

113. 107.5 ⎫
114. 3.2   ⎬ review Section A.7

APPENDIX B
Answers to Exercises

NOTE: COMPUTATIONAL ACCURACY

Whenever necessary, round numbers to two digits to the right of the decimal point. (See Section 7 of Appendix A for information about rounding numbers.) If your preliminary computations are based on numbers with more than two digits to the right of the decimal point—possibly because you're using the full capacity of a calculator—*slight* differences might appear occasionally between your answers and those in this appendix.

When answers are obtained directly from a table, such as the standard normal table (Table A in Appendix C), the complete tabular entry, for instance, .0571, is listed, along with the customary answer, rounded two digits to the right of the decimal point.

CHAPTER 1

1. (a) qualitative   (i) ranked
  (b) quantitative   (j) quantitative
  (c) quantitative   (k) quantitative
  (d) qualitative   (l) quantitative
  (e) quantitative   (m) qualitative
  (f) qualitative   (n) ranked
  (g) ranked   (o) qualitative
  (h) qualitative   (p) qualitative

2. (d) academic rank of college teachers

3. (a) interval-ratio   (i) interval-ratio
  (b) ordinal   (j) ordinal
  (c) interval-ratio   (k) nominal
  (d) nominal   (l) interval-ratio?
  (e) interval-ratio?   (m) interval-ratio
  (f) ordinal   (n) interval-ratio?
  (g) nominal   (o) nominal
  (h) interval-ratio?   (p) ordinal
         (q) interval-ratio

4. (a) false   (g) false
  (b) true   (h) false
  (c) false   (i) false
  (d) false   (j) false
  (e) false   (k) true
  (f) true   (l) false (nominal not represented)
         (m) true

CHAPTER 2

1. (a)

SECONDS	FREQUENCY
123	I
⋮	
114	I
113	
112	I
111	
110	I
⋮	
105	I
⋮	
100	II
99	I
98	II
97	
96	I
95	II
94	I
93	
92	
91	I
90	IIII
89	II
88	I
87	
86	
85	I
84	I
83	I
82	I
81	
80	II
79	III
⋮	
75	I
⋮	
71	I
⋮	
62	I
61	
60	I

259

(b) A majority of students hold their breath between approximately 80 and 100 seconds (or some variation on this statement).

(c) Using stems of size ten:

STEM	LEAF
12	3
11	024
10	500
9	158609400580
8	350428099
7	15999
6	20

(d) Calculating width of the class interval (Formula 2.1)

$$i = \frac{123 - 60}{15} = \frac{63}{15} = 4.2$$

Round 4.2 off to a more convenient value of i, say 5.

TIME	FREQUENCY
120–124	1
115–119	0
110–114	3
105–109	1
100–104	2
95–99	6
90–94	6
85–89	4
80–84	5
75–79	4
70–74	1
65–69	0
60–64	2
	Total 35

(e) The stem and leaf arrangement (c) and the grouped frequency distribution (d) are particularly helpful. Both show the heavy concentration of observations in the 80s and 90s.

2. (a)

RATING	FREQUENCY
10	|
9	||
8	|||
7	|||
6	||||
5	||
4	|
3	|||
2	||
1	|
0	|||

(b) No well-defined pattern.

3. Not all observations can be assigned to one and only one interval (because of gap between 20–22 and 25–30 and overlap between 25–30 and 30–34). All intervals aren't equal in width (25–30 vs. 30–34). All class intervals don't have both boundaries (35–above).

	INTERVAL WIDTH	TWO LOWEST CLASS INTERVALS	ONE HIGHEST CLASS INTERVAL
(a)	2	28–29; 30–31	64–65
(b)	30	150–179; 180–209	570–599
(c)	10	20–29; 30–39	160–169
(d)	150	600–749; 750–899	2850–2999
(e)	400	13,600–13,999; 14,000–14,399	19,200–19,599
(f)	3	0–2; 3–5	42–44
(g)	0.030	0.420–0.449; 0.450–0.479	0.960–0.989
(h)	25	850–874; 875–899	1200–1224

5. (a)

STEM	LEAF
19	03
18	00
17	98, 89, 54, 88
16	44, 30, 23, 05
15	
14	25, 70
13	
12	08
11	
10	59, 26
9	61, 23, 98
8	93, 21, 47, 80, 89
7	89, 63, 56, 77, 52, 43, 76, 41, 71
6	54, 77, 79, 60, 32, 38
5	62, 43, 55, 76, 93, 67
4	21, 29, 23, 31
3	78
2	33

(b)

MILES BETWEEN BREAKDOWNS	FREQUENCY
1900–1999	1
1800–1899	1
1700–1799	4
1600–1699	4
1500–1599	0
1400–1499	2
1300–1399	0
1200–1299	1
1100–1199	0
1000–1099	2
900–999	3
800–899	5
700–799	9
600–699	6
500–599	6
400–499	4
300–399	1
200–299	1
	Total 50

(c) Most buses average less than 1000 miles between breakdowns. However, a substantial minority of buses average more than 1500 miles between breakdowns.

6. (a)

MEDITATORS		NONMEDITATORS	
3.8	5 0	3.8	0 9
3.7	8 5 1	3.7	
3,6	3 8	3.6	7
3.5	6 7 6 6	3.5	9
3.4		3.4	
3.3	0 0	3.3	7
3.2	5	3.2	5 0
3.1	0	3.1	0
3.0	0	3.0	0
2.9	5 5 0	2.9	0 0
2.8		2.8	0 3 6
2.7	5 5	2.7	5 6
2.6		2.6	7 5 6 7
2.5	8	2.5	0 8
2.4	5	2.4	
2.3		2.3	4
2.2	5 5	2.2	
		2.1	0 0

(b)

MEDITATORS	FREQUENCY	NONMEDITATORS	FREQUENCY
3.80–3.89	2	3.80–3.89	2
3.70–3.79	3	3.70–3.79	0
3.60–3.69	2	3.60–3.69	1
3.50–3.59	4	3.50–3.59	1
3.40–3.49	0	3.40–3.49	0
3.30–3.39	2	3.30–3.39	1
3.20–3.29	1	3.20–3.29	2
3.10–3.19	1	3.10–3.19	1
3.00–3.09	1	3.00–3.09	1
2.90–2.99	3	2.90–2.99	2
2.80–2.89	0	2.80–2.89	3
2.70–2.79	2	2.70–2.79	2
2.60–2.69	0	2.60–2.69	4
2.50–2.59	1	2.50–2.59	2
2.40–2.49	1	2.40–2.49	0
2.30–2.39	0	2.30–2.39	1
2.20–2.29	2	2.20–2.29	0
		2.10–2.19	2
	Total 25		Total 25

(c) Slight tendency for meditators to have higher GPAs than nonmeditators.

9.

GRE	(a) RELATIVE FREQUENCY		(b) CUMULATIVE FREQUENCY	(c) CUMULATIVE RELATIVE FREQ.	
	Proportion	Percent (%)		Proportion	Percent (%)
720–739	.01	1	200	1.00	100
700–719	.01	1	198	.99	99
680–699	.05	5	196	.98	98
660–679	.07	7	186	.93	93
640–659	.15	15	172	.86	86
620–639	.17	17	142	.71	71
600–619	.16	16	108	.54	54
580–599	.13	13	76	.38	38
560–579	.10	10	50	.25	25
540–559	.08	8	30	.15	15
520–539	.05	5	15	.08	8
500–519	.02	2	5	.03	3
480–499	.01	1	1	.01	1
	Total 1.01	101%			

10.

NO. OF CHILDREN	(a) PERCENT (%)	(b) CUMULATIVE FREQUENCY	(c) CUMULATIVE RELATIVE FREQUENCY (%)
4 or more	10	51,237,000	100
3	11	46,113,300	90
2	17	40,477,230	79
1	18	31,766,940	62
0	44	22,544,280	44
	Total 100%		

CITY		TOWN	
Age	**Proportion**	**Age**	**Proportion**
90–99	.00	90–99	.02
80–89	.00	80–89	.07
70–79	.03	70–79	.07
60–69	.09	60–69	.10
50–59	.12	50–59	.12
40–49	.15	40–49	.13
30–39	.17	30–39	.15
20–29	.16	20–29	.16
10–19	.13	10–19	.09
0–9	.14	0–9	.09
	Total .99		Total 1.00

Proportionately more elderly people live in town, and proportionately more children live in the city.

12.

RATINGS	(a) FREQUENCY	(c) PERCENT (%)	(d) CUMULATIVE FREQUENCY
X	2	10	20
R	7	35	18
PG	8	40	11
G	3	15	3
	Total 20	Total 100%	

(b) Majority of films are rated R and PG.

13.

TYPE	(a) FREQUENCY	(c) RELATIVE FREQUENCY	
		Proportion	Percent (%)
O	14	.47	47
A	13	.43	43
B	2	.07	7
AB	1	.03	3
	Total 30	1.00	100%

(b) Most Types are O and A.

(d) Not appropriate, since blood types can't be ordered from least to most.

CHAPTER 3

1.

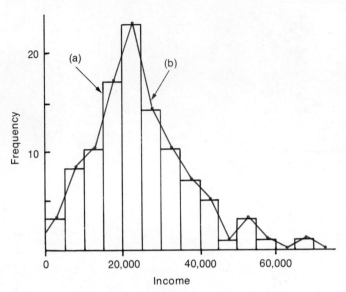

Note: Ordinarily, either (a) or (b) only would be shown. When closing the left flank of (b), imagine extending a line to the midpoint of the first unoccupied interval (-5000 to -1) on the left, but stop the line at the vertical axis, as shown.

(c) A majority of incomes range from approximately \$10,000 to \$30,000, with a few in excess of \$50,000.

2.

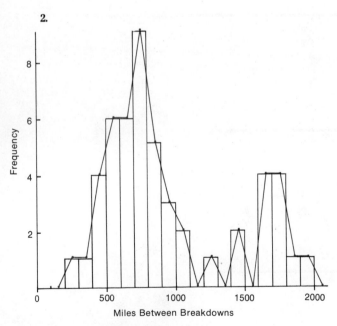

3.

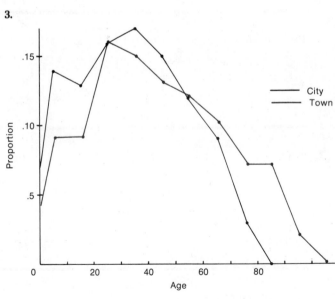

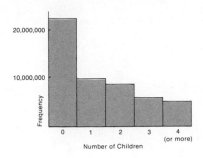

6. (a)

(b)

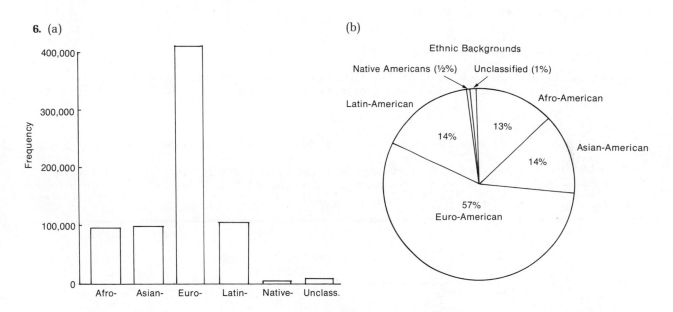

(c) Euro-Americans constitute a slight majority, with substantial minorities equally distributed among Afro-Americans, Asian-Americans, and Latin-Americans.

7. (a) Widths of two rightmost bars aren't the same as those of two leftmost bars.
 (b) Histogram is more appropriate for quantitative data.
 (c) Height of vertical axis is too small relative to width of horizontal axis—causing the histogram to be squashed.
 (d) Poorly selected frequency scale along the vertical axis—causing the histogram to be squashed.
 (e) Bars have unequal widths. No wiggly line along vertical axis indicating break between 0 and 50.

CHAPTER 4

1. (a) mode   = 5
   median = 5
   mean $= \dfrac{117}{15} = 7.8$

(b) Positively skewed, since the mean exceeds the median.

2. mode   = 21.4
   median $= \dfrac{21.4 + 20.9}{2} = 21.15$
   mean $= \dfrac{127.3}{6} = 21.22$

3. mode   = 112, 120, 125 (multimodal)
   median = 125
   mean   = 129.51

4. mode   = 63
   median = 63
   mean   = 61.09

5. mode   = Hawaii (H)
   Impossible to find median since these qualitative data can't be ordered.

6. mode = Euro-American
   Impossible to find median since these qualitative data can't be ordered.

7. mode   = PG $\Big\{$ Note: Possible to find median since these qualitative data can be ordered from least to most
   median = PG $\quad$ restrictive. The PG rating contains the two middlemost observations (the 10th and 11th) in the set of 20 observations—ordered from least to most restrictive.

8. (a) No—median exceeds mean   (c) Yes
   (b) Yes—mean exceeds median   (d) No

9. (a) . . . it has a smaller sampling variability.
   (b) . . . it's more appropriate for data that reflect merely order (ordinal measurement).

10. Two different averages are being used to describe the central tendency in a skewed distribution of teacher's salaries. The school board is probably using the mean while the teachers' association is probably using the median.

11. modal interval = 0
   median interval = 1   Note: The cumulative percent first exceeds 50% in the median interval of 1.

12. modal interval = 110–119 $\Big\{$ Note: The cumulative frequency first exceeds 27 (the middlemost ranked
   median interval = 120–129 $\quad$ observation) in the median interval of 120–129.

13. (a)   (b)

RANKS	RANKS	RANKS	RANKS
NEGATIVE EVENTS (*)	POSITIVE EVENTS	TWENTIETH CENTURY EVENTS (√)	EARLIER EVENTS
1	2	3	1
3	4	4	2
5	9	5	10
6 (median)	11 median =	6	13 median =
7 = 7	13 $\dfrac{13 + 15}{2} = 14$	7	16 $\dfrac{10 + 13}{2} = 11.5$
8	15	8 (median)	18
10	16	9 = 9	
12	17	11	
14	18	12	
	19	14	
		15	
		17	
		19	

1. b, c   **3.** $S = \sqrt{1.5} = 1.22$

4. Yes. Adding 10 to each of the four observations doesn't produce any change in variability among these observations. It does not, for instance, change the value of the range.
$$S = \sqrt{1.5} = 1.22$$

5. No. Adding 10 to only one of the four observations does produce a change in variability among the observations. It does, for instance, change the value of the range.
$$S = \sqrt{10.25} = 3.20$$

6. $S = \sqrt{10} = 3.16$

7. $S = \sqrt{\dfrac{(53)(917,246) - (6864)^2}{(53)^2}} = 23.10$

8. (a) False   (b) False   (c) True   (d) False   (e) True

9. Corporation A—if you want to be reasonably certain of a salary in the vicinity of $30,000. Corporation B—if you're willing to gamble in order to earn appreciably more than $30,000.

10. No. Negative deviations become positive—because of squaring—and, therefore, the sum of all squared deviations always equals zero or some positive number.

11. Intermediate variability.

CHAPTER 6

1. (a) .0918 = .09   (e) 2200
　　(b) .0013 = .00   (f) 2200 + (2.33)(600) = 3598
　　(c) .2586 = .26   (g) 2200 + (−1.41)(600) = 1354
　　(d) 2200 + (−1.28)(600) = 1432

2. (a) .0475 = .05   (e) 100 + (2.05)(15) = 130.75
　　(b) .1151 = .12   (f) 100 + (−1.28)(15) = 80.8
　　(c) .4514 = .45   (g) 100 + (±1.96)(15) = 129.4 and 70.6
　　(d) .0076 = .01   (h) 100 + (±2.57)(15) = 138.55 and 61.45 (or 138.70 and 61.30 given $z = \pm2.58$)

3. (a) .9599 = .96   (e) 83 + (−1.65)(20) = 50 (or 50.2 given $z = -1.64$)
　　(b) < .00003 = .00   (f) 83 + (0.84)(20) = 99.8
　　(c) .7021 = .70   (g) 83 + (± 0.25)(20) = 88 and 78
　　(d) .2896 = .29   (h) 83 + (±2.33)(20) = 129.6 and 36.4

4. (a) .1587 = .16   (c) .0336 = .03
　　(b) .0099 = .01   (d) 3.20 + (1.28)(.30) = 3.58

5. The heights of Bill Walton and the midget both deviate five standard deviations from the mean.

7. A score of 64 deviates seven standard deviations above the mean in distribution (a) —more than in any other distribution.

8.

	$\bar{X} = 0;\ S = 1$	$\bar{X} = 50;\ S = 10$	$\bar{X} = 100;\ S = 15$	$\bar{X} = 500,\ S = 100$	PERCENTILE RANK
(a)	0.80	58	112	580	78.81
(b)	−1.67	33.3	74.95	333	4.75
(c)	1.80	68	127	680	96.41
(d)	−1.85	31.5	72.25	315	3.22

Note: Ordinarily, percentile ranks are reported to the nearest whole number.

9. Conversion to standard scores preserves the relative standings of scores within the distribution.

10. (a) 91.92   (c) 6.68
     (b) $100 + (1.48)(15) = 122.2$ (d) $500 + (.25)(100) = 525$

CHAPTER 7

1. (a) Positive relationship since high values of X tend to be paired with high values of Y, while low values of X tend to be paired with low values of Y.

   (b)

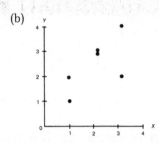

Scatterplot doesn't show any pronounced curvilinear trend.

    (c) $r = \dfrac{(6)(33) - (12)(15)}{\sqrt{[(6)(28) - 144]\,[(6)(43) - 225]}} = .64$

2. (a) Negative relationship since high values of X are regularly paired with low values of Y, and vice versa.

   (b)

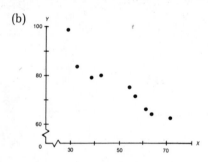

Scatterplot doesn't show a pronounced curvilinear trend.

    (c) $r = \dfrac{(9)(33,304) - (451)(687)}{\sqrt{[(9)(24,357) - 203,401][(9)(53,299) - 471,969]}} = -.91$

3. (a) $r = \dfrac{(5)(344) - (35)(60)}{\sqrt{[5(325) - 1225]\,[5(800) - 3600]}} = -.95$

   (b) $r_s = 1 - \dfrac{(6)(38)}{(5)(5^2 - 1)} = -.90$

4. (a) $r = \dfrac{(5)(420) - (35)(60)}{\sqrt{[(5)(325) - 1225][(5)(800) - 3600]}} = 0$

   (b) $r_s = 1 - \dfrac{(6)(20)}{(5)(5^2 - 1)} = 0$

5. (a) 2   (b) above   (c) $1\frac{1}{2}$ (d) below   (e) $\frac{1}{3}$ (f) above

6. (a) 2   (b) below   (c) 1½   (d) above   (e) ⅓   (f) below

7. No. The value of r depends only on the pattern among pairs of z scores, which, themselves, show no traces of the units of measurement for the original X and Y observations.

8. No. A value of r can't be interpreted as a percent.

9. No. Correlation doesn't demonstrate a cause-effect relationship. It could be, for example, that both the seven "golden rules" of behavior and a longer life are effects of a common cause—leading a low-keyed, orderly life.

10. $r = \dfrac{(9)(169) - (45)(45)}{\sqrt{[(9)(285) - 2025][(9)(283) - 2025]}}$

$= -.95$ (which should be reported as a value of r_s, corrected for ties).

CHAPTER 8

1. (a) $b = \left(\dfrac{4}{2}\right)(.70) = 1.40;\ a = 8 - (1.40)(13) = -10.20$

$Y' = 1.40(X) - 10.20$

(b) $Y' = (1.40)(15) - 10.20$
$= 10.80$

(c) $Y' = (1.40)(11) - 10.20$
$= 5.20$

(d) $S_{Y \cdot X} = (4)\left[\sqrt{1 - (.70)^2}\right]$
$= 2.86$

2. (a) $b = \left(\dfrac{6}{3}\right)(-.80) = -1.60$

$a = 60 - (-1.60)(5) = 68$
$Y' = -1.60\ (X) + 68$

(b) $S_{Y \cdot X} = (6)\left[\sqrt{1 - (-.80)^2}\right]$
$= 3.60$

(c) $Y' = (-1.60)(8) + 68$
$= 55.20$

(d) 68% between 51.60 and 58.80
95% between 48.14 and 62.26

(e) $Y' = (-1.60)(0) + 68 = 68$

(f) 68% between 64.40 and 71.60
95% between 60.94 and 75.06

3. (a)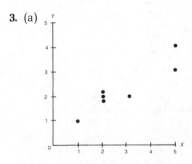

No pronounced curvilinearity.

(c) $S_{Y \cdot X} = (2.33)\left[\sqrt{1 - (.92)^2}\right] = .91$

(d) $Y' = (.56)(2) + .69$
$= 1.81$ (or 2)
$Y' = (.56)(5) + .69$
$= 3.49$ (or 3)

(b) $r = \dfrac{(7)(54) - (20)(16)}{\sqrt{[(7)(72) - 400][(7)(42) - 256]}}$

$= .92$

$\bar{X} = \dfrac{20}{7} = 2.86 \qquad \bar{Y} = \dfrac{16}{7} = 2.29$

$S_X = \sqrt{\dfrac{(7)(72) - 400}{7}} = 3.85$

$S_Y = \sqrt{\dfrac{(7)(42) - 256}{7}} = 2.33$

$b = \left(\dfrac{2.33}{3.85}\right)(.92) - .56$

$a = 2.29 - (.56)(2.86) = .69$

$Y' = .56(X) + .69$

5. (a) $z'_Y = 1.50$ (b) $z'_Y = -.75$
(c) Sam: $z'_Y = .50$ (d) Sam: $z'_Y = -1.00$
     Tim: $z'_Y = -.25$ Tim: $z'_Y = .50$
(e) pair of predictions in (c) when $r = .25$
(f) pair of predictions in (a) and (b) when $r = .75$.

6. Reversing the direction of prediction changes the values of both b and a in the prediction equation, that is,

$$b = \left(\frac{3.85}{2.33}\right)(.92) = 1.52$$

$$a = 2.86 - (1.52)(2.29) = -.62$$
$$Y' = 1.52\,(X) - .62$$

CHAPTER 9

1. (a) Yes   (c) Yes   (e) No   (g) Yes   (i) Yes
(b) Yes   (d) No   (f) Yes   (h) No

2. (g) all lab rats, similar to those used, that could undergo the same experiment.
(i) all possible tosses of a coin.

3. (a) false   (b) true   (c) false   (d) true

4. (a) There are many ways. For instance, on the assumption that the classroom contains a maximum of nine rows of seats and that each row contains a maximum of 9 seats, you could proceed as follows: Consult the tables of random numbers, using the first digit of each five-digit random number to identify the row (previously labeled 1, 2, 3, and so on), and the second digit of the same random number to locate a particular student's seat within that row. Repeat this process until 5 students have been identified. (If the classroom is larger, use additional digits so that every student can be sampled.)
(b) Once again, there are many ways. For instance, if a student directory exists, use the initial digits of each random number to identify the page number of the directory and the final digits to count down the list of names on that page. Repeat this process until 40 students have been identified.

5. (a) Yes   (b) Yes   (c) No   (Subjects are more likely to be assigned to control group.)
(d) Yes

6. (a) For instance, focus on the first two digits of each random number. Assign a subject to group 1 if the first two digits are between 00 and 24; to group 2 if the first two digits are between 25 and 49; to group 3 if the first two digits are between 50 and 74; and to group 4 if the first two digits are between 75 and 99.
(b) Random assignment could involve sets of 4 subjects. The first subject can be *randomly* assigned to any of the four experimental conditions; the second subject can be *randomly* assigned to any of the three remaining conditions; the third subject can be *randomly* assigned to either of the two remaining conditions; and the fourth subject is *automatically* assigned to the one remaining condition. Repeat this process for each set of four subjects.

7. (a) $(\frac{1}{2})(\frac{1}{2}) = \frac{1}{4}$ (b) $(\frac{1}{2})(\frac{1}{2}) = \frac{1}{4}$ (c) $\frac{1}{4} + \frac{1}{4} = \frac{2}{4}$

8. (a) $(\frac{1}{2})(\frac{1}{2})(\frac{1}{2}) = \frac{1}{8}$ (c) $\frac{1}{8} + \frac{1}{8} = \frac{2}{8}$
(b) $(\frac{1}{2})(\frac{1}{2})(\frac{1}{2}) = \frac{1}{8}$ (d) $1 - (\frac{1}{8} + \frac{1}{8}) = \frac{6}{8}$

9. (a) $(.05 + .02 + .01 + .01) + (.07 + .03 + .02 + .01) = .22$
(b) $(.22)(.22) = .05$

10. (a) .0250   (d) $.4750 + .4750 = .9500 = .95$
(b) .0250   (e) $.0051 + .0051 = .0102 = .01$
(c) $.0250 + .0250 = .0500 = .05$ (f) $.0005 + .0005 = .0010 = .001$

1. The sampling distribution of the mean is the probability distribution of means for all possible samples of a given size from some population.

2. (a)

(1) 2,2	(6) 4,2	(11) 6,2	(16) 8,2	(21) 10,2
(2) 2,4	(7) 4,4	(12) 6,4	(17) 8,4	(22) 10,4
(3) 2,6	(8) 4,6	(13) 6,6	(18) 8,6	(23) 10,6
(4) 2,8	(9) 4,8	(14) 6,8	(19) 8,8	(24) 10,8
(5) 2,10	(10) 4,10	(15) 6,10	(20) 8,10	(25) 10,10

(b)

$\overline{X}$	PROBABILITY
10	1/25
9	2/25
8	3/25
7	4/25
6	5/25
5	4/25
4	3/25
3	2/25
2	1/25

(c) $\mu_{\overline{X}} = 6$ (since $\mu_{\overline{X}} = \mu$)

(d) $\sigma_{\overline{X}} = \dfrac{2.83}{\sqrt{2}} = 2.00 \left(\text{since } \sigma_{\overline{X}} = \dfrac{\sigma}{\sqrt{n}}\right)$

(e) Population is shaped like a rectangle, while sampling distribution is shaped like a triangle—peaked in the middle with tapered flanks.

(f) $\frac{3}{25} + \frac{2}{25} + \frac{1}{25} = \frac{6}{25}$

3. (a) μ (b) $\mu_{\overline{X}}$ (c) $\overline{X}$ (d) $\sigma_{\overline{X}}$ (e) S (f) σ

4. Samples are not exact replicas of populations, and therefore, most sample means fail to equal the population mean—even though the mean of all possible sample means equals the population mean.

5. (a) The shape of the sampling distribution approximates a normal curve, since the sample size of 144 satisfies the requirements of the central limit theorem.
(b) The standard error of the mean, $\sigma_{\overline{X}}$, equals .67, since

$$\sigma_{\overline{X}} = \frac{\sigma}{\sqrt{n}} = \frac{8}{\sqrt{144}} = \frac{8}{12} = .67$$

(c) The mean of the sampling distribution, $\mu_{\overline{X}}$, equals 21, since $\mu_{\overline{X}} = \mu$.

CHAPTER 11

1. *Problem:*

Does the average salary of all female APA members (with a PhD and full time teaching appointment) equal $21,500?

Statistical Hypotheses: *Statistical Test:*
                                  z test

$H_0: \mu = 21,500$
$H_1: \mu \neq 21,500$

Decision Rule:
Reject H_0 at the .05 level of significance if $z \geq 1.96$ or $z \leq -1.96$.

Calculations: *Decision:*

Given $\overline{X} = 20,300$; $\sigma_{\overline{X}} = \dfrac{3000}{\sqrt{100}} = \dfrac{3000}{10} = 300$ Reject H_0.

$z = \dfrac{20,300 - 21,500}{300} = -4.00$

Interpretation:

The average salary of female APA members is less than $21,500 (and by implication, less than their male counterparts).

2. *Problem:*

Does the mean IQ of all students within the district equal 100?

Statistical Hypotheses: *Statistical Test:*

H_0: $\mu = 100$ z test
H_1: $\mu \neq 100$

Decision Rule:
Reject H_0 at the .05 level of significance if $z \geq 1.96$ or $z \leq -1.96$.

Calculations: *Decision:*

Given $\overline{X} = 105$; $\sigma_{\overline{X}} = \dfrac{15}{\sqrt{25}} = \dfrac{15}{5} = 3$ Retain H_0

$$z = \frac{105 - 100}{3} = \frac{5}{3} = 1.67$$

Interpretation:
No evidence that the mean IQ of all students differs from 100.

3. *Problem:*

Does the mean weight of all "2 pound" boxes of candy (produced by the most recent shift) deviate from 33 ounces?

Statistical Hypotheses: *Statistical Test:*

H_0: $\mu = 33$ z test
H_1: $\mu \neq 33$

Decision Rule:
Reject H_0 at .05 level of significance if $z \geq 1.96$ or $z \leq -1.96$.

Calculations: *Decision:*

Given $\overline{X} = 33.09$; $\sigma_{\overline{X}} = \dfrac{.30}{\sqrt{36}} = .05$ Retain H_0.

$$z = \frac{33.09 - 33}{.05} = \frac{.09}{.05} = 1.80$$

Interpretation:
No evidence that mean weight deviates from 33 ounces.

4. *Problem:*

Does the mean commuting distance of all Chicago area workers differ from 13 miles?

Statistical Hypotheses: *Statistical Test:*

H_0: $\mu = 13$ z test
H_1: $\mu \neq 13$

Decision Rule:
Reject H_0 at .05 level of significance if $z \geq 1.96$ or $z \leq -1.96$.

Calculations: *Decision:*

Given $\overline{X} = 14.5$; $\sigma_{\overline{X}} = \dfrac{13}{\sqrt{169}} = \dfrac{13}{13} = 1$ Retain H_0.

$$z = \frac{14.5 - 13}{1} = 1.5$$

Interpretation:
No evidence that mean commuting distance for all Chicago area workers differs from the national average of 13 miles.

5. This is a meaningless test. The value of z always would equal 0 (since the value of the sample mean always would coincide with the hypothesized value), and therefore, H_0 always would be retained.

6. Retaining the null hypothesis is a weak conclusion. The null hypothesis *could* be true but so could null hypotheses having other values similar to that tested.

7. (a) $H_1: \mu > 0.54$ (b) $H_1: \mu < 23$ (c) $H_1: \mu > 134$

8. (a) probable
   (b) retained
   (c) improbable
   (d) rejected
   (e) critical
   (f) level of significance
   (g) α
   (h) z
   (i) central limit theorem
   (j) population standard deviation
   (k) one-tailed
   (l) directional
   (m) two-tailed
   (n) nondirectional
   (o) true
   (p) .05
   (q) .01
   (r) .001

CHAPTER 12

1. (a) Correct decision (True H_0 is retained)
      Type I error
      Correct decision (False H_0 is rejected)
      Type II error

   (b)

STATUS OF H_0

DECISION	True H_0 (Innocent)	False H_0 (Guilty)
Retain H_0 (Release)	**Correct Decision:** Innocent defendant is released	**Type II Error:** Guilty defendant is released (Miss)
Reject H_0 (Sentence)	**Type I Error:** Innocent defendant is sentenced (False Alarm)	**Correct Decision:** Guilty defendant is sentenced

2. Type I error occurs when a true H_0 is rejected.
   Type II error occurs when a false H_0 is retained.

3. (a) level of significance
   (b) true
   (c) hypothesized
   (d) sample size
   (e) decrease
   (f) II
   (g) difference
   (h) standard deviation
   (i) small, medium, large
   (j) α
   (k) β
   (l) effect

4. (a) Given a critical z of 1.65, first find the equivalent IQ score, that is,

$$X = \mu_{hyp} + (1.65)(\sigma_{\overline{X}}) = 100 + (1.65)(2.5) = 104.13$$

Next, convert this IQ score to a new z score, expressed relative to the *true* population mean of 110, that is,

$$z = \frac{104.13 - \mu_{true}}{\sigma_{\overline{X}}} = \frac{104.13 - 110}{2.5} = -2.35$$

Finally, verify that the area below a z of -2.35 equals .0057 or approximately .01.

(b) First, express the critical z as an IQ score and convert to a new z score, expressed relative to the true population mean of 103.

$$z = \frac{104.13 - 103}{2.5} = .45$$

Then, verify that the area below a z of .45 equals .6736 or approximately .67.

5. (a) No. Given that H_0 has been retained, a type II error, but not a type I error, could have been committed.
(b) No. Given that H_0 has been rejected, a type I error, but not a type II error, could have been committed.

6. Values of d, α, and β, and whether the test has one or two tails.

7. (a) Conclude that local population mean differs from the national average of 429 when, in fact, it doesn't.
(b) Conclude that there's a lack of evidence that the local population mean differs from the national average of 429 when, in fact, it does.
(c) 22 SAT points—from (.2) (110)
(d) 198—from Table 12.2

8. a_1, b_1, c_2, d_2, e_2

9. (a) 54   (b) 28   (c) 294

10. (a) Because of small sample size, only very large effects will be detected.
(b) Because of large sample size, even small, unimportant effects will be detected.

CHAPTER 13

1. *Problem:*

Does the mean weight for all packages of ground beef drop below the specified weight of 16 ounces?

Statistical Hypotheses: *Statistical Test:*
H_0: $\mu = 16$ t test
H_1: $\mu < 16$
Decision Rule:
Reject H_0 at the .05 level of significance if $t \leq 1.833$, given $df = 10 - 1 = 9$.
Calculations: *Decision:*
$$\overline{X} = \frac{147}{10} = 14.7 \qquad s = \sqrt{\frac{10(2167) - (147)^2}{10(10-1)}} = 0.82 \qquad \text{Reject } H_0$$

$$s_{\overline{X}} = \frac{.82}{\sqrt{10}} = 0.26 \qquad t = \frac{14.7 - 16}{.26} = -5.00$$

Interpretation:
The mean weight for all packages drops below the specified weight of 16 ounces.

2. *Problem:*

On the average, do all library patrons borrow books for longer or shorter periods than the currently specified loan period of 21 days?

Statistical Hypotheses: *Statistical Test:*
H_0: $\mu = 21$ t test
H_1: $\mu \neq 21$
Decision Rule:
Reject H_0 at the .05 level of significance if $t \geq 2.365$ or $t \leq -2.365$, given $df = 8 - 1 = 7$.

Calculations:

$$\overline{X} = \frac{142}{8} = 17.75; \qquad s = \sqrt{\frac{8(2652) - (142)^2}{8(8-1)}} = 4.33$$

$$s_{\overline{X}} = \frac{4.33}{\sqrt{8}} = 1.53 \qquad t = \frac{17.75 - 21}{1.53} = -2.12$$

Decision:
   Retain H_0.

Interpretation:

No evidence that, on the average, all library patrons borrow books for longer or shorter periods than 21 days.

3. Problem:

Does the mean number of trials to criterion differ from 32 trials for lab rats who receive a slight electrical shock just prior to each trial in a water maze?

Statistical Hypotheses: Statistical Test:
   H_0: $\mu = 32$ t test
   H_1: $\mu \neq 32$

Decision Rule:
   Reject H_0 at .05 level of significance if $t \geq 2.447$ or $t \leq -2.447$, given $df = 7 - 1 = 6$.

Calculations:

$$\overline{X} = \frac{244}{7} = 34.86 \qquad s = \sqrt{\frac{7(8561) - (244)^2}{7(7-1)}} = 3.05$$

$$s_{\overline{X}} = \frac{3.05}{\sqrt{7}} = 1.15 \qquad t = \frac{34.86 - 32}{1.15} = 2.49$$

Decision:
   Reject H_0.

Interpretation:

When given a slight electrical shock, lab rats require, on the average, more than 32 trials to reach criterion in a water maze.

4. Problem:

On the average, does a particular brand of steel-belted radial tires provide more than 50,000 miles of wear?

Statistical Hypotheses: Statistical Test
   H_0: $\mu = 50,000$ t test
   H_1: $\mu > 50,000$

Decision Rule:
   Reject H_0 at .01 level of significance if $t \geq 2.457$, given $df = 36 - 1 = 35$ (read as 30 in Table B).

Calculations:

$$s_{\overline{X}} = \frac{2500}{\sqrt{36}} = 416.67 \qquad t = \frac{52,100 - 50,000}{416.67} = 5.04$$

Decision:
   Reject H_0.

Interpretation:

On the average, this brand of tires lasts longer than 50,000 miles.

5. Problem:

Does the consumption of alcohol affect the average amount of dream time (90 minutes) of young adults?

Statistical Hypotheses: Statistical Test
   H_0: $\mu = 90$ t test
   H_1: $\mu \neq 90$

Decision Rule:
   Reject H_0 at .05 level of significance if $t \geq 2.052$ or $t \leq -2.052$, given $df = 28 - 1 = 27$.

Calculations:

$$s_{\overline{X}} = \frac{9}{\sqrt{28}} = 1.70 \qquad t = \frac{88 - 90}{1.70} = -1.18$$

Decision:
   Retain H_0.

Interpretation:
No evidence that consumption of alcohol affects the average amount of dream time of young adults.

6. The true level of significance is larger than .05 because of the greater area in the tails of the t distribution.

7. The degrees of freedom would equal n. No degrees of freedom are lost since *all* n of these deviations in the sample are free to vary about the population mean.

8. *Problem:*

For the population of California taxpayers, is there a relationship between educational level and annual income?

Statistical Hypotheses: *Statistical Test:*
     H_0: $\rho = 0$ t test
     H_1: $\rho \neq 0$
Decision Rule:
Reject H_0 at .05 level of significance if $t \geq 2.060$ or $t \leq -2.060$, given $df = 27 - 2 = 25$.

Calculations: *Decision:*

$$t = \frac{.43(\sqrt{27 - 2})}{\sqrt{1 - (.43)^2}} = 2.38$$
                                      Reject H_0.

Interpretation:
For the population of California taxpayers, there is a relationship (positive) between educational level and annual income.

9. *Problem:*

In a large statistics class, is there a relationship between test score on a statistics exam and amount of time spent taking the exam?

Statistical Hypotheses: *Statistical Test:*
     H_0: $\rho = 0$ t test
     H_1: $\rho \neq 0$
Decision Rule:
Reject H_0 at .01 level if $t \geq 2.750$ or $t \leq -2.750$, given $df = 38 - 2 = 36$ (read as 30 in Table B).

Calculations: *Decision:*

$$t = \frac{-.24(\sqrt{38 - 2})}{\sqrt{1 - (-.24)^2}} = -1.48$$
                                      Retain H_0.

Interpretation:
No evidence that, in a large statistics class, there is a relationship between test score and amount of time spent taking the test.

CHAPTER 14

1. *Problem:*

Is there a difference, on the average, between compliance scores of subjects in a committee and those of solitary subjects?

Statistical Hypotheses: *Statistical Test:*
     H_0: $\mu_1 - \mu_2 = 0$ t test
     H_1: $\mu_1 - \mu_2 \neq 0$
Decision Rule:
Reject H_0 at .05 level of significance if $t \geq 2.228$ or $t \leq -2.228$, given $df = 6 + 6 - 2 = 10$.

Calculations:

$$\bar{X}_1 = \frac{56}{6} = 9.33 \qquad \bar{X}_2 = \frac{42}{6} = 7.00$$

$$s_1^2 = \frac{6(770) - (56)^2}{6(6-1)} = 49.47 \qquad s_2^2 = \frac{6(418) - (42)^2}{6(6-1)} = 24.80$$

$$s_p^2 = \frac{(6-1)(49.47) + (6-1)(24.80)}{(6-1) + (6-1)} = 37.14$$

$$s_{\bar{X}_1 - \bar{X}_2} = \sqrt{\frac{37.14}{6} + \frac{37.14}{6}} = 3.52$$

$$t = \frac{(9.33 - 7.00) - 0}{3.52} = 0.66$$

Decision:
Retain H_0.

Interpretation:

No evidence that compliance scores differ for subjects in committee and solitary conditions.

2. (a) *Problem:*

Is there a difference, on the average, between the puzzle-solving times required by subjects who are told that the puzzle is difficult and by subjects who are told that the puzzle is easy?

Statistical Hypotheses: *Statistical Test:*
   H_0: $\mu_1 - \mu_2 = 0$ t test
   H_1: $\mu_1 - \mu_2 \neq 0$

Decision Rule:

Reject H_0 at .05 level of significance if $t \geq 2.101$ or $t \leq -2.101$, given $df = 10 + 10 - 2 = 18$.

Calculations:

$$\bar{X}_1 = \frac{138}{10} = 13.8 \qquad \bar{X}_2 = \frac{90}{10} = 9.0$$

$$s_1^2 = \frac{10(2608) - (138)^2}{10(10-1)} = 78.18 \qquad s_2^2 = \frac{10(1036) - (90)^2}{10(10-1)} = 25.11$$

$$s_p^2 = \frac{(10-1)(78.18) + (10-1)(25.11)}{(10-1) + (10-1)} = 51.65$$

$$s_{\bar{X}_1 - \bar{X}_2} = \sqrt{\frac{51.65}{10} + \frac{51.65}{10}} = 3.21$$

$$t = \frac{(13.8 - 9.0) - 0}{3.21} = 1.50$$

Decision:
Retain H_0.

Interpretation:

No evidence that puzzle-solving times differ for subjects who are told that the puzzle is difficult and for those who are told that the puzzle is easy.
(b) 64 subjects in each group.

3. (a) *Problem:*

Do daily doses of vitamin C increase the mean IQ score of high school students?

Statistical Hypotheses: *Statistical Test:*
   H_0: $\mu_1 - \mu_2 = 0$ t test
   H_1: $\mu_1 - \mu_2 > 0$

Decision Rule:

Reject H_0 at .01 level of significance if $t \geq 2.390$, given $df = 35 + 35 - 2 = 68$ (read as 60 in Table B).

Calculations:

$$t = \frac{(110 - 100) - 0}{1.80} = 1.11$$

Decision:

Retain H_0.

Interpretation:

No evidence that daily doses of vitamin C increase the mean IQ scores of high school students.

(b) There is a lack of evidence that vitamin C increases IQ scores ($t(68) = 1.11$, $p > .05$).

(c) Approximately speaking, a sample size of 35 per group is adequate to detect only a large effect size, given a one-tailed test with $\alpha = .01$ and $\beta = .20$.

4. (a) *Problem:*

Is the mean performance of college students in an introductory biology course affected by grading policy?

Statistical Hypotheses:	*Statistical Test:*
H_0: $\mu_1 - \mu_2 = 0$	t test
H_1: $\mu_1 - \mu_2 \neq 0$	

Decision Rule:

Reject H_0 at .05 level of significance if $t \geq 2.000$ or $t \leq -2.000$, given $df = 40 + 40 - 2 = 78$ (read as 60 in Table B).

Calculations: *Decision:*

$$t = \frac{(86.2 - 81.6) - 0}{1.50} = 3.07$$

Reject H_0.

Interpretation:

Introductory biology students have higher achievement scores, on the average, when awarded letter grades rather than a simple pass/fail.

(b) Compared with a simple pass/fail, letter grades are associated with higher achievement scores among introductory biology students ($t(78) = 3.07$, $p < .05$).

(c) Because of self-selection, groups might differ with respect to any one or several uncontrolled variables, such as motivation, aptitude, and so on, in addition to the difference in grading policy. Hence, any observed difference between the mean achievement scores for these two groups couldn't be attributed solely to the difference in grading policy.

5. (a) *Problem:*

Does alcoholic consumption cause an increase in mean performance errors on a driving simulator?

Statistical Hypotheses:	*Statistical Test:*
H_0: $\mu_1 - \mu_2 = 0$	t test
H_1: $\mu_1 - \mu_2 > 0$	

Decision Rule:

Reject H_0 at .05 level of significance if $t \geq 1.671$, given $df = 60 + 60 - 2 = 118$ (read as 60 in Table B).

Calculations: *Decision:*

$$t = \frac{26.4 - 18.6}{2.4} = 3.25$$

Reject H_0.

Interpretation:

Alcoholic consumption causes an increase in mean performance errors on a driving simulator.

(b) Alcoholic consumption causes an increase in errors on a driving simulator ($t(118) = 3.25$, $p < .05$).

(c) Approximately speaking, a sample size of 60 is adequate to detect a medium effect size, given a one-tailed test with $\alpha = .05$ and $\beta = .20$.

6. (a) A—since it is more likely to detect only a larger effect (having more practical importance).

(b) B

(c) The probability of a type I error is the same (.05) for both A and B.

(d) No. A would be better if you wished to detect *only* a large effect, whereas B would be better if you wished to detect a medium effect (as well as a large effect).

1. (a) *Problem:*

When school children are matched for home environment, does vitamin C reduce the mean frequency of common colds?

Statistical Hypotheses: *Statistical Test:*
   H_0: $\mu_D = 0$ t test
   H_1: $\mu_D < 0$
Decision Rule:
   Reject H_0 at .05 level of significance if $t \leq -1.833$, given $df = 10 - 1 = 9$.
Calculations: *Decision:*

$$\bar{D} = \frac{-15}{10} = -1.5 \qquad s_D = \sqrt{\frac{10(37) - (-15)^2}{10(10 - 1)}} = 1.27 \qquad \text{Reject } H_0.$$

$$s_{\bar{D}} = \frac{1.27}{\sqrt{10}} = 0.40 \qquad t = \frac{-1.5 - 0}{.40} = -3.75$$

Interpretation:

When school children are matched for home environment, vitamin C reduces the mean frequency of common colds.

(b) Roughly speaking, 10 pairs of children are adequate to detect a large effect size, given a one-tailed test with $\alpha = .05$ and $\beta = .20$ (and an estimate of .60 for ρ).

(c) Thirty-six pairs of children should be used.

2. (a) *Problem:*

Does "ABC" meditation cause an increase in mean GPAs for students—given that pairs of students are originally matched for their GPAs?

Statistical Hypotheses: *Statistical Test:*
   H_0: $\mu_D = 0$ t test
   H_1: $\mu_D > 0$
Decision Rule:
   Reject H_0 at .01 level of significance if $t > 3.143$, given $df = 7 - 1 = 6$.
Calculations: *Decision:*

$$\bar{D} = \frac{1.56}{7} = .22 \qquad s_D = \sqrt{\frac{7(.50) - (1.56)^2}{7(7 - 1)}} = .16 \qquad \text{Reject } H_0.$$

$$s_{\bar{D}} = \frac{.16}{\sqrt{7}} = .06 \qquad t = \frac{.22 - 0}{.06} = 3.67$$

Interpretation:

"ABC" meditation causes an increase in mean GPAs when pairs of students are matched for their original GPAs.

(b) Seven pairs of observations are adequate to detect only a *very* large effect (not listed in Table H). Given a one-tailed test with $\alpha = .05$, $\beta = .20$, and ρ estimated as .50, there should be at least 12 pairs of observations to detect a large effect size. This number reflects the addition of one pair of subjects because the recommended procedure underestimates required sample sizes, particularly small sample sizes.

3. (a) *Problem:*

Is there a decline in the mean daily cigarette consumption of heavy smokers who see an anti-smoking film?

Statistical Hypotheses: *Statistical Test:*
   H_0: $\mu_D = 0$ t test
   H_1: $\mu_D > 0$
Decision Rule:
   Reject H_0 at .05 level of significance if $t \geq 1.895$, given $df = 8 - 1 = 7$.

Calculations:

$$\bar{D} = \frac{16}{8} = 2 \qquad s_D = \sqrt{\frac{8(54) - (16)^2}{8(8-1)}} = 1.77$$

$$s_{\bar{D}} = \frac{1.77}{\sqrt{8}} = .63 \qquad t = \frac{2-0}{.63} = 3.17$$

Decision:

Reject H_0.

Interpretation:

There is a decline in the mean daily cigarette consumption of heavy smokers who see an anti-smoking film.

(b) A better design would incorporate repeated measurements for a control group of heavy smokers who don't see the anti-smoking film. Difference scores for the experimental subjects (who see the film) and control subjects (who don't see the film) then could be analyzed with a *t* test for two independent groups.

4. (a) *Problem:*

On the average, does a gas additive improve gas mileage?

Statistical Hypotheses: *Statistical Test:*

H_0: $\mu_D = 0$ *t* test

H_1: $\mu_D > 0$

Decision Rule:

Reject H_0 at .05 level of significance if $t \geq 1.699$, given $df = 30 - 1 = 29$.

Calculations: *Decision:*

$$t = \frac{2.12 - 0}{1.50} = 1.41$$

Retain H_0.

Interpretation:

No evidence that, on the average, a gas additive increases gas mileage.

(b) Roughly speaking, thirty pairs are adequate to detect a large effect size, given a one-tailed test with $\alpha = .05$, $\beta = .05$, and ρ estimated as .20.

(c) A random half of all pairs of tests should begin with the additive, while the other half of the pairs should end with the additive. Counterbalancing eliminates any possible bias due to the order of testing.

5. On the assumption that matching involves a relevant variable, the value of the standard error term will be inflated by use of the formula for two independent samples. This inflated standard error term will reduce the value of *t* and, therefore, cause an increase in the probability of a type II error.

6. (a) Reduces the standard error term. This, in turn, reduces the probability of a type II error.

(b) When compared with the *df* for two independent samples, the *df* for two dependent samples is reduced by one half. Taken by itself, a loss of degrees of freedom translates into a less precise analysis.

(c) If it has a considerable effect on the variable being measured.

(d) Contamination of experimental and control conditions.

CHAPTER 16

1. (a) 3.82   (b) $3.82 \pm (2.000) \left(\dfrac{.4}{\sqrt{64}} \right)$

                      lower limit = 3.72
                      upper limit = 3.92

(c) We can claim, with 95% confidence, that the interval between 3.72 and 3.92 includes the true *population mean* reading score for the fourth graders. All of these values suggest that, on the average, the fourth graders are underachieving.

2. (a) $8.46 - 8.20 = .26$ (b) $.26 \pm (2.021)(.09)$
                                       lower limit = .08
                                       upper limit = .44

(c) We can claim, with 95% confidence, that the interval between .08 and .44 includes the true *population mean difference*, $\mu_1 - \mu_2$. All of these values suggest that the special reading program has some positive effect.

3. (a) -15.4 (b) $-15.4 \pm (2.064)\left(\dfrac{12}{\sqrt{25}}\right)$

lower limit $= -20.34$
upper limit $= -10.46$

(c) We can claim, with 95% confidence, that the interval between -20.34 and -10.46 includes the true *population mean for all difference scores*. All of these values suggest that the special weight-control program has an effect in the desired direction.

4. $14.7 \pm (2.262)\left(\dfrac{.26}{\sqrt{10}}\right)$ **5.** $34.86 \pm (2.447)(1.15)$

lower limit $= 14.52$ lower limit $= 32.04$
upper limit $= 14.88$ upper limit $= 37.68$

6. (a) $110 - 108 \pm (2.660)(1.80)$
lower limit -2.79
upper limit 6.79

(b) We can claim, with 99% confidence, that the interval between -2.79 and 6.79 includes the true *population mean difference*. These results are ambiguous. Negative differences between population means (down to -2.79) suggest that vitamin C decreases IQ scores, while positive differences between population means (up to 6.79) suggest that vitamin C increases IQ scores.

7. $86.2 - 81.6 \pm (2.000)(1.50)$
lower limit $= 1.6$
upper limit $= 7.6$

8. (a) $2 \pm (2.365)(.63)$
lower limit $= 0.51$
upper limit $= 3.49$

(b) We can claim, with 95% confidence, that the interval between 0.51 and 3.49 includes the true *population mean for all difference scores*. All of these values suggest that a decline in smoking occurs with exposure to the anti-smoking film.

9. (a) 3   (b) 3   (c) 4   (d) 3   (e) 5   (f) 4

CHAPTER 17

1.

	TYPE OF VARIABILITY	
	Between Groups	Within Groups
(a)	No	No
(b)	Yes	No
(c)	No	Yes
(d)	Yes	Yes

2. (a) *Problem:*

On the average, are subjects' alarm reactions to potentially dangerous smoke affected by crowds of zero, two, and four confederates?

Statistical Hypotheses:
H_0: $\mu_0 = \mu_2 = \mu_4$
H_1: H_0 is not true
Decision Rule:
Reject H_0 at .01 level of significance if $F \geq 10.92$, given $df_{between} = 2$ and $df_{within} = 6$.
Calculations:
$F = .50$ (See 2(b) for more information about this calculation.)
Decision:
Retain H_0.
Interpretation:
No evidence that mean alarm reactions are affected by crowd size.

Statistical Test:
F test

(b) SOURCE	SS	df	MS	F
Between	6	2	3	0.50
Within	36	6	6	
Total	42	8		

3. (a) *Problem:*

On the average, are subjects' alarm reactions to potentially dangerous smoke affected by crowds of zero, two, and four confederates?

Statistical Hypotheses:
H_0: $\mu_0 = \mu_2 = \mu_4$
H_1: H_0 is not true
Decision Rule:
Reject H_0 at .05 level if $F \geq 4.26$, given $df_{between} = 2$ and $df_{within} = 9$.
Calculations:
$F = 13.70$ (See 3(b) for information.)
Decision:
Reject H_0.
Interpretation:
Mean alarm reactions are affected by crowd size.

Statistical Test:
F test

(b) SOURCE	SS	df	MS	F
Between	106.05	2	53.03	13.70*
Within	34.87	9	3.87	
Total	140.92	11		

*Significant at .05 level.

(c) Since H_0 has been rejected, Scheffé's test may be used.

SCHEFFÉ'S TEST

POPULATION COMPARISON	OBSERVED (MEAN DIFFERENCE)	SCHEFFÉ'S CRITICAL VALUE
$\mu_0 - \mu_2$	−2.73	±4.18
$\mu_0 - \mu_4$	−6.90*	±3.85
$\mu_2 - \mu_4$	−4.17	±4.37

*Significant at .05 level.

Mean alarm reaction of subjects in crowds of zero confederates is shorter than that of subjects in crowds of four confederates. There is no evidence, however, that the mean alarm reaction of subjects in crowds of two confederates differs either from that of subjects in crowds of zero confederates or from that of subjects in crowds of four confederates.

4. (a) *Problem:*

Are subjects' mean alarm reactions to potentially dangerous smoke affected by crowds of zero, two, four, eight, and twelve confederates?

Statistical Hypotheses: *Statistical test;*
H_0: $\mu_0 = \mu_2 = \mu_4 = \mu_8 = \mu_{12}$ F test
H_1: H_0 is not true

Decision rule:

Reject H_0 at the .05 level of significance if $F \geq 3.06$, given $df_{between} = 4$ and $df_{within} = 15$.

Calculations:

$F = 6.21$ (See 4(b) for more information.)

Decision:

Reject H_0.

Interpretation:

Mean alarm reactions are affected by crowd size.

(b)

SOURCE	SS	df	MS	F
Between	552.80	4	138.20	6.21*
Within	333.75	15	22.25	
Total	886.55	19		

*Significant at .05 level.

(c) SCHEFFÉ'S TEST

POPULATION COMPARISON	OBSERVED MEAN DIFFERENCE	SCHEFFÉ'S CRITICAL VALUE
$\mu_0 - \mu_2$	−1.00	±11.67
$\mu_0 - \mu_4$	−2.00	±11.67
$\mu_0 - \mu_8$	−9.00	±11.67
$\mu_0 - \mu_{12}$	−13.50*	±11.67
$\mu_2 - \mu_4$	−1.00	±11.67
$\mu_2 - \mu_8$	−8.00	±11.67
$\mu_2 - \mu_{12}$	−12.50*	±11.67
$\mu_4 - \mu_8$	−7.00	±11.67
$\mu_4 - \mu_{12}$	−11.50	±11.67
$\mu_8 - \mu_{12}$	−4.50	±11.67

*Significant at .05 level.

The mean alarm reaction of subjects in crowds of twelve confederates is longer than that of subjects in crowds of zero confederates, as well as that of subjects in crowds of two confederates. There is no evidence, however, of a significance difference between the means for any other pair of groups.

5. (a) 4   (b) 21   (c) A and B   (d) C   (e) A (Larger F makes null hypothesis more suspect.)   (f) D (Because of the small number of subjects, only a large treatment effect would have been detected.) (g) C (Because of the relatively large number of subjects, even a fairly small treatment effect should have been detected.)

6. (a) *Problem:*

On the average, are weight changes affected by the type of reduction program: diet, exercise, or behavioral modification?

Statistical Hypotheses: *Statistical Test:*
H_0: $\mu_{diet} = \mu_{exercise} = \mu_{behavioral\ modification}$ F test
H_1: H_0 is not true

Decision Rule:

Reject H_0 at .05 level of significance if $F \geq 3.49$, given $df_{between} = 2$ and $df_{within} = 20$.

Calculations:

$F = 2.67$ (See 6(b) for more information.)

Decision:

Retain H_0.

Interpretation:

No evidence that the type of reduction program affects the mean amount of weight loss.

(b) SOURCE	SS	df	MS	F
Between	116.41	2	58.21	2.67
Within	435.50	20	21.78	
Total	551.91	22		

(c) No evidence that the three different treatment programs produce different mean weight losses [$F(2,20) = 2.67$, $p > .05$].

(d) Scheffé's test may not be used.

7. (a) *Problem:*

On the average, is aggressive behavior influenced by sleep deprivation?

Statistical Hypotheses: *Statistical Test:*

H_0: $\mu_0 = \mu_{24} = \mu_{48} = \mu_{72}$ F test

H_1: H_0 is not true

Decision Rule:

Reject H_0 at .05 level of significance if $F \geq 2.95$, given $df_{between} = 3$ and $df_{within} = 28$.

Calculations:

$F = 10.84$ (See 7(b) for more information.)

Decision:

Reject H_0.

Interpretation

Mean aggression score is influenced by sleep deprivation.

(b) SOURCE	SS	df	MS	F
Between	154.12	3	51.37	10.84*
Within	132.75	28	4.74	
Total	286.87	31		

*Significant at .05 level.

(c) There is evidence that the mean aggression score is influenced by sleep deprivation [$F(3,28) = 10.84$, $p < .05$].

(d) Scheffé's test may be used.

POPULATION COMPARISON	OBSERVED MEAN DIFFERENCE	SCHEFFÉ'S CRITICAL VALUE
$\mu_0 - \mu_{24}$	−1.50	±3.24
$\mu_0 - \mu_{48}$	−3.37*	±3.24
$\mu_0 - \mu_{72}$	−5.87*	±3.24
$\mu_{24} - \mu_{48}$	−1.87	±3.24
$\mu_{24} - \mu_{72}$	−4.37*	±3.24
$\mu_{48} - \mu_{72}$	−2.50	±3.24

*Significant at .05 level.

Subjects deprived of sleep for 48 or 72 hours have higher mean aggression scores than those deprived of sleep for 0 hours. Furthermore, subjects deprived of sleep for 72 hours have higher mean aggression scores than those deprived of sleep for 24 hours. No other differences are significant.

CHAPTER 18

1.

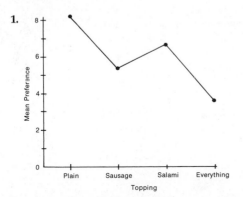

Interpretation: H_0 for topping is suspect.

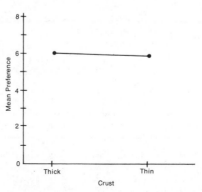

Interpretation: H_0 for crust is *not* suspect.

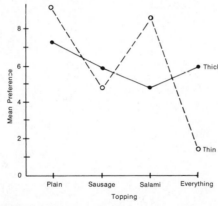

Interpretation: H_0 for interaction is suspect.

2. (a)

SOURCE	df
Food deprivation (F)	3
Reward magnitude (R)	2
F × R	6
Error	84
Total	95

(b)

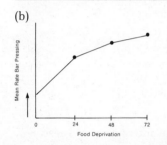

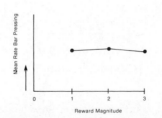

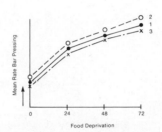

3. (a) Column has 2 levels. Row has 3 levels. There are 6 groups with 11 subjects in each group.

SOURCE	SS	df	MS	F
Column	790	1	790	3.22
Row	326	2	163	0.67
Interaction	1887	2	943.50	3.85*
Within	14702	60	245.03	
Total	17705	65		

*Significant at .05 level.

(b) Treatment A has 3 levels. Treatment B has 3 levels. There are 9 groups with 10 subjects in each group.

SOURCE	SS	df	MS	F
Treatment A	142	2	71	1.14
Treatment B	480	2	240	3.86*
A × B	209	4	52.25	0.84
Error	5030	81	62.10	
Total	5861	89		

*Significant at .05 level.

4. (a)

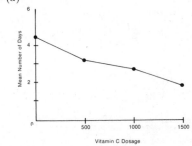 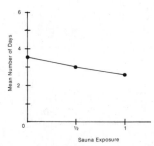

Interpretation: H_0 for vitamin C dosage is suspect.

Interpretation: H_0 for sauna exposure is suspect (although not as much as H_0 for vitamin C dosage).

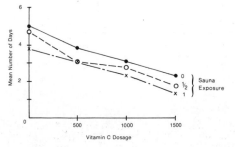

Interpretation: H_0 for interaction is not suspect.

(b) *Problem:*

Do vitamin C dosage and sauna exposure, as well as the interaction of these two factors, affect the mean number of days of discomfort due to colds?

Statistical Hypotheses:

H_0: no treatment effect due to columns or vitamin C dosage (or $\mu_0 = \mu_{500} = \mu_{1000} = \mu_{1500}$)

H_0: no treatment effect due to rows or sauna exposure (or $\mu_0 = \mu_{1/2} = \mu_1$)

H_0: no interaction

H_1: H_0 is not true

Statistical Tests:

F test

Decision Rule:

Reject H_0 at the .05 level of significance if $F_{column} \geq 3.01$, given 3 and 24 degrees of freedom; if $F_{row} \geq 3.40$, given 2 and 24 degrees of freedom; and if $F_{interaction} \geq 2.51$, given 6 and 24 degrees of freedom.

Calculations:

$$\left. \begin{array}{l} F_{column} = 17.52 \\ F_{row} = 3.95 \\ F_{interaction} = 0.25 \end{array} \right\} \text{(See 4(c) for more information.)}$$

Decision:

Reject H_0 for column (vitamin C) and H_0 for row (sauna).

Interpretation:

Both vitamin C dosage and sauna exposure affect the mean number of days of discomfort due to colds. There is no evidence, however, of an interaction between these two factors.

(c) SOURCE	SS	df	MS	F
Column (Vitamin C)	33.64	3	11.21	17.52*
Row (Sauna)	5.05	2	2.53	3.95*
Interaction	0.95	6	0.16	0.25
Error	15.33	24	0.64	
Total	54.97	35		

*Significant at .05 level.

5. (a)

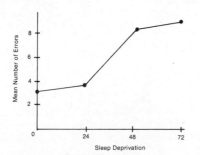

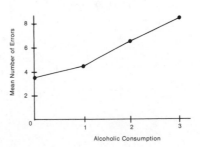

Interpretation: H_0 for sleep deprivation is suspect.

Interpretation: H_0 for alcoholic consumption is suspect.

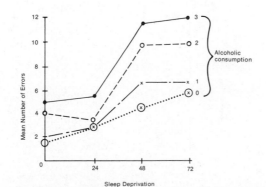

Interpretation: H_0 for interaction is not suspect.

(b) *Problem:*

Do sleep deprivation and alcoholic consumption, as well as the interaction of these two factors, affect the mean number of driving errors?

Statistical Hypotheses:

H_0: no treatment effect due to columns or sleep deprivation (or $\mu_0 = \mu_{24} = \mu_{48} = \mu_{72}$)

H_0: no treatment effect due to rows or alcoholic consumption (or $\mu_0 = \mu_1 = \mu_2 = \mu_3$)

H_0: no interaction

H_1: H_0 is not true

Statistical Tests:

F test

Decision Rule:

Reject H_0 at the .05 level if F_{column} or $F_{\text{row}} \geq 3.24$, given 3 and 16 degrees of freedom, and if $F_{\text{interaction}} \geq 2.54$, given 9 and 16 degrees of freedom.

Calculations:

$$\left.\begin{array}{r} F_{\text{column}} = 14.53 \\ F_{\text{row}} = 9.16 \\ F_{\text{interaction}} = 0.53 \end{array}\right\} \text{(See 5(c) for more information.)}$$

Decision:

Reject H_0 for sleep deprivation and for alcoholic consumption.

Interpretation:

Both sleep deprivation and alcoholic consumption affect the mean number of driving errors. There is no evidence, however, that these two factors interact.

(c)

SOURCE	SS	df	MS	F
Column (Sleep deprivation)	182.62	3	60.87	14.53*
Row (Alcoholic consumption)	115.12	3	38.37	9.16*
Interaction	20.13	9	2.24	0.53
Error	67.00	16	4.19	
Total	384.87	31		

*Significant at .05 level.

6. (a)

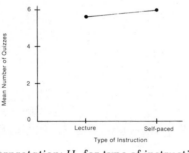

Interpretation: H_0 for type of instruction is not suspect.

Interpretation: H_0 for grading policy is not suspect.

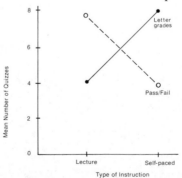

Interpretation: H_0 for interaction is suspect.

(b) *Problem:*
Do the type of instruction and the grading policy, as well as the interaction of these two factors, affect the mean number of quizzes successfully completed during the semester?

Statistical Hypotheses:
H_0: no treatment effect due to columns or type of instruction (or $\mu_{\text{lecture}} = \mu_{\text{self-paced}}$)
H_0: no treatment effect due to rows or grading policy (or $\mu_{\text{letter grades}} = \mu_{\text{pass/fail}}$)
H_0: no interaction
H_1: H_0 is not true

Statistical Tests:
F test

Decision Rule:
Reject H_0 at .01 level of significance if F_{column} or F_{row} or $F_{\text{interaction}} \geq 8.10$, given 1 and 20 degrees of freedom.

Calculations:

$$\left. \begin{array}{l} F_{\text{column}} = 0.03 \\ F_{\text{row}} = 0.12 \\ F_{\text{interaction}} = 13.01 \end{array} \right\} \quad \text{(See 6(c) for more information.)}$$

Decision:
Reject H_0 for interaction.

Interpretation:
There is evidence of an interaction. Given that instruction is based on lectures, more quizzes are successfully completed, on the average, when the grading policy is pass/fail. However, given that instruction is self-paced, more quizzes are successfully passed, on the average, when letter grades are awarded.

(c)	SOURCE	SS	df	MS	F
	Column (Instruction)	0.16	1	0.16	0.03
	Row (Grades)	0.66	1	0.66	0.12
	Interaction	73.51	1	73.51	13.01*
	Error	113.00	20	5.65	
	Total	187.33	23		

*Significant at .01 level.

CHAPTER 19

1. *Problem:*

The attribute most desired by a population of college students is equally distributed among various possibilities.

Statistical Hypotheses:
H_0: $P_{\text{love}} = P_{\text{wealth}} = P_{\text{power}} = P_{\text{health}} = P_{\text{fame}} = P_{\text{family happiness}}$
H_1: H_0 is not true

Statistical Test:
χ^2 test

Decision Rule:
Reject H_0 at .05 level of significance if $\chi^2 \geq 11.07$, given $df = 5$.

Calculations:
$$\chi^2 = \frac{(65-50)^2}{50} + \frac{(55-50)^2}{50} + \frac{(45-50)^2}{50} + \frac{(60-50)^2}{50} + \frac{(40-50)^2}{50} + \frac{(35-50)^2}{50}$$

$$= 14$$

Decision:
Reject H_0.

Interpretation:
The attribute most desired by a population of college students is not equally distributed among various possibilities.

2. *Problem:*

Are crimes more likely to be committed on certain days of the week?

Statistical Hypotheses:
H_0: $P_{Mon} = P_{Tue} = P_{Wed} = P_{Thu} = P_{Fri} = P_{Sat} = P_{Sun}$
H_1: H_0 is not true

Statistical Test:
χ^2 test

Decision Rule:
Reject H_0 at the .01 level of significance if $\chi^2 \geq 16.81$, given $df = 6$.

Calculations:

$$\chi^2 = \frac{(17-20)^2}{20} + \frac{(21-20)^2}{20} + \frac{(22-20)^2}{20} + \frac{(18-20)^2}{20} + \frac{(23-20)^2}{20} + \frac{(24-20)^2}{20} + \frac{(15-2}{20}$$
$$= 3.4$$

Decision:
Retain H_0.

Interpretation:
No evidence that crimes are more likely to be committed on certain days of the week.

3. *Problem:*

Is a particular coin unbiased?

Statistical Hypotheses:
H_0: $P_{heads} = P_{tails}$ (or $P_{heads} = .50$)
H_1: H_0 is not true

Statistical Test:
χ^2 test (with Yates' correction)

Decision Rule:
Reject H_0 at the .05 level of significance if $\chi^2 \geq 3.84$, given $df = 1$.

Calculations:

$$\chi^2 = \frac{(|30-25|-.5)^2}{25} + \frac{(|20-25|-.5)^2}{25} = 1.62$$

Decision:
Retain H_0.

Interpretation:
No evidence that the coin is biased.

4. *Problem:*

Is there a tendency for students to report their weights to the nearest five or ten pounds (rather than to the nearest pound)?

Statistical Hypotheses:
H_0: $P = .20$ (where P represents the population proportion of weights reported to the nearest five or ten pounds)
H_1: H_0 is not true

Statistical Test:
χ^2 test (with Yates' correction)

Decision Rule:
Reject H_0 at .05 level if $\chi^2 \geq 3.84$, given $df = 1$.

Calculations:

$$\chi^2 = \frac{(|27-10.6|-.5)^2}{10.6} + \frac{(|26-42.4|-.5)^2}{42.4} = 29.81$$

Decision:
Reject H_0.

Interpretation:
Weights tend to be reported to the nearest five or ten pounds.

| | OBSERVED PROPORTIONS | | | | EXPECTED PROPORTION |
	Red	Blond	Brown	Black	Total
Rash	.33	.50	.67	.67	.60

Preliminary interpretation: H_0 is suspect (because of the discrepancies between observed and expected proportions).

(b) *Problem:*

Is hair color related to susceptibility to poison oak?

Statistical Hypotheses:

H_0: Hair color and susceptibility to poison oak are not related

H_1: H_0 is false

Statistical Test:

χ^2 test

Decision:

Reject H_0 at .01 level if $\chi^2 \geq 11.34$, given $df = (2 - 1)(4 - 1) = 3$.

Calculations:

$$\chi^2 = \frac{(10 - 18)^2}{18} + \frac{(30 - 36)^2}{36} + \frac{(60 - 54)^2}{54} + \frac{(80 - 72)^2}{72} + \frac{(20 - 12)^2}{12} + \frac{(30 - 24)^2}{24}$$

$$+ \frac{(30 - 36)^2}{36} + \frac{(40 - 48)^2}{48} = 15.28$$

Decision:

Reject H_0.

Interpretation:

There is a relationship between hair color and susceptibility to poison oak.

6. (a)

| | OBSERVED PROPORTIONS | | | | | EXPECTED PROPORTIONS |
	Buddhist	Jewish	Protestant	Roman Catholic	Other	Total
Democrat	.60	.60	.40	.60	.20	.40
Republican	.20	.20	.40	.20	.10	.20
Independent	.20	.20	.20	.20	.20	.20
Other	.00	.00	.00	.00	.50	.20

Preliminary interpretation: H_0 is suspect (because of large discrepancies between observed and expected proportions in the last row and, to a lesser extent, in the first row).

(b) DEGREES OF FREEDOM (✔)

✔	✔	✔	✔	X
✔	✔	✔	✔	X
✔	✔	✔	✔	X
X	X	X	X	X

$df = (5 - 1)(4 - 1) = 12$

(c) *Problem:*

Is the religious preference of students related to their political affiliation?

Statistical Hypotheses:

H_0: Religious preference and political affiliation are not related.

H_1: H_0 is false

Statistical Test:

χ^2 test

Decision Rule:

Reject H_0 at .05 level if $\chi^2 \geq 21.03$, given $df = (4 - 1)(5 - 1) = 12$.

Calculations:

$$\chi^2 = \frac{(30 - 20)^2}{20} + \frac{(30 - 20)^2}{20} + \ldots + \frac{(100 - 40)^2}{40} = 220$$

Decision:

Reject H_0.

Interpretation:

There is a relationship between the religious preference of students and their political affiliation.

7. *Problem:*

Has the prevalence of marijuana smoking, as reported by college students, changed between the early 1970s and the late 1970s?

Statistical Hypotheses:

H_0: There is no relationship between the reported prevalence of marijuana smoking and two different generations of college students

H_1: H_0 is false

Statistical Test:

χ^2 test

Decision:

Reject H_0 at .05 level if $\chi^2 \geq 3.84$, given $df = 1$.

Calculations:

$$\chi^2 = \frac{(62 - 65)^2}{65} + \frac{(55 - 52)^2}{52} + \frac{(38 - 35)^2}{35} + \frac{(25 - 28)^2}{28} = 0.89$$

Decision:

Retain H_0.

Interpretation:

No evidence of a change in the reported prevalence of marijuana smoking between two different generations of college students during the early 1970s and late 1970s.

8. *Problem:*

Does the distribution of blood types of college students comply with that described (for the U.S. population) in a blood bank bulletin?

Statistical Hypotheses:

H_0: $P_O = .44$; $P_A = .42$; $P_{B \text{ and } AB} = .14$

H_1: H_0 is false

Statistical Test:

χ^2 test

Decision Rule:

Reject H_0 at .01 level if $\chi^2 \geq 9.21$, given $df = 3 - 1 = 2$.

Calculations:

$$\chi^2 = \frac{(27 - 26.4)^2}{26.4} + \frac{(24 - 25.2)^2}{25.2} + \frac{(9 - 8.4)^2}{8.4} = 0.11$$

Decision:

Retain H_0.

Interpretation:

No evidence that the distribution of blood types among college students departs from that described in the bulletin.

1. *Problem:*

Do encounter groups with aggressive leaders (1) produce more or less growth (in members) than encounter groups with supportive leaders (2)?

Statistical Hypotheses:
H_0: Population distribution 1 = Population distribution 2
H_1: Population distribution 1 ≠ Population distribution 2
Statistical Test:
U Test
Decision Rule:
Reject H_0 at .05 level if $U \leq 5$, given $n_1 = 6$ and $n_2 = 6$.

Calculations:	*Decision:*
$U_1 = 31$	Reject H_0.
$U_2 = 5$	
$U = 5$	

Interpretation:
Encounter groups with aggressive leaders produce less growth than those with supportive leaders.

2. (a) Each group of numbers includes one very large value, suggesting that the underlying populations might not be normally distributed, as required by the *t* test.
(b) *Problem:*
Does the type of penalty—attendance at a traffic school (1) or performance of volunteer work (2)—affect the number of moving violations during the subsequent five-year period?

Statistical Hypotheses:
H_0: Population distribution 1 = Population distribution 2
H_1: Population distribution 1 ≠ Population distribution 2
Statistical Test:
U test
Decision Rule:
Reject H_0 at the .05 level if $U \leq 17$, given $n_1 = 10$ and $n_2 = 8$.

Calculations:	*Decision:*
$U_1 = 45$	Retain H_0.
$U_2 = 35$	
$U = 35$	

Interpretation:
No evidence that the type of penalty affects the number of moving violations during the subsequent five-year period.

3. (a) Two very large difference scores suggest that the underlying population of difference scores might not be normally distributed, as required by the *t* test.
(b) *Problem:*
Does an anti-smoking workshop cause a decline in cigarette smoking?

Statistical Hypotheses:
H_0: Population distribution 1 = Population distribution 2
H_1: Population distribution 1 > Population distribution 2
Note— the directional H_1 assumes both population distributions have roughly similar (positively skewed) shapes.
Statistical Test:
T test
Decision Rule:
Reject H_0 at .05 level (directional test) if $T \leq 5$, given $n = 8$.

Calculations:	*Decision:*
$R_+ = 30$	Retain H_0.
$R_- = 6$	
$T = 6$	

Interpretation:
    No evidence that anti-smoking workshop causes a decline in smoking.

4. *Problem:*

    Does our attitude toward other people tend to reflect our perception of their attitude toward us?

Statistical Hypotheses:
    H_0: Population distribution 1 = Population distribution 2
    H_1: Population distribution 1 > Population distribution 2
    Note— the directional H_1 assumes both population distributions have roughly similar (symmetrical) shapes.
Statistical Test:
    T test
Decision Rule:
    Reject H_0 at .01 level (directional test) if $T \leq 7$, given $n = 11$.
Calculations: *Decision:*
    $R_+ = 63.5$ Reject H_0.
    $R_- = 2.5$
    $T = 2.5$
Interpretation:
    Our attitude toward other people tends to reflect our perception of their attitude toward us.

5. (a) Radically different ranges for X- and G-rated films suggests that the variances of the underlying populations might not be equal, as required by the F test.
    (b) *Problem:*
        Are motion picture ratings associated with the number of violent or sexually explicit scenes in films?

Statistical Hypotheses:
    H_0: Population dist. X = Population dist. R = Population dist. PG = Population dist. G
    H_1: H_0 is not true
Statistical Test:
    H test
Decision Rule:
    Reject H_0 at .05 level if $H \geq 7.82$, given $df = 3$.
Calculations:

$$H = \frac{12}{20(20 + 1)}\left[\frac{(82.5)^2}{5} + \frac{(64.5)^2}{5} + \frac{(48)^2}{5} + \frac{(15)^2}{5}\right] - 3(20 + 1) = 14.12$$

Decision:
    Reject H_0.
Interpretation:
    Motion picture ratings are associated with the number of violent or sexually explicit scenes in films.

6. (a) One or two very low scores in each sample suggest that the underlying populations might not be normally distributed, as required by the F test in ANOVA.
    (b) *Problem:*
        Does the type of background music tend to influence the scores of college students on a reading comprehension test?

Statistical Hypotheses:
    H_0: Population dist. 1 = Population dist. 2 = Population dist. 3
    H_1: H_0 is false
Statistical Test:
    H test
Decision Rule:
    Reject H_0 at .05 level if $H \geq 5.99$, given $df = 2$.

$$H = \frac{12}{30(30 + 1)}\left[\frac{(153)^2}{10} + \frac{(152)^2}{10} + \frac{(160)^2}{10}\right] - 3(30 + 1) = 0.05$$

Decision:
    Retain H_0.
Interpretation:
    No evidence that type of background music influences reading comprehension scores.

7. Problem:

    Is there a tendency for compliance scores of subjects in committee (1) to differ from those of subjects in a solitary (2) condition?

Statistical Hypotheses:
    H_0: Population distribution 1 = Population distribution 2
    H_1: Population distribution 1 ≠ Population distribution 2
Statistical Test:
    U test
Decision Rule:
    Reject H_0 at the .05 level if $U \leq 5$, given $n_1 = 6$ and $n_2 = 6$.
Calculations: *Decision:*
    $U_1 = 14.5$ Retain H_0.
    $U_2 = 21.5$
    $U = 14.5$
Interpretation:
    No evidence that compliance scores differ between subjects in the two conditions.

8. *Problem:*

    Does "ABC" meditation increase GPAs for students, given that pairs of students are originally matched for their GPAs?

Statistical Hypotheses:
    H_0: Population distribution X_1 = Population distribution X_2
    H_1: Population distribution X_1 > Population distribution X_2
    Note— the directional test assumes that both population distributions have roughly
          similar (symmetrical) shapes.
Statistical Test:
    T test
Decision Rule:
    Reject H_0 at .01 level (directional test) if $T = 0$, given $n = 7$.
Calculations: *Decision:*
    $R_+ = 27$ Retain H_0.
    $R_- = 1$
    $T = 1$
Interpretation:
    No evidence that "ABC" meditation increases GPAs of students—even when students are matched for their original GPAs.

9. *Problem:*

    Is there a tendency for weight changes to be affected by the type of weight reduction program?

Statistical Hypotheses:
    H_0: Population distribution (diet) = Population distribution (exercise)
          = Population distribution (behavioral modification)
    H_1: H_0 is false
Statistical Test:
    H test

Decision Rule:

Reject H_0 at .05 level if $H \geq 5.99$, given $df = 2$.

Calculations:

$$H = \frac{12}{23(23 + 1)}\left[\frac{(81.5)^2}{8} + \frac{(53.5)^2}{6} + \frac{(141)^2}{9}\right] - 3(23 + 1) = 4.44$$

Decision:

Retain H_0.

Interpretation:

No evidence that weight changes are affected by the type of weight reduction program.

10. When its assumptions are satisfied, the F test is less likely than the H test to produce a type II error.

APPENDIX C
Tables

A. AREAS UNDER STANDARD NORMAL CURVE FOR VALUES OF z

B. CRITICAL VALUES OF t

C. CRITICAL VALUES OF F

D. CRITICAL VALUES OF χ^2

E. CRITICAL VALUES OF MANN-WHITNEY U

F. CRITICAL VALUES OF WILCOXON T

G. RANDOM NUMBERS

H. APPROXIMATE SAMPLE SIZE FOR HYPOTHESIS TESTS OF POPULATION MEANS

TABLE A
AREAS UNDER STANDARD NORMAL CURVE FOR VALUES OF Z[a]

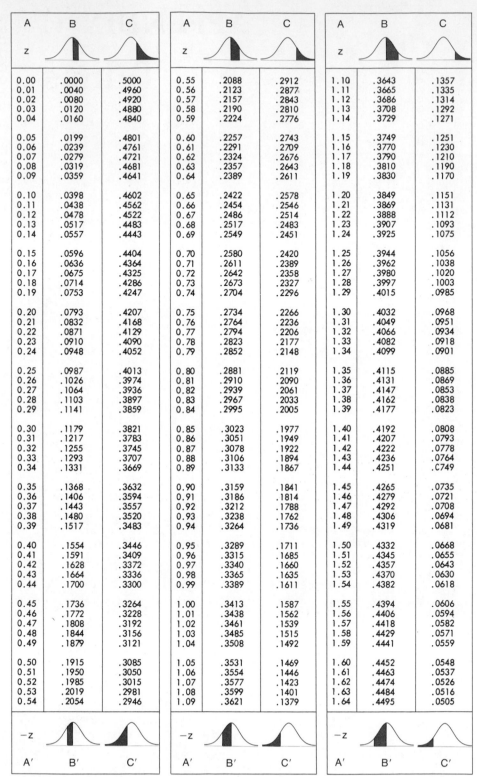

A z	B	C	A z	B	C	A z	B	C
0.00	.0000	.5000	0.55	.2088	.2912	1.10	.3643	.1357
0.01	.0040	.4960	0.56	.2123	.2877	1.11	.3665	.1335
0.02	.0080	.4920	0.57	.2157	.2843	1.12	.3686	.1314
0.03	.0120	.4880	0.58	.2190	.2810	1.13	.3708	.1292
0.04	.0160	.4840	0.59	.2224	.2776	1.14	.3729	.1271
0.05	.0199	.4801	0.60	.2257	.2743	1.15	.3749	.1251
0.06	.0239	.4761	0.61	.2291	.2709	1.16	.3770	.1230
0.07	.0279	.4721	0.62	.2324	.2676	1.17	.3790	.1210
0.08	.0319	.4681	0.63	.2357	.2643	1.18	.3810	.1190
0.09	.0359	.4641	0.64	.2389	.2611	1.19	.3830	.1170
0.10	.0398	.4602	0.65	.2422	.2578	1.20	.3849	.1151
0.11	.0438	.4562	0.66	.2454	.2546	1.21	.3869	.1131
0.12	.0478	.4522	0.67	.2486	.2514	1.22	.3888	.1112
0.13	.0517	.4483	0.68	.2517	.2483	1.23	.3907	.1093
0.14	.0557	.4443	0.69	.2549	.2451	1.24	.3925	.1075
0.15	.0596	.4404	0.70	.2580	.2420	1.25	.3944	.1056
0.16	.0636	.4364	0.71	.2611	.2389	1.26	.3962	.1038
0.17	.0675	.4325	0.72	.2642	.2358	1.27	.3980	.1020
0.18	.0714	.4286	0.73	.2673	.2327	1.28	.3997	.1003
0.19	.0753	.4247	0.74	.2704	.2296	1.29	.4015	.0985
0.20	.0793	.4207	0.75	.2734	.2266	1.30	.4032	.0968
0.21	.0832	.4168	0.76	.2764	.2236	1.31	.4049	.0951
0.22	.0871	.4129	0.77	.2794	.2206	1.32	.4066	.0934
0.23	.0910	.4090	0.78	.2823	.2177	1.33	.4082	.0918
0.24	.0948	.4052	0.79	.2852	.2148	1.34	.4099	.0901
0.25	.0987	.4013	0.80	.2881	.2119	1.35	.4115	.0885
0.26	.1026	.3974	0.81	.2910	.2090	1.36	.4131	.0869
0.27	.1064	.3936	0.82	.2939	.2061	1.37	.4147	.0853
0.28	.1103	.3897	0.83	.2967	.2033	1.38	.4162	.0838
0.29	.1141	.3859	0.84	.2995	.2005	1.39	.4177	.0823
0.30	.1179	.3821	0.85	.3023	.1977	1.40	.4192	.0808
0.31	.1217	.3783	0.86	.3051	.1949	1.41	.4207	.0793
0.32	.1255	.3745	0.87	.3078	.1922	1.42	.4222	.0778
0.33	.1293	.3707	0.88	.3106	.1894	1.43	.4236	.0764
0.34	.1331	.3669	0.89	.3133	.1867	1.44	.4251	.0749
0.35	.1368	.3632	0.90	.3159	.1841	1.45	.4265	.0735
0.36	.1406	.3594	0.91	.3186	.1814	1.46	.4279	.0721
0.37	.1443	.3557	0.92	.3212	.1788	1.47	.4292	.0708
0.38	.1480	.3520	0.93	.3238	.1762	1.48	.4306	.0694
0.39	.1517	.3483	0.94	.3264	.1736	1.49	.4319	.0681
0.40	.1554	.3446	0.95	.3289	.1711	1.50	.4332	.0668
0.41	.1591	.3409	0.96	.3315	.1685	1.51	.4345	.0655
0.42	.1628	.3372	0.97	.3340	.1660	1.52	.4357	.0643
0.43	.1664	.3336	0.98	.3365	.1635	1.53	.4370	.0630
0.44	.1700	.3300	0.99	.3389	.1611	1.54	.4382	.0618
0.45	.1736	.3264	1.00	.3413	.1587	1.55	.4394	.0606
0.46	.1772	.3228	1.01	.3438	.1562	1.56	.4406	.0594
0.47	.1808	.3192	1.02	.3461	.1539	1.57	.4418	.0582
0.48	.1844	.3156	1.03	.3485	.1515	1.58	.4429	.0571
0.49	.1879	.3121	1.04	.3508	.1492	1.59	.4441	.0559
0.50	.1915	.3085	1.05	.3531	.1469	1.60	.4452	.0548
0.51	.1950	.3050	1.06	.3554	.1446	1.61	.4463	.0537
0.52	.1985	.3015	1.07	.3577	.1423	1.62	.4474	.0526
0.53	.2019	.2981	1.08	.3599	.1401	1.63	.4484	.0516
0.54	.2054	.2946	1.09	.3621	.1379	1.64	.4495	.0505
−z			−z			−z		
A'	B'	C'	A'	B'	C'	A'	B'	C'

Table A is taken from Table IIi of R.A. Fisher and F. Yates, *Statistical Tables for Biological, Agricultural and Medical Research,* published by Longman Group Ltd., London (previously published by Oliver and Boyd, Edinburgh), and by permission of the authors and publishers.

[a]Discussed in Section 6.4.

TABLE A (*Continued*)
AREAS UNDER STANDARD NORMAL CURVE FOR VALUES OF Z

A z	B	C
1.65	.4505	.0495
1.66	.4515	.0485
1.67	.4525	.0475
1.68	.4535	.0465
1.69	.4545	.0455
1.70	.4554	.0446
1.71	.4564	.0436
1.72	.4573	.0427
1.73	.4582	.0418
1.74	.4591	.0409
1.75	.4599	.0401
1.76	.4608	.0392
1.77	.4616	.0384
1.78	.4625	.0375
1.79	.4633	.0367
1.80	.4641	.0359
1.81	.4649	.0351
1.82	.4656	.0344
1.83	.4664	.0336
1.84	.4671	.0329
1.85	.4678	.0322
1.86	.4686	.0314
1.87	.4693	.0307
1.88	.4699	.0301
1.89	.4706	.0294
1.90	.4713	.0287
1.91	.4719	.0281
1.92	.4726	.0274
1.93	.4732	.0268
1.94	.4738	.0262
1.95	.4744	.0256
1.96	.4750	.0250
1.97	.4756	.0244
1.98	.4761	.0239
1.99	.4767	.0233
2.00	.4772	.0228
2.01	.4778	.0222
2.02	.4783	.0217
2.03	.4788	.0212
2.04	.4793	.0207
2.05	.4798	.0202
2.06	.4803	.0197
2.07	.4808	.0192
2.08	.4812	.0188
2.09	.4817	.0183
2.10	.4821	.0179
2.11	.4826	.0174
2.12	.4830	.0170
2.13	.4834	.0166
2.14	.4838	.0162
2.15	.4842	.0158
2.16	.4846	.0154
2.17	.4850	.0150
2.18	.4854	.0146
2.19	.4857	.0143
2.20	.4861	.0139
2.21	.4864	.0136

A z	B	C
2.22	.4868	.0132
2.23	.4871	.0129
2.24	.4875	.0125
2.25	.4878	.0122
2.26	.4881	.0119
2.27	.4884	.0116
2.28	.4887	.0113
2.29	.4890	.0110
2.30	.4893	.0107
2.31	.4896	.0104
2.32	.4898	.0102
2.33	.4901	.0099
2.34	.4904	.0096
2.35	.4906	.0094
2.36	.4909	.0091
2.37	.4911	.0089
2.38	.4913	.0087
2.39	.4916	.0084
2.40	.4918	.0082
2.41	.4920	.0080
2.42	.4922	.0078
2.43	.4925	.0075
2.44	.4927	.0073
2.45	.4929	.0071
2.46	.4931	.0069
2.47	.4932	.0068
2.48	.4934	.0066
2.49	.4936	.0064
2.50	.4938	.0062
2.51	.4940	.0060
2.52	.4941	.0059
2.53	.4943	.0057
2.54	.4945	.0055
2.55	.4946	.0054
2.56	.4948	.0052
2.57	.4949	.0051
2.58	.4951	.0049
2.59	.4952	.0048
2.60	.4953	.0047
2.61	.4955	.0045
2.62	.4956	.0044
2.63	.4957	.0043
2.64	.4959	.0041
2.65	.4960	.0040
2.66	.4961	.0039
2.67	.4962	.0038
2.68	.4963	.0037
2.69	.4964	.0036
2.70	.4965	.0035
2.71	.4966	.0034
2.72	.4967	.0033
2.73	.4968	.0032
2.74	.4969	.0031
2.75	.4970	.0030
2.76	.4971	.0029
2.77	.4972	.0028
2.78	.4973	.0027

A z	B	C
2.79	.4974	.0026
2.80	.4974	.0026
2.81	.4975	.0025
2.82	.4976	.0024
2.83	.4977	.0023
2.84	.4977	.0023
2.85	.4978	.0022
2.86	.4979	.0021
2.87	.4979	.0021
2.88	.4980	.0020
2.89	.4981	.0019
2.90	.4981	.0019
2.91	.4982	.0018
2.92	.4982	.0018
2.93	.4983	.0017
2.94	.4984	.0016
2.95	.4984	.0016
2.96	.4985	.0015
2.97	.4985	.0015
2.98	.4986	.0014
2.99	.4986	.0014
3.00	.4987	.0013
3.01	.4987	.0013
3.02	.4987	.0013
3.03	.4988	.0012
3.04	.4988	.0012
3.05	.4989	.0011
3.06	.4989	.0011
3.07	.4989	.0011
3.08	.4990	.0010
3.09	.4990	.0010
3.10	.4990	.0010
3.11	.4991	.0009
3.12	.4991	.0009
3.13	.4991	.0009
3.14	.4992	.0008
3.15	.4992	.0008
3.16	.4992	.0008
3.17	.4992	.0008
3.18	.4993	.0007
3.19	.4993	.0007
3.20	.4993	.0007
3.21	.4993	.0007
3.22	.4994	.0006
3.23	.4994	.0006
3.24	.4994	.0006
3.25	.4994	.0006
3.30	.4995	.0005
3.35	.4996	.0004
3.40	.4997	.0003
3.45	.4997	.0003
3.50	.4998	.0002
3.60	.4998	.0002
3.70	.4999	.0001
3.80	.4999	.0001
3.90	.49995	.00005
4.00	.49997	.00003

−z		
A′	B′	C′

TABLE B.
CRITICAL VALUES OF t[a]

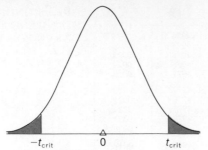

Two-tailed or Nondirectional Test

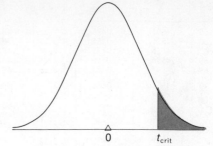

One-tailed or Directional Test

LEVEL OF SIGNIFICANCE

df	.05*	.01**	.001	df	.05	.01	.001
1	12.706	63.657	636.62	1	6.314	31.821	318.31
2	4.303	9.925	31.598	2	2.920	6.965	22.326
3	3.182	5.841	12.924	3	2.353	4.541	10.213
4	2.776	4.604	8.610	4	2.132	3.747	7.173
5	2.571	4.032	6.869	5	2.015	3.365	5.893
6	2.447	3.707	5.959	6	1.943	3.143	5.208
7	2.365	3.499	5.408	7	1.895	2.998	4.785
8	2.306	3.355	5.041	8	1.860	2.896	4.501
9	2.262	3.250	4.781	9	1.833	2.821	4.297
10	2.228	3.169	4.587	10	1.812	2.764	4.144
11	2.201	3.106	4.437	11	1.796	2.718	4.025
12	2.179	3.055	4.318	12	1.782	2.681	3.930
13	2.160	3.012	4.221	13	1.771	2.650	3.852
14	2.145	2.977	4.140	14	1.761	2.624	3.787
15	2.131	2.947	4.073	15	1.753	2.602	3.733
16	2.120	2.921	4.015	16	1.746	2.583	3.686
17	2.110	2.898	3.965	17	1.740	2.567	3.646
18	2.101	2.878	3.922	18	1.734	2.552	3.610
19	2.093	2.861	3.883	19	1.729	2.539	3.579
20	2.086	2.845	3.850	20	1.725	2.528	3.552
21	2.080	2.831	3.819	21	1.721	2.518	3.527
22	2.074	2.819	3.792	22	1.717	2.508	3.505
23	2.069	2.807	3.767	23	1.714	2.500	3.485
24	2.064	2.797	3.745	24	1.711	2.492	3.467
25	2.060	2.787	3.725	25	1.708	2.485	3.450
26	2.056	2.779	3.707	26	1.706	2.479	3.435
27	2.052	2.771	3.690	27	1.703	2.473	3.421
28	2.048	2.763	3.674	28	1.701	2.467	3.408
29	2.045	2.756	3.659	29	1.699	2.462	3.396
30	2.042	2.750	3.646	30	1.697	2.457	3.385
40	2.021	2.704	3.551	40	1.684	2.423	3.307
60	2.000	2.660	3.460	60	1.671	2.390	3.232
120	1.980	2.617	3.373	120	1.658	2.358	3.160
∞	1.960	2.576	3.291	∞	1.645	2.326	3.090

Table B is taken from Table 12 of E. Pearson and H. Hartley (Eds.), *Biometrika Tables for Statisticians,* Vol 1, 3rd ed., University Press, Cambridge, 1966, with permission of the Biometrika Trustees.

[a]Discussed in Section 13.5.

*95% level of confidence.

**99% level of confidence.

TABLE C
CRITICAL VALUES OF F[a]

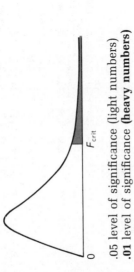

F_{crit}

.05 level of significance (light numbers)
.01 level of significance (**heavy numbers**)

DEGREES OF FREEDOM IN NUMERATOR

DEGREES OF FREEDOM IN DENOMINATOR	1	2	3	4	5	6	7	8	9	10	11	12	14	16	20	24	30	40	50	75	100	200	500	∞
1	161 **4,052**	200 **4,999**	216 **5,403**	225 **5,625**	230 **5,764**	234 **5,859**	237 **5,928**	239 **5,981**	241 **6,022**	242 **6,056**	243 **6,082**	244 **6,106**	245 **6,142**	246 **6,169**	248 **6,208**	249 **6,234**	250 **6,258**	251 **6,286**	252 **6,302**	253 **6,323**	253 **6,334**	254 **6,352**	254 **6,361**	254 **6,366**
2	18.51 **98.49**	19.00 **99.00**	19.16 **99.17**	19.25 **99.25**	19.30 **99.30**	19.33 **99.33**	19.36 **99.34**	19.37 **99.36**	19.38 **99.38**	19.39 **99.40**	19.40 **99.41**	19.41 **99.42**	19.42 **99.43**	19.43 **99.44**	19.44 **99.45**	19.45 **99.46**	19.46 **99.47**	19.47 **99.48**	19.47 **99.48**	19.48 **99.49**	19.49 **99.49**	19.49 **99.49**	19.50 **99.50**	19.50 **99.50**
3	10.13 **34.12**	9.55 **30.82**	9.28 **29.46**	9.12 **28.71**	9.01 **28.24**	8.94 **27.91**	8.88 **27.67**	8.84 **27.49**	8.81 **27.34**	8.78 **27.23**	8.76 **27.13**	8.74 **27.05**	8.71 **26.92**	8.69 **26.83**	8.66 **26.69**	8.64 **26.60**	8.62 **26.50**	8.60 **26.41**	8.58 **26.35**	8.57 **26.27**	8.56 **26.23**	8.54 **26.18**	8.54 **26.14**	8.53 **26.12**
4	7.71 **21.20**	6.94 **18.00**	6.59 **16.69**	6.39 **15.98**	6.26 **15.52**	6.16 **15.21**	6.09 **14.98**	6.04 **14.80**	6.00 **14.66**	5.96 **14.54**	5.93 **14.45**	5.91 **14.37**	5.87 **14.24**	5.84 **14.15**	5.80 **14.02**	5.77 **13.93**	5.74 **13.83**	5.71 **13.74**	5.70 **13.69**	5.68 **13.61**	5.66 **13.57**	5.65 **13.52**	5.64 **13.48**	5.63 **13.46**
5	6.61 **16.26**	5.79 **13.27**	5.41 **12.06**	5.19 **11.39**	5.05 **10.97**	4.95 **10.67**	4.88 **10.45**	4.82 **10.27**	4.78 **10.15**	4.74 **10.05**	4.70 **9.96**	4.68 **9.89**	4.64 **9.77**	4.60 **9.68**	4.56 **9.55**	4.53 **9.47**	4.50 **9.38**	4.46 **9.29**	4.44 **9.24**	4.42 **9.17**	4.40 **9.13**	4.38 **9.07**	4.37 **9.04**	4.36 **9.02**
6	5.99 **13.74**	5.14 **10.92**	4.76 **9.78**	4.53 **9.15**	4.39 **8.75**	4.28 **8.47**	4.21 **8.26**	4.15 **8.10**	4.10 **7.98**	4.06 **7.87**	4.03 **7.79**	4.00 **7.72**	3.96 **7.60**	3.92 **7.52**	3.87 **7.39**	3.84 **7.31**	3.81 **7.23**	3.77 **7.14**	3.75 **7.09**	3.72 **7.02**	3.71 **6.99**	3.69 **6.94**	3.68 **6.90**	3.67 **6.88**
7	5.59 **12.25**	4.74 **9.55**	4.35 **8.45**	4.12 **7.85**	3.97 **7.46**	3.87 **7.19**	3.79 **7.00**	3.73 **6.84**	3.68 **6.71**	3.63 **6.62**	3.60 **6.54**	3.57 **6.47**	3.52 **6.35**	3.49 **6.27**	3.44 **6.15**	3.41 **6.07**	3.38 **5.98**	3.34 **5.90**	3.32 **5.85**	3.29 **5.78**	3.28 **5.75**	3.25 **5.70**	3.24 **5.67**	3.23 **5.65**
8	5.32 **11.26**	4.46 **8.65**	4.07 **7.59**	3.84 **7.01**	3.69 **6.63**	3.58 **6.37**	3.50 **6.19**	3.44 **6.03**	3.39 **5.91**	3.34 **5.82**	3.31 **5.74**	3.28 **5.67**	3.23 **5.56**	3.20 **5.48**	3.15 **5.36**	3.12 **5.28**	3.08 **5.20**	3.05 **5.11**	3.03 **5.06**	3.00 **5.00**	2.98 **4.96**	2.96 **4.91**	2.94 **4.88**	2.93 **4.86**
9	5.12 **10.56**	4.26 **8.02**	3.86 **6.99**	3.63 **6.42**	3.48 **6.06**	3.37 **5.80**	3.29 **5.62**	3.23 **5.47**	3.18 **5.35**	3.13 **5.26**	3.10 **5.18**	3.07 **5.11**	3.02 **5.00**	2.98 **4.92**	2.93 **4.80**	2.90 **4.73**	2.86 **4.64**	2.82 **4.56**	2.80 **4.51**	2.77 **4.45**	2.76 **4.41**	2.73 **4.36**	2.72 **4.33**	2.71 **4.31**

Reprinted by permission from George W. Snedecor and William G. Cochran, *Statistical Methods*, 6th ed. Copyright © 1967 by The Iowa State University Press, Ames, Iowa.
[a]Discussed in Section 17.9.

Table continued on following page

TABLE C (*Continued*)
CRITICAL VALUES OF F

DEGREES OF FREEDOM IN NUMERATOR

DEGREES OF FREEDOM IN DENOMINATOR	1	2	3	4	5	6	7	8	9	10	11	12	14	16	20	24	30	40	50	75	100	200	500	∞
10	4.96 / 10.04	4.10 / 7.56	3.71 / 6.55	3.48 / 5.99	3.33 / 5.64	3.22 / 5.39	3.14 / 5.21	3.07 / 5.06	3.02 / 4.95	2.97 / 4.85	2.94 / 4.78	2.91 / 4.71	2.86 / 4.60	2.82 / 4.52	2.77 / 4.41	2.74 / 4.33	2.70 / 4.25	2.67 / 4.17	2.64 / 4.12	2.61 / 4.05	2.59 / 4.01	2.56 / 3.96	2.55 / 3.93	2.54 / 3.91
11	4.84 / 9.65	3.98 / 7.20	3.59 / 6.22	3.36 / 5.67	3.20 / 5.32	3.09 / 5.07	3.01 / 4.88	2.95 / 4.74	2.90 / 4.63	2.86 / 4.54	2.82 / 4.46	2.79 / 4.40	2.74 / 4.29	2.70 / 4.21	2.65 / 4.10	2.61 / 4.02	2.57 / 3.94	2.53 / 3.86	2.50 / 3.80	2.47 / 3.74	2.45 / 3.70	2.42 / 3.66	2.41 / 3.62	2.40 / 3.60
12	4.75 / 9.33	3.88 / 6.93	3.49 / 5.95	3.26 / 5.41	3.11 / 5.06	3.00 / 4.82	2.92 / 4.65	2.85 / 4.50	2.80 / 4.39	2.76 / 4.30	2.72 / 4.22	2.69 / 4.16	2.64 / 4.05	2.60 / 3.98	2.54 / 3.86	2.50 / 3.78	2.46 / 3.70	2.42 / 3.61	2.40 / 3.56	2.36 / 3.49	2.35 / 3.46	2.32 / 3.41	2.31 / 3.38	2.30 / 3.36
13	4.67 / 9.07	3.80 / 6.70	3.41 / 5.74	3.18 / 5.20	3.02 / 4.86	2.92 / 4.62	2.84 / 4.44	2.77 / 4.30	2.72 / 4.19	2.67 / 4.10	2.63 / 4.02	2.60 / 3.96	2.55 / 3.85	2.51 / 3.78	2.46 / 3.67	2.42 / 3.59	2.38 / 3.51	2.34 / 3.42	2.32 / 3.37	2.28 / 3.30	2.26 / 3.27	2.24 / 3.21	2.22 / 3.18	2.21 / 3.16
14	4.60 / 8.86	3.74 / 6.51	3.34 / 5.56	3.11 / 5.03	2.96 / 4.69	2.85 / 4.46	2.77 / 4.28	2.70 / 4.14	2.65 / 4.03	2.60 / 3.94	2.56 / 3.86	2.53 / 3.80	2.48 / 3.70	2.44 / 3.62	2.39 / 3.51	2.35 / 3.43	2.31 / 3.34	2.27 / 3.26	2.24 / 3.21	2.21 / 3.14	2.19 / 3.11	2.16 / 3.06	2.14 / 3.02	2.13 / 3.00
15	4.54 / 8.68	3.68 / 6.36	3.29 / 5.42	3.06 / 4.89	2.90 / 4.56	2.79 / 4.32	2.70 / 4.14	2.64 / 4.00	2.59 / 3.89	2.55 / 3.80	2.51 / 3.73	2.48 / 3.67	2.43 / 3.56	2.39 / 3.48	2.33 / 3.36	2.29 / 3.29	2.25 / 3.20	2.21 / 3.12	2.18 / 3.07	2.15 / 3.00	2.12 / 2.97	2.10 / 2.92	2.08 / 2.89	2.07 / 2.87
16	4.49 / 8.53	3.63 / 6.23	3.24 / 5.29	3.01 / 4.77	2.85 / 4.44	2.74 / 4.20	2.66 / 4.03	2.59 / 3.89	2.54 / 3.78	2.49 / 3.69	2.45 / 3.61	2.42 / 3.55	2.37 / 3.45	2.33 / 3.37	2.28 / 3.25	2.24 / 3.18	2.20 / 3.10	2.16 / 3.01	2.13 / 2.96	2.09 / 2.89	2.07 / 2.86	2.04 / 2.80	2.02 / 2.77	2.01 / 2.75
17	4.45 / 8.40	3.59 / 6.11	3.20 / 5.18	2.96 / 4.67	2.81 / 4.34	2.70 / 4.10	2.62 / 3.93	2.55 / 3.79	2.50 / 3.68	2.45 / 3.59	2.41 / 3.52	2.38 / 3.45	2.33 / 3.35	2.29 / 3.27	2.23 / 3.16	2.19 / 3.08	2.15 / 3.00	2.11 / 2.92	2.08 / 2.86	2.04 / 2.79	2.02 / 2.76	1.99 / 2.70	1.97 / 2.67	1.96 / 2.65
18	4.41 / 8.28	3.55 / 6.01	3.16 / 5.09	2.93 / 4.58	2.77 / 4.25	2.66 / 4.01	2.58 / 3.85	2.51 / 3.71	2.46 / 3.60	2.41 / 3.51	2.37 / 3.44	2.34 / 3.37	2.29 / 3.27	2.25 / 3.19	2.19 / 3.07	2.15 / 3.00	2.11 / 2.91	2.07 / 2.83	2.04 / 2.78	2.00 / 2.71	1.98 / 2.68	1.95 / 2.62	1.93 / 2.59	1.92 / 2.57
19	4.38 / 8.18	3.52 / 5.93	3.13 / 5.01	2.90 / 4.50	2.74 / 4.17	2.63 / 3.94	2.55 / 3.77	2.48 / 3.63	2.43 / 3.52	2.38 / 3.43	2.34 / 3.36	2.31 / 3.30	2.26 / 3.19	2.21 / 3.12	2.15 / 3.00	2.11 / 2.92	2.07 / 2.84	2.02 / 2.76	2.00 / 2.70	1.96 / 2.63	1.94 / 2.60	1.91 / 2.54	1.90 / 2.51	1.88 / 2.49
20	4.35 / 8.10	3.49 / 5.85	3.10 / 4.94	2.87 / 4.43	2.71 / 4.10	2.60 / 3.87	2.52 / 3.71	2.45 / 3.56	2.40 / 3.45	2.35 / 3.37	2.31 / 3.30	2.28 / 3.23	2.23 / 3.13	2.18 / 3.05	2.12 / 2.94	2.08 / 2.86	2.04 / 2.77	1.99 / 2.69	1.96 / 2.63	1.92 / 2.56	1.90 / 2.53	1.87 / 2.47	1.85 / 2.44	1.84 / 2.42
21	4.32 / 8.02	3.47 / 5.78	3.07 / 4.87	2.84 / 4.37	2.68 / 4.04	2.57 / 3.81	2.49 / 3.65	2.42 / 3.51	2.37 / 3.40	2.32 / 3.31	2.28 / 3.24	2.25 / 3.17	2.20 / 3.07	2.15 / 2.99	2.09 / 2.88	2.05 / 2.80	2.00 / 2.72	1.96 / 2.63	1.93 / 2.58	1.89 / 2.51	1.87 / 2.47	1.84 / 2.42	1.82 / 2.38	1.81 / 2.36
22	4.30 / 7.94	3.44 / 5.72	3.05 / 4.82	2.82 / 4.31	2.66 / 3.99	2.55 / 3.76	2.47 / 3.59	2.40 / 3.45	2.35 / 3.35	2.30 / 3.26	2.26 / 3.18	2.23 / 3.12	2.18 / 3.02	2.13 / 2.94	2.07 / 2.83	2.03 / 2.75	1.98 / 2.67	1.93 / 2.58	1.91 / 2.53	1.87 / 2.46	1.84 / 2.42	1.81 / 2.37	1.80 / 2.33	1.78 / 2.31
23	4.28 / 7.88	3.42 / 5.66	3.03 / 4.76	2.80 / 4.26	2.64 / 3.94	2.53 / 3.71	2.45 / 3.54	2.38 / 3.41	2.32 / 3.30	2.28 / 3.21	2.24 / 3.14	2.20 / 3.07	2.14 / 2.97	2.10 / 2.89	2.04 / 2.78	2.00 / 2.70	1.96 / 2.62	1.91 / 2.53	1.88 / 2.48	1.84 / 2.41	1.82 / 2.37	1.79 / 2.32	1.77 / 2.28	1.76 / 2.26

DEGREES OF FREEDOM IN DENOMINATOR

	1	2	3	4	5	6	7	8	9	10	11	12	14	16	20	24	30	40	50	75	100	200	500	∞
24	4.26 / 7.82	3.40 / 5.61	3.01 / 4.72	2.78 / 4.22	2.62 / 3.90	2.51 / 3.67	2.43 / 3.50	2.36 / 3.36	2.30 / 3.25	2.26 / 3.17	2.22 / 3.09	2.18 / 3.03	2.13 / 2.93	2.09 / 2.85	2.02 / 2.74	1.98 / 2.66	1.94 / 2.58	1.89 / 2.49	1.86 / 2.44	1.82 / 2.36	1.80 / 2.33	1.76 / 2.27	1.74 / 2.23	1.73 / 2.21
25	4.24 / 7.77	3.38 / 5.57	2.99 / 4.68	2.76 / 4.18	2.60 / 3.86	2.49 / 3.63	2.41 / 3.46	2.34 / 3.32	2.28 / 3.21	2.24 / 3.13	2.20 / 3.05	2.16 / 2.99	2.11 / 2.89	2.06 / 2.81	2.00 / 2.70	1.96 / 2.62	1.92 / 2.54	1.87 / 2.45	1.84 / 2.40	1.80 / 2.32	1.77 / 2.29	1.74 / 2.23	1.72 / 2.19	1.71 / 2.17
26	4.22 / 7.72	3.37 / 5.53	2.98 / 4.64	2.74 / 4.14	2.59 / 3.82	2.47 / 3.59	2.39 / 3.42	2.32 / 3.29	2.27 / 3.17	2.22 / 3.09	2.18 / 3.02	2.15 / 2.96	2.10 / 2.86	2.05 / 2.77	1.99 / 2.66	1.95 / 2.58	1.90 / 2.50	1.85 / 2.41	1.82 / 2.36	1.78 / 2.28	1.76 / 2.25	1.72 / 2.19	1.70 / 2.15	1.69 / 2.13
27	4.21 / 7.68	3.35 / 5.49	2.96 / 4.60	2.73 / 4.11	2.57 / 3.79	2.46 / 3.56	2.37 / 3.39	2.30 / 3.26	2.25 / 3.14	2.20 / 3.06	2.16 / 2.98	2.13 / 2.93	2.08 / 2.83	2.03 / 2.74	1.97 / 2.63	1.93 / 2.55	1.88 / 2.47	1.84 / 2.38	1.80 / 2.33	1.76 / 2.25	1.74 / 2.21	1.71 / 2.16	1.68 / 2.12	1.67 / 2.10
28	4.20 / 7.64	3.34 / 5.45	2.95 / 4.57	2.71 / 4.07	2.56 / 3.76	2.44 / 3.53	2.36 / 3.36	2.29 / 3.23	2.24 / 3.11	2.19 / 3.03	2.15 / 2.95	2.12 / 2.90	2.06 / 2.80	2.02 / 2.71	1.96 / 2.60	1.91 / 2.52	1.87 / 2.44	1.81 / 2.35	1.78 / 2.30	1.75 / 2.22	1.72 / 2.18	1.69 / 2.13	1.67 / 2.09	1.65 / 2.06
29	4.18 / 7.60	3.33 / 5.42	2.93 / 4.54	2.70 / 4.04	2.54 / 3.73	2.43 / 3.50	2.35 / 3.33	2.28 / 3.20	2.22 / 3.08	2.18 / 3.00	2.14 / 2.92	2.10 / 2.87	2.05 / 2.77	2.00 / 2.68	1.94 / 2.57	1.90 / 2.49	1.85 / 2.41	1.80 / 2.32	1.77 / 2.27	1.73 / 2.19	1.71 / 2.15	1.68 / 2.10	1.65 / 2.06	1.64 / 2.03
30	4.17 / 7.56	3.32 / 5.39	2.92 / 4.51	2.69 / 4.02	2.53 / 3.70	2.42 / 3.47	2.34 / 3.30	2.27 / 3.17	2.21 / 3.06	2.16 / 2.98	2.12 / 2.90	2.09 / 2.84	2.04 / 2.74	1.99 / 2.66	1.93 / 2.55	1.89 / 2.47	1.84 / 2.38	1.79 / 2.29	1.76 / 2.24	1.72 / 2.16	1.69 / 2.13	1.66 / 2.07	1.64 / 2.03	1.62 / 2.01
32	4.15 / 7.50	3.30 / 5.34	2.90 / 4.46	2.67 / 3.97	2.51 / 3.66	2.40 / 3.42	2.32 / 3.25	2.25 / 3.12	2.19 / 3.01	2.14 / 2.94	2.10 / 2.86	2.07 / 2.80	2.02 / 2.70	1.97 / 2.62	1.91 / 2.51	1.86 / 2.42	1.82 / 2.34	1.76 / 2.25	1.74 / 2.20	1.69 / 2.12	1.67 / 2.08	1.64 / 2.02	1.61 / 1.98	1.59 / 1.96
34	4.13 / 7.44	3.28 / 5.29	2.88 / 4.42	2.65 / 3.93	2.49 / 3.61	2.38 / 3.38	2.30 / 3.21	2.23 / 3.08	2.17 / 2.97	2.12 / 2.89	2.08 / 2.82	2.05 / 2.76	2.00 / 2.66	1.95 / 2.58	1.89 / 2.47	1.84 / 2.38	1.80 / 2.30	1.74 / 2.21	1.71 / 2.15	1.67 / 2.08	1.64 / 2.04	1.61 / 1.98	1.59 / 1.94	1.57 / 1.91
36	4.11 / 7.39	3.26 / 5.25	2.86 / 4.38	2.63 / 3.89	2.48 / 3.58	2.36 / 3.35	2.28 / 3.18	2.21 / 3.04	2.15 / 2.94	2.10 / 2.86	2.06 / 2.78	2.03 / 2.72	1.98 / 2.62	1.93 / 2.54	1.87 / 2.43	1.82 / 2.35	1.78 / 2.26	1.72 / 2.17	1.69 / 2.12	1.65 / 2.04	1.62 / 2.00	1.59 / 1.94	1.56 / 1.90	1.55 / 1.87
38	4.10 / 7.35	3.25 / 5.21	2.85 / 4.34	2.62 / 3.86	2.46 / 3.54	2.35 / 3.32	2.26 / 3.15	2.19 / 3.02	2.14 / 2.91	2.09 / 2.82	2.05 / 2.75	2.02 / 2.69	1.96 / 2.59	1.92 / 2.51	1.85 / 2.40	1.80 / 2.32	1.76 / 2.22	1.71 / 2.14	1.67 / 2.08	1.63 / 2.00	1.60 / 1.97	1.57 / 1.90	1.54 / 1.86	1.53 / 1.84
40	4.08 / 7.31	3.23 / 5.18	2.84 / 4.31	2.61 / 3.83	2.45 / 3.51	2.34 / 3.29	2.25 / 3.12	2.18 / 2.99	2.12 / 2.88	2.07 / 2.80	2.04 / 2.73	2.00 / 2.66	1.95 / 2.56	1.90 / 2.49	1.84 / 2.37	1.79 / 2.29	1.74 / 2.20	1.69 / 2.11	1.66 / 2.05	1.61 / 1.97	1.59 / 1.94	1.55 / 1.88	1.53 / 1.84	1.51 / 1.81
42	4.07 / 7.27	3.22 / 5.15	2.83 / 4.29	2.59 / 3.80	2.44 / 3.49	2.32 / 3.26	2.24 / 3.10	2.17 / 2.96	2.11 / 2.86	2.06 / 2.77	2.02 / 2.70	1.99 / 2.64	1.94 / 2.54	1.89 / 2.46	1.82 / 2.35	1.78 / 2.26	1.73 / 2.17	1.68 / 2.08	1.64 / 2.02	1.60 / 1.94	1.57 / 1.91	1.54 / 1.85	1.51 / 1.80	1.49 / 1.78
44	4.06 / 7.24	3.21 / 5.12	2.82 / 4.26	2.58 / 3.78	2.43 / 3.46	2.31 / 3.24	2.23 / 3.07	2.16 / 2.94	2.10 / 2.84	2.05 / 2.75	2.01 / 2.68	1.98 / 2.62	1.92 / 2.52	1.88 / 2.44	1.81 / 2.32	1.76 / 2.24	1.72 / 2.15	1.66 / 2.06	1.63 / 2.00	1.58 / 1.92	1.56 / 1.88	1.52 / 1.82	1.50 / 1.78	1.48 / 1.75
46	4.05 / 7.21	3.20 / 5.10	2.81 / 4.24	2.57 / 3.76	2.42 / 3.44	2.30 / 3.22	2.22 / 3.05	2.14 / 2.92	2.09 / 2.82	2.04 / 2.73	2.00 / 2.66	1.97 / 2.60	1.91 / 2.50	1.87 / 2.42	1.80 / 2.30	1.75 / 2.22	1.71 / 2.13	1.65 / 2.04	1.62 / 1.98	1.57 / 1.90	1.54 / 1.86	1.51 / 1.80	1.48 / 1.76	1.46 / 1.72
48	4.04 / 7.19	3.19 / 5.08	2.80 / 4.22	2.56 / 3.74	2.41 / 3.42	2.30 / 3.20	2.21 / 3.04	2.14 / 2.90	2.08 / 2.80	2.03 / 2.71	1.99 / 2.64	1.96 / 2.58	1.90 / 2.48	1.86 / 2.40	1.79 / 2.28	1.74 / 2.20	1.70 / 2.11	1.64 / 2.02	1.61 / 1.96	1.56 / 1.88	1.53 / 1.84	1.50 / 1.78	1.47 / 1.73	1.45 / 1.70

Table continued on following page

TABLE C (*Continued*)
CRITICAL VALUES OF F

DEGREES OF FREEDOM IN NUMERATOR

DEGREES OF FREEDOM IN DENOMINATOR	1	2	3	4	5	6	7	8	9	10	11	12	14	16	20	24	30	40	50	75	100	200	500	∞
50	4.03 **7.17**	3.18 **5.06**	2.79 **4.20**	2.56 **3.72**	2.40 **3.41**	2.29 **3.18**	2.20 **3.02**	2.13 **2.88**	2.07 **2.78**	2.02 **2.70**	1.98 **2.62**	1.95 **2.56**	1.90 **2.46**	1.85 **2.39**	1.78 **2.26**	1.74 **2.18**	1.69 **2.10**	1.63 **2.00**	1.60 **1.94**	1.55 **1.86**	1.52 **1.82**	1.48 **1.76**	1.46 **1.71**	1.44 **1.68**
55	4.02 **7.12**	3.17 **5.01**	2.78 **4.16**	2.54 **3.68**	2.38 **3.37**	2.27 **3.15**	2.18 **2.98**	2.11 **2.85**	2.05 **2.75**	2.00 **2.66**	1.97 **2.59**	1.93 **2.53**	1.88 **2.43**	1.83 **2.35**	1.76 **2.23**	1.72 **2.15**	1.67 **2.06**	1.61 **1.96**	1.58 **1.90**	1.52 **1.82**	1.50 **1.78**	1.46 **1.71**	1.43 **1.66**	1.41 **1.64**
60	4.00 **7.08**	3.15 **4.98**	2.76 **4.13**	2.52 **3.65**	2.37 **3.34**	2.25 **3.12**	2.17 **2.95**	2.10 **2.82**	2.04 **2.72**	1.99 **2.63**	1.95 **2.56**	1.92 **2.50**	1.86 **2.40**	1.81 **2.32**	1.75 **2.20**	1.70 **2.12**	1.65 **2.03**	1.59 **1.93**	1.56 **1.87**	1.50 **1.79**	1.48 **1.74**	1.44 **1.68**	1.41 **1.63**	1.39 **1.60**
65	3.99 **7.04**	3.14 **4.95**	2.75 **4.10**	2.51 **3.62**	2.36 **3.31**	2.24 **3.09**	2.15 **2.93**	2.08 **2.79**	2.02 **2.70**	1.98 **2.61**	1.94 **2.54**	1.90 **2.47**	1.85 **2.37**	1.80 **2.30**	1.73 **2.18**	1.68 **2.09**	1.63 **2.00**	1.57 **1.90**	1.54 **1.84**	1.49 **1.76**	1.46 **1.71**	1.42 **1.64**	1.39 **1.60**	1.37 **1.56**
70	3.98 **7.01**	3.13 **4.92**	2.74 **4.08**	2.50 **3.60**	2.35 **3.29**	2.23 **3.07**	2.14 **2.91**	2.07 **2.77**	2.01 **2.67**	1.97 **2.59**	1.93 **2.51**	1.89 **2.45**	1.84 **2.35**	1.79 **2.28**	1.72 **2.15**	1.67 **2.07**	1.62 **1.98**	1.56 **1.88**	1.53 **1.82**	1.47 **1.74**	1.45 **1.69**	1.40 **1.62**	1.37 **1.56**	1.35 **1.53**
80	3.96 **6.96**	3.11 **4.88**	2.72 **4.04**	2.48 **3.56**	2.33 **3.25**	2.21 **3.04**	2.12 **2.87**	2.05 **2.74**	1.99 **2.64**	1.95 **2.55**	1.91 **2.48**	1.88 **2.41**	1.82 **2.32**	1.77 **2.24**	1.70 **2.11**	1.65 **2.03**	1.60 **1.94**	1.54 **1.84**	1.51 **1.78**	1.45 **1.70**	1.42 **1.65**	1.38 **1.57**	1.35 **1.52**	1.32 **1.49**
100	3.94 **6.90**	3.09 **4.82**	2.70 **3.98**	2.46 **3.51**	2.30 **3.20**	2.19 **2.99**	2.10 **2.82**	2.03 **2.69**	1.97 **2.59**	1.92 **2.51**	1.88 **2.43**	1.85 **2.36**	1.79 **2.26**	1.75 **2.19**	1.68 **2.06**	1.63 **1.98**	1.57 **1.89**	1.51 **1.79**	1.48 **1.73**	1.42 **1.64**	1.39 **1.59**	1.34 **1.51**	1.30 **1.46**	1.28 **1.43**
125	3.92 **6.84**	3.07 **4.78**	2.68 **3.94**	2.44 **3.47**	2.29 **3.17**	2.17 **2.95**	2.08 **2.79**	2.01 **2.65**	1.95 **2.56**	1.90 **2.47**	1.86 **2.40**	1.83 **2.33**	1.77 **2.23**	1.72 **2.15**	1.65 **2.03**	1.60 **1.94**	1.55 **1.85**	1.49 **1.75**	1.45 **1.68**	1.39 **1.59**	1.36 **1.54**	1.31 **1.46**	1.27 **1.40**	1.25 **1.37**
150	3.91 **6.81**	3.06 **4.75**	2.67 **3.91**	2.43 **3.44**	2.27 **3.14**	2.16 **2.92**	2.07 **2.76**	2.00 **2.62**	1.94 **2.53**	1.89 **2.44**	1.85 **2.37**	1.82 **2.30**	1.76 **2.20**	1.71 **2.12**	1.64 **2.00**	1.59 **1.91**	1.54 **1.83**	1.47 **1.72**	1.44 **1.66**	1.37 **1.56**	1.34 **1.51**	1.29 **1.43**	1.25 **1.37**	1.22 **1.33**
200	3.89 **6.76**	3.04 **4.71**	2.65 **3.88**	2.41 **3.41**	2.26 **3.11**	2.14 **2.90**	2.05 **2.73**	1.98 **2.60**	1.92 **2.50**	1.87 **2.41**	1.83 **2.34**	1.80 **2.28**	1.74 **2.17**	1.69 **2.09**	1.62 **1.97**	1.57 **1.88**	1.52 **1.79**	1.45 **1.69**	1.42 **1.62**	1.35 **1.53**	1.32 **1.48**	1.26 **1.39**	1.22 **1.33**	1.19 **1.28**
400	3.86 **6.70**	3.02 **4.66**	2.62 **3.83**	2.39 **3.36**	2.23 **3.06**	2.12 **2.85**	2.03 **2.69**	1.96 **2.55**	1.90 **2.46**	1.85 **2.37**	1.81 **2.29**	1.78 **2.23**	1.72 **2.12**	1.67 **2.04**	1.60 **1.92**	1.54 **1.84**	1.49 **1.74**	1.42 **1.64**	1.38 **1.57**	1.32 **1.47**	1.28 **1.42**	1.22 **1.32**	1.16 **1.24**	1.13 **1.19**
1000	3.85 **6.66**	3.00 **4.62**	2.61 **3.80**	2.38 **3.34**	2.22 **3.04**	2.10 **2.82**	2.02 **2.66**	1.95 **2.53**	1.89 **2.43**	1.84 **2.34**	1.80 **2.26**	1.76 **2.20**	1.70 **2.09**	1.65 **2.01**	1.58 **1.89**	1.53 **1.81**	1.47 **1.71**	1.41 **1.61**	1.36 **1.54**	1.30 **1.44**	1.26 **1.38**	1.19 **1.28**	1.13 **1.19**	1.08 **1.11**
∞	3.84 **6.64**	2.99 **4.60**	2.60 **3.78**	2.37 **3.32**	2.21 **3.02**	2.09 **2.80**	2.01 **2.64**	1.94 **2.51**	1.88 **2.41**	1.83 **2.32**	1.79 **2.24**	1.75 **2.18**	1.69 **2.07**	1.64 **1.99**	1.57 **1.87**	1.52 **1.79**	1.46 **1.69**	1.40 **1.59**	1.35 **1.52**	1.28 **1.41**	1.24 **1.36**	1.17 **1.25**	1.11 **1.15**	1.00 **1.00**

CRITICAL VALUES OF χ^{2a}

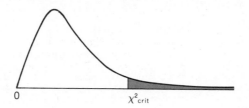

LEVEL OF SIGNIFICANCE

df	.10	.05	.01	.001
1	2.71	3.84	6.64	10.83
2	4.60	5.99	9.21	13.82
3	6.25	7.82	11.34	16.27
4	7.78	9.49	13.28	18.47
5	9.24	11.07	15.09	20.52
6	10.64	12.59	16.81	22.46
7	12.02	14.07	18.48	24.32
8	13.36	15.51	20.09	26.12
9	14.68	16.92	21.67	27.88
10	15.99	18.31	23.21	29.59
11	17.28	19.68	24.72	31.26
12	18.55	21.03	26.22	32.91
13	19.81	22.36	27.69	34.53
14	21.06	23.68	29.14	36.12
15	22.31	25.00	30.58	37.70
16	23.54	26.30	32.00	39.25
17	24.77	27.59	33.41	40.79
18	25.99	28.87	34.80	42.31
19	27.20	30.14	36.19	43.82
20	28.41	31.41	37.57	45.32
21	29.62	32.67	38.93	46.80
22	30.81	33.92	40.29	48.27
23	32.01	35.17	41.64	49.73
24	33.20	36.42	42.98	51.18
25	34.38	37.65	44.31	52.62
26	35.56	38.88	45.64	54.05
27	36.74	40.11	46.96	55.48
28	37.92	41.34	48.28	56.89
29	39.09	42.56	49.59	58.30
30	40.26	43.77	50.89	59.70
40	51.80	55.76	63.69	73.40
50	63.17	67.50	76.15	86.66
60	74.40	79.08	88.38	99.61
70	85.53	90.53	100.42	112.32

Table D is taken from Table IV of R.A. Fisher and F. Yates, *Statistical Tables for Biological, Agricultural and Medical Research,* published by Longman Group, Ltd., London (previously published by Oliver and Boyd, Edinburgh), and by permission of the authors and publishers.

[a]Discussed in Section 19.3.

TABLE E
CRITICAL VALUES OF MANN-WHITNEY U[a]

NONDIRECTIONAL TEST
.05 Level of Significance (light numbers)
.01 Level of Significance (heavy numbers)

n_2 \ n_1	1	2	3	4	5	6	7	8	9	10	11	12	13	14	15	16	17	18	19	20
1	—	—	—	—	—	—	—	—	—	—	—	—	—	—	—	—	—	—	—	—
	—	—	—	—	—	—	—	—	—	—	—	—	—	—	—	—	—	—	—	—
2	—	—	—	—	—	—	—	0	0	0	0	1	1	1	1	1	2	2	2	2
	—	—	—	—	—	—	—	—	—	—	—	—	—	—	—	—	—	—	**0**	**0**
3	—	—	—	—	0	1	1	2	2	3	3	4	4	5	5	6	6	7	7	8
	—	—	—	—	—	—	—	**0**	**0**	**0**	**1**	**1**	**1**	**2**	**2**	**2**	**2**	**2**	**3**	**3**
4	—	—	—	0	1	2	3	4	4	5	6	7	8	9	10	11	11	12	13	13
	—	—	—	—	—	**0**	**0**	**1**	**1**	**2**	**2**	**3**	**3**	**4**	**5**	**5**	**6**	**6**	**7**	**8**
5	—	—	0	1	2	3	5	6	7	8	9	11	12	13	14	15	17	18	19	20
	—	—	—	—	**0**	**1**	**1**	**2**	**3**	**4**	**5**	**6**	**7**	**7**	**8**	**9**	**10**	**11**	**12**	**13**
6	—	—	1	2	3	5	6	8	10	11	13	14	16	17	19	21	22	24	25	27
	—	—	—	**0**	**1**	**2**	**3**	**4**	**5**	**6**	**7**	**9**	**10**	**11**	**12**	**13**	**15**	**16**	**17**	**18**
7	—	—	1	3	5	6	8	10	12	14	16	18	20	22	24	26	28	30	32	34
	—	—	—	**0**	**1**	**3**	**4**	**6**	**7**	**9**	**10**	**12**	**13**	**15**	**16**	**18**	**19**	**21**	**22**	**24**
8	—	0	2	4	6	8	10	13	15	17	19	22	24	26	29	31	34	36	38	41
	—	—	**1**	**2**	**4**	**6**	**7**	**9**	**11**	**13**	**15**	**17**	**18**	**20**	**22**	**24**	**26**	**28**	**30**
9	—	0	2	4	7	10	12	15	17	20	23	26	28	31	34	37	39	42	45	48
	—	—	**0**	**1**	**3**	**5**	**7**	**9**	**11**	**13**	**16**	**18**	**20**	**22**	**24**	**27**	**29**	**31**	**33**	**36**
10	—	0	3	5	8	11	14	17	20	23	26	29	33	36	39	42	45	48	52	55
	—	—	**0**	**2**	**4**	**6**	**9**	**11**	**13**	**16**	**18**	**21**	**24**	**26**	**29**	**31**	**34**	**37**	**39**	**42**
11	—	0	3	6	9	13	16	19	23	26	30	33	37	40	44	47	51	55	58	62
	—	—	**0**	**2**	**5**	**7**	**10**	**13**	**16**	**18**	**21**	**24**	**27**	**30**	**33**	**36**	**39**	**42**	**45**	**48**
12	—	1	4	7	11	14	18	22	26	29	33	37	41	45	49	53	57	61	65	69
	—	—	**1**	**3**	**6**	**9**	**12**	**15**	**18**	**21**	**24**	**27**	**31**	**34**	**37**	**41**	**44**	**47**	**51**	**54**
13	—	1	4	8	12	16	20	24	28	33	37	41	45	50	54	59	63	67	72	76
	—	—	**1**	**3**	**7**	**10**	**13**	**17**	**20**	**24**	**27**	**31**	**34**	**38**	**42**	**45**	**49**	**53**	**56**	**60**
14	—	1	5	9	13	17	22	26	31	36	40	45	50	55	59	64	67	74	78	83
	—	—	**1**	**4**	**7**	**11**	**15**	**18**	**22**	**26**	**30**	**34**	**38**	**42**	**46**	**50**	**54**	**58**	**63**	**67**
15	—	1	5	10	14	19	24	29	34	39	44	49	54	59	64	70	75	80	85	90
	—	—	**2**	**5**	**8**	**12**	**16**	**20**	**24**	**29**	**33**	**37**	**42**	**46**	**51**	**55**	**60**	**64**	**69**	**73**
16	—	1	6	11	15	21	26	31	37	42	47	53	59	64	70	75	81	86	92	98
	—	—	**2**	**5**	**9**	**13**	**18**	**22**	**27**	**31**	**36**	**41**	**45**	**50**	**55**	**60**	**65**	**70**	**74**	**79**
17	—	2	6	11	17	22	28	34	39	45	51	57	63	67	75	81	87	93	99	105
	—	—	**2**	**6**	**10**	**15**	**19**	**24**	**29**	**34**	**39**	**44**	**49**	**54**	**60**	**65**	**70**	**75**	**81**	**86**
18	—	2	7	12	18	24	30	36	42	48	55	61	67	74	80	86	93	99	106	112
	—	—	**2**	**6**	**11**	**16**	**21**	**26**	**31**	**37**	**42**	**47**	**53**	**58**	**64**	**70**	**75**	**81**	**87**	**92**
19	—	2	7	13	19	25	32	38	45	52	58	65	72	78	85	92	99	106	113	119
	—	**0**	**3**	**7**	**12**	**17**	**22**	**28**	**33**	**39**	**45**	**51**	**56**	**63**	**69**	**74**	**81**	**87**	**93**	**99**
20	—	2	8	13	20	27	34	41	48	55	62	69	76	83	90	98	105	112	119	127
	—	**0**	**3**	**8**	**13**	**18**	**24**	**30**	**36**	**42**	**48**	**54**	**60**	**67**	**73**	**79**	**86**	**92**	**99**	**105**

Table E is taken from the *Bulletin of the Institute of Educational Research* Vol. 1, No. 2, Indiana University, with permission of the publishers.

[a]Discussed in Section 20.4. To be significant, the observed U must equal or be *less than* the value shown in the table. Dashes in the table indicate that no decision is possible at the specified level of significance.

DIRECTIONAL TEST
.05 Level of Significance (light numbers)
.01 Level of Significance (heavy numbers)

n_2 \ n_1	1	2	3	4	5	6	7	8	9	10	11	12	13	14	15	16	17	18	19	20
1	—	—	—	—	—	—	—	—	—	—	—	—	—	—	—	—	—	—	0	0
	—	—	—	—	—	—	—	—	—	—	—	—	—	—	—	—	—	—	—	—
2	—	—	—	—	0	0	0	1	1	1	1	2	2	2	3	3	3	4	4	4
	—	—	—	—	—	—	—	—	—	—	—	—	**0**	**0**	**0**	**0**	**0**	**0**	**1**	**1**
3	—	—	0	0	1	2	2	3	3	4	5	5	6	7	7	8	9	9	10	11
	—	—	—	—	—	**0**	**0**	**1**	**1**	**1**	**2**	**2**	**2**	**3**	**3**	**4**	**4**	**4**	**4**	**5**
4	—	—	0	1	2	3	4	5	6	7	8	9	10	11	12	14	15	16	17	18
	—	—	—	—	**0**	**1**	**1**	**2**	**3**	**3**	**4**	**5**	**5**	**6**	**7**	**7**	**8**	**9**	**9**	**10**
5	—	0	1	2	4	5	6	8	9	11	12	13	15	16	18	19	20	22	23	25
	—	—	—	**0**	**1**	**2**	**3**	**4**	**5**	**6**	**7**	**8**	**9**	**10**	**11**	**12**	**13**	**14**	**15**	**16**
6	—	0	2	3	5	7	8	10	12	14	16	17	19	21	23	25	26	28	30	32
	—	—	—	**1**	**2**	**3**	**4**	**6**	**7**	**8**	**9**	**11**	**12**	**13**	**15**	**16**	**18**	**19**	**20**	**22**
7	—	0	2	4	6	8	11	13	15	17	19	21	24	26	28	30	33	35	37	39
	—	—	**0**	**1**	**3**	**4**	**6**	**7**	**9**	**11**	**12**	**14**	**16**	**17**	**19**	**21**	**23**	**24**	**26**	**28**
8	—	1	3	5	8	10	13	15	18	20	23	26	28	31	33	36	39	41	44	47
	—	—	**0**	**2**	**4**	**6**	**7**	**9**	**11**	**13**	**15**	**17**	**20**	**22**	**24**	**26**	**28**	**30**	**32**	**34**
9	—	1	3	6	9	12	15	18	21	24	27	30	33	36	39	42	45	48	51	54
	—	—	**1**	**3**	**5**	**7**	**9**	**11**	**14**	**16**	**18**	**21**	**23**	**26**	**28**	**31**	**33**	**36**	**38**	**40**
10	—	1	4	7	11	14	17	20	24	27	31	34	37	41	44	48	51	55	58	62
	—	—	**1**	**3**	**6**	**8**	**11**	**13**	**16**	**19**	**22**	**24**	**27**	**30**	**33**	**36**	**38**	**41**	**44**	**47**
11	—	1	5	8	12	16	19	23	27	31	34	38	42	46	50	54	57	61	65	69
	—	—	**1**	**4**	**7**	**9**	**12**	**15**	**18**	**22**	**25**	**28**	**31**	**34**	**37**	**41**	**44**	**47**	**50**	**53**
12	—	2	5	9	13	17	21	26	30	34	38	42	47	51	55	60	64	68	72	77
	—	—	**2**	**5**	**8**	**11**	**14**	**17**	**21**	**24**	**28**	**31**	**35**	**38**	**42**	**46**	**49**	**53**	**56**	**60**
13	—	2	6	10	15	19	24	28	33	37	42	47	51	56	61	65	70	75	80	84
	—	**0**	**2**	**5**	**9**	**12**	**16**	**20**	**23**	**27**	**31**	**35**	**39**	**43**	**47**	**51**	**55**	**59**	**63**	**67**
14	—	2	7	11	16	21	26	31	36	41	46	51	56	61	66	71	77	82	87	92
	—	**0**	**2**	**6**	**10**	**13**	**17**	**22**	**26**	**30**	**34**	**38**	**43**	**47**	**51**	**56**	**60**	**65**	**69**	**73**
15	—	3	7	12	18	23	28	33	39	44	50	55	61	66	72	77	83	88	94	100
	—	**0**	**3**	**7**	**11**	**15**	**19**	**24**	**28**	**33**	**37**	**42**	**47**	**51**	**56**	**61**	**66**	**70**	**75**	**80**
16	—	3	8	14	19	25	30	36	42	48	54	60	65	71	77	83	89	95	101	107
	—	**0**	**3**	**7**	**12**	**16**	**21**	**26**	**31**	**36**	**41**	**46**	**51**	**56**	**61**	**66**	**71**	**76**	**82**	**87**
17	—	3	9	15	20	26	33	39	45	51	57	64	70	77	83	89	96	102	109	115
	—	**0**	**4**	**8**	**13**	**18**	**23**	**28**	**33**	**38**	**44**	**49**	**55**	**60**	**66**	**71**	**77**	**82**	**88**	**93**
18	—	4	9	16	22	28	35	41	48	55	61	68	75	82	88	95	102	109	116	123
	—	**0**	**4**	**9**	**14**	**19**	**24**	**30**	**36**	**41**	**47**	**53**	**59**	**65**	**70**	**76**	**82**	**88**	**94**	**100**
19	0	4	10	17	23	30	37	44	51	58	65	72	80	87	94	101	109	116	123	130
	—	**1**	**4**	**9**	**15**	**20**	**26**	**32**	**38**	**44**	**50**	**56**	**63**	**69**	**75**	**82**	**88**	**94**	**101**	**107**
20	0	4	11	18	25	32	39	47	54	62	69	77	84	92	100	107	115	123	130	138
	—	**1**	**5**	**10**	**16**	**22**	**28**	**34**	**40**	**47**	**53**	**60**	**67**	**73**	**80**	**87**	**93**	**100**	**107**	**114**

TABLE F
CRITICAL VALUES OF WILCOXON T[a]

LEVEL OF SIGNIFICANCE

	Nondirectional Test						Directional Test				
	.05	.01		.05	.01		.05	.01		.05	.01
n			**n**			**n**			**n**		
5	—	—	**28**	116	91	**5**	0	—	**28**	130	101
6	0	—	**29**	126	100	**6**	2	—	**29**	140	110
7	2	—	**30**	137	109	**7**	3	0	**30**	151	120
8	3	0	**31**	147	118	**8**	5	1	**31**	163	130
9	5	1	**32**	159	128	**9**	8	3	**32**	175	140
10	8	3	**33**	170	138	**10**	10	5	**33**	187	151
11	10	5	**34**	182	148	**11**	13	7	**34**	200	162
12	13	7	**35**	195	159	**12**	17	9	**35**	213	173
13	17	9	**36**	208	171	**13**	21	12	**36**	227	185
14	21	12	**37**	221	182	**14**	25	15	**37**	241	198
15	25	15	**38**	235	194	**15**	30	19	**38**	256	211
16	29	19	**39**	249	207	**16**	35	23	**39**	271	224
17	34	23	**40**	264	220	**17**	41	27	**40**	286	238
18	40	27	**41**	279	233	**18**	47	32	**41**	302	252
19	46	32	**42**	294	247	**19**	53	37	**42**	319	266
20	52	37	**43**	310	261	**20**	60	43	**43**	336	281
21	58	42	**44**	327	276	**21**	67	49	**44**	353	296
22	65	48	**45**	343	291	**22**	75	55	**45**	371	312
23	73	54	**46**	361	307	**23**	83	62	**46**	389	328
24	81	61	**47**	378	322	**24**	91	69	**47**	407	345
25	89	68	**48**	396	339	**25**	100	76	**48**	426	362
26	98	75	**49**	415	355	**26**	110	84	**49**	446	379
27	107	83	**50**	434	373	**27**	119	92	**50**	466	397

Table F is taken from F. Wilcoxon and R.A. Wilcox, *Some Rapid Approximate Statistical Procedures*, Lederle Laboratories, New York, 1964, with permission of the publishers.

[a]Discussed in Section 20.11. To be significant, the observed T must equal or be *less than* the value shown in the table. Dashes in the table indicate that no decision is possible at the specified level of significance.

TABLE G
RANDOM NUMBERS[a]

Row number										
00000	10097	32533	76520	13586	34673	54876	80959	09117	39292	74945
00001	37542	04805	64894	74296	24805	24037	20636	10402	00822	91665
00002	08422	68953	19645	09303	23209	02560	15953	34764	35080	33606
00003	99019	02529	09376	70715	38311	31165	88676	74397	04436	27659
00004	12807	99970	80157	36147	64032	36653	98951	16877	12171	76833
00005	66065	74717	34072	76850	36697	36170	65813	39885	11199	29170
00006	31060	10805	45571	82406	35303	42614	86799	07439	23403	09732
00007	85269	77602	02051	65692	68665	74818	73053	85247	18623	88579
00008	63573	32135	05325	47048	90553	57548	28468	28709	83491	25624
00009	73796	45753	03529	64778	35808	34282	60935	20344	35273	88435
00010	98520	17767	14905	68607	22109	40558	60970	93433	50500	73998
00011	11805	05431	39808	27732	50725	68248	29405	24201	52775	67851
00012	83452	99634	06288	98033	13746	70078	18475	40610	68711	77817
00013	88685	40200	86507	58401	36766	67951	90364	76493	29609	11062
00014	99594	67348	87517	64969	91826	08928	93785	61368	23478	34113
00015	65481	17674	17468	50950	58047	76974	73039	57186	40218	16544
00016	80124	35635	17727	08015	45318	22374	21115	78253	14385	53763
00017	74350	99817	77402	77214	43236	00210	45521	64237	96286	02655
00018	69916	26803	66252	29148	36936	87203	76621	13990	94400	56418
00019	09893	20505	14225	68514	46427	56788	96297	78822	54382	14598
00020	91499	14523	68479	27686	46162	83554	94750	89923	37089	20048
00021	80336	94598	26940	36858	70297	34135	53140	33340	42050	82341
00022	44104	81949	85157	47954	32979	26575	57600	40881	22222	06413
00023	12550	73742	11100	02040	12860	74697	96644	89439	28707	25815
00024	63606	49329	16505	34484	40219	52563	43651	77082	07207	31790
00025	61196	90446	26457	47774	51924	33729	65394	59593	42582	60527
00026	15474	45266	95270	79953	59367	83848	82396	10118	33211	59466
00027	94557	28573	67897	54387	54622	44431	91190	42592	92927	45973
00028	42481	16213	97344	08721	16868	48767	03071	12059	25701	46670
00029	23523	78317	73208	89837	68935	91416	26252	29663	05522	82562
00030	04493	52494	75246	33824	45862	51025	61962	79335	65337	12472
00031	00549	97654	64051	88159	96119	63896	54692	82391	23287	29529
00032	35963	15307	26898	09354	33351	35462	77974	50024	90103	39333
00033	59808	08391	45427	26842	83609	49700	13021	24892	78565	20106
00034	46058	85236	01390	92286	77281	44077	93910	83647	70617	42941
00035	32179	00597	87379	25241	05567	07007	86743	17157	85394	11838
00036	69234	61406	20117	45204	15956	60000	18743	92423	97118	96338
00037	19565	41430	01758	75379	40419	21585	66674	36806	84962	85207
00038	45155	14938	19476	07246	43667	94543	59047	90033	20826	69541
00039	94864	31994	36168	10851	34888	81553	01540	35456	05014	51176
00040	98086	24826	45240	28404	44999	08896	39094	73407	35441	31880
00041	33185	16232	41941	50949	89435	48581	88695	41994	37548	73043
00042	80951	00406	96382	70774	20151	23387	25016	25298	94624	61171
00043	79752	49140	71961	28296	69861	02591	74852	20539	00387	59579
00044	18633	32537	98145	06571	31010	24674	05455	61427	77938	91936
00045	74029	43902	77557	32270	97790	17119	52527	58021	80814	51748
00046	54178	45611	80993	37143	05335	12969	56127	19255	36040	90324
00047	11664	49883	52079	84827	59381	71539	09973	33440	88461	23356
00048	48324	77928	31249	64710	02295	36870	32307	57546	15020	09994
00049	69074	94138	87637	91976	35584	04401	10518	21615	01848	76938
00050	09188	20097	32825	39527	04220	86304	83389	87374	64278	58044
00051	90045	85497	51981	50654	94938	81997	91870	76150	68476	64659
00052	73189	50207	47677	26269	62290	64464	27124	67018	41361	82760
00053	75768	76490	20971	87749	90429	12272	95375	05871	93823	43178
00054	54016	44056	66281	31003	00682	27398	20714	53295	07706	17813
00055	08358	69910	78542	42785	13661	58873	04618	97553	31223	08420
00056	28306	03264	81333	10591	40510	07893	32604	60475	94119	01840
00057	53840	86233	81594	13628	51215	90290	28466	68795	77762	20791
00058	91757	53741	61613	62669	50263	90212	55781	76514	83483	47055
00059	89415	92694	00397	58391	12607	17646	48949	72306	94541	37408

Table G is taken from The Rand Corporation, *A Million Random Digits with 100,000 Normal Deviates*, The Free Press, Glencoe, Ill., 1955, with permission of the Rand Corporation.
[a]Discussed in Section 9.5.

TABLE H
APPROXIMATE SAMPLE SIZE FOR HYPOTHESIS TEST ABOUT POPULATION MEANS (z AND t TESTS)

SINGLE SAMPLE[a]

Effect Size	.05 (α) .20 (β)	.01 (α) .20 (β)	.05 (α) .05 (β)	.01 (α) .05 (β)
		Two-tailed Test		
d				
0.2 (small)	198	294	326	447
0.5 (medium)	34	51	54	75
0.8 (large)	15	22	23	32
		One-tailed Test		
d				
0.2 (small)	156	253	272	396
0.5 (medium)	27	43	45	66
0.8 (large)	12	19	19	28

TWO SAMPLES[b]

Effect Size	.05 (α) .20 (β)	.01 (α) .20 (β)	.05 (α) .05 (β)	.01 (α) .05 (β)
		Two-tailed Test		
d				
0.2 (small)	393	586	651	892
0.5 (medium)	64	96	105	144
0.8 (large)	26	39	42	57
		One-tailed Test		
d				
0.2 (small)	310	503	542	790
0.5 (medium)	51	82	88	128
0.8 (large)	21	33	35	51

Table H is taken from J. Cohen, *Statistical Power Analysis for the Behavioral Sciences*, rev. ed., Academic Press, New York, 1977, with permission of the author and publishers.

[a]Discussed in Section 12.6.

[b]Discussed in Section 14.12 for two independent samples and in Section 15.5 for two dependent samples. Values in this table refer to the size for *each* of the two samples.

Index

Addition rule. See *Probability*.
Alpha (α) error. See *Type I error*.
Alternative hypothesis, 124–126. See also *Hypothesis*.
Analysis of variance, 190–220
  alternative hypothesis, 194, 202, 211
  ANOVA tables, 199–200, 214–215
  assumptions, 202–203, 217
  degrees of freedom, 196–197, 212–214
  F ratio, 192, 198, 210–211, 214
  F test, 193–194, 202, 211
  interaction, 208–209, 215–216
  meaning of, 190–193
  mean squares, 195, 214
  multiple comparisons, 200–202, 217
  null hypothesis, 190, 207–211
  one-factor case, 190–192
  published reports, 200
  Scheffé's test, 201–202
  sum of squares, 195–197, 212–213
  tables, 199
  two-factor case, 207–211
  variability between groups, 192
  variability within groups, 192
  variance estimates, 194–195
ANOVA, 190
Arithmetic mean. See *Mean*.
Arithmetic operations, 253–258
Average(s), 35–45
  and skewed distributions, 37–39
  common usage, 42–43
  for grouped data, 41
  for qualitative data, 41
  for quantitative data, 35–37
  for ranked data, 41–42
  mean, 36–37
  median, 36
  mode, 35–36

Bar graph, 26
Beta (β) error. See *Type II error*.
Bimodal distribution, 36
Bradley, J., 248

Camilli, G., 233
Central limit theorem, 112–114
Central tendency, measures of, 35
Chi square, alternative hypothesis, 225
  degrees of freedom, 222–223, 231
  expected frequencies, 221–222, 229–230, 233
  expected proportions, 228–229
  formula, 222
  null hypothesis, 221–222
  one degree of freedom, 225–227, 232–233
  one-variable case, 221
  precautions, 233
  published reports, 232
  tables, 222–223
  two-variable case, 221
  Yates' correction, 226
Class intervals, 13–16
Cohen, J., 139
Confidence, level of, 181–182
Confidence interval, 177–189
  and effect of sample size, 182
  compared to hypothesis test, 186–187
  for difference between population
     means, dependent samples, 186
    independent samples, 184–185
  for single population mean, 178–181, 182–183
  interpretation of, 181

Correlation, 71–83
  and cause-effect, 79
  and prediction, 90–91
  coefficient, Pearson r, 75–79
    Spearman r_s, 79–81
  hypothesis test for, 150–151
  meaning of, 71–73, 78–79
  scatterplots for, 73–75
Counterbalancing, 173
Critical z scores, 122
  table of, 127
Cumulative frequency distribution, 17
Curvilinear relationship, 75

Data, overview, 5–6
  qualitative, 4–5
  quantitative, 2
  ranked, 4
Decision rule, 122–123
Degrees of freedom, 147–150
  in analysis of variance, 196–197, 212–214
  in chi square, 222–223, 231
  in one sample, 145, 147–150
  in two samples, 157, 168
Dependent samples, 172–173
Descriptive statistics, 1, 9–93
  definition of, 1
Difference, between population means, 155
  between sample means, 154
Difference score, 168–169
Directional and nondirectional tests, 124–126
Distribution, bimodal, 36
  normal, 54–56
  sampling, 107–110
  skewed, negatively, 39–40
    positively, 39–40
Distribution-free tests, 248

Effect, 138
Effect size, 138–139, 159–161, 310
Error, prediction, 84–86
  random, 192
  type I, 133
  type II, 133
Estimate, interval. See Confidence interval.
  point, 178, 184
Expected frequency, 221–222, 229–230

F Test, 193–194, 202, 211. See also Analysis of variance.
Frequency distribution, cumulative, 17
  for qualitative data, 17–18
  for quantitative data, 11–17
    grouped, 13
    ungrouped, 11
  for ranked data, 18
  guidelines for construction of, 13–15
Frequency polygon, 24–25

Grouped data, 13

H (Kruskall-Wallis) Test, 245–247
  and ties, 247–248
  as replacement for F, 203, 246, 248
  decision rule, 246
  tables, 246
Histogram, 23
Homogeneity of variance, assumption of, 156
Homoscedasticity, assumption of, 91
Hopkins, K.D., 233
Huff, D., 28
Hypothesis, alternative, 122, 124–126
  choice of, 126
  defined, 122
  directional, 124–125
  nondirectional, 125–126
  null, defined, 121
    secondary status of, 161–162
  research, 161–162
Hypothesis tests, 116–176, 190–251
  compared to confidence intervals, 186–187
  for qualitative data. See Chi square.
  for quantitative data. See F, t, and z tests.
  for ranked data. See H, T, and U tests.
  less structured approach to, 163
  meaning of, 118–119
  published reports of, 162–163

Independent samples, 161
Inferential statistics, 1–2, 95–251
  definition of, 1–2
Interaction, 208–209, 215–216
Interval estimate. See Confidence interval.
Interval-ratio measurement, 2–4, 42

Keppel, G., 202, 217
Kruskall-Wallis test. See H test.

Least squares prediction, 86–88
Level of measurement, interval-ratio, 2–4, 42
  nominal, 5, 221
  ordinal, 4, 42, 236
  overview, 5–6
Level of significance, as type I error, 134
  choice of, 126–127
  defined, 123
Linear relationship, 75

Mann-Whitney test. See U Test.
Matching subjects, 172

Mean, and skewed distributions, 37–39
  for quantitative data, 36–37
  for ranked data, 42
  of difference scores, 169
  of population, 109–111
  of sample, 109–111
  of sampling distribution of mean, 109–111
  sampling distribution of, 107–110
  special property of, 39–40
  standard error of, 111–112
Mean squares, 195, 197–198, 214
Measurement, level of. See *Level of measurement.*
Median, for grouped data, 41
  for qualitative data, 41
  for quantitative data, 36
  for ranked data, 42
Minium, E.W., 182, 202
Mode, and bimodal distributions, 36
  and multimodal distributions, 36
  for grouped data, 41
  for qualitative data, 41
  for quantitative data, 35–36
Multiple comparisons, 200–202, 217
  with Scheffé's test, 201–202
Multiplication rule. See *Probability.*

Negatively skewed distribution, 39–40
Negative relationship, 73, 74, 75
Nominal measurement, 5, 221
Nondirectional and directional tests, 124–126
Nonparametric tests, 248
Normal distribution, 54–70
  and central limit theorem, 112–114
  and common standard scores, 67
  compared to *t* distribution, 145–146
  general properties, 54–56
  problems, finding areas, 59–63
    finding scores, 63–65
  standard, 57
  tables, 57–58
Null hypothesis, 121. See also *Hypothesis.*

Observed frequency, 222, 229
One-tailed and two-tailed tests, 124–126
Ordinal measurement, 4, 42, 236

Parameter, 248
Parametric test, 248
Pearson, Karl, 75
Pearson r, 75. See also *Correlation.*
  and population correlation coefficient, 150
Percentile ranks, 67–68
Pie chart, 26
Point estimate, for difference between population means, 184
  for single population mean, 178

Pooled variance estimate, 157
Population, 98
  hypothetical, 98
  mean, 109–111
  mean of difference scores, 169
  real, 98
  standard deviation, 109, 111, 112, 156
  variance, 156
Positively skewed distribution, 39–40
Positive relationship, 71–72, 74, 75
Prediction, 84–93
  and correlation, 90
  and regression toward the mean, 89–90
  assumptions, 91
  equation, 86–88
  error, 84–86, 88–89
  standard error of, 88–89
Probability, and addition rule, 103
  and multiplication rule, 103
  as area under curve, 103–104
  defined, 102

Qualitative data, averages for, 41
  defined, 4–5
  frequency distribution for, 17–18
  graphs for, 23–25
  hypothesis test for, 224, 226, 232
  measures of variability for, 51–52
Quantitative data, averages for, 35–41
  correlation coefficient for, 75–79
  defined, 2
  frequency distribution for, 11–17
  graphs for, 26–28
  hypothesis tests for, 120, 146, 150, 159, 169, 194, 211
  measures of variability for, 46–51

Random assignment of subjects, 101–102
Random error, 192, 210
Random numbers, tables of, 100, 102, 309
Random sample, and hypothetical populations, 101
  defined, 99
  fish-bowl method for, 99–100
  tables of random numbers for, 100, 102, 309
Range, 46–47
Ranked data, averages for, 41–42
  correlation coefficient for, 79–81
  defined, 4
  frequency distributions for, 18
  hypothesis tests for, 240, 243, 247
  measures of variability for, 51–52
Regression. See *Prediction.*
Regression toward the mean, 89–90
Relationships between variables, curvilinear, 75
  linear, 74
  negative, 73, 74, 75
  positive, 71–72, 74, 75
Relative frequency distribution, 16

Research hypothesis, 161–162
Rounding, rule for, 257

Sample(s), 98
    all possible, 107–109
    mean, 109–111
    mean of difference scores, 168
    standard deviation, 144
    variance, 157
Sample size, and probability of type II error,
        136–137
    selection of, for confidence interval, 182
        for one sample, 137–140, 143
        for two dependent samples, 170–171
        for two independent samples, 159–
            161, 162
Sampling distribution, of difference be-
        tween means, dependent samples,
        168
    independent samples, 154–157
    of F, 199, 301
    of mean, defined, 107–109, 114
        hypothesized, 116–119
        hypothesized and true, 134–135
        mean, 111
        shape, 112–114
        standard error, 111–112, 144
    of t, 145–146, 147
    of z, 119–120
    of χ^2, 224
Sampling variability, of correlation coeffi-
        cient, 15, 78–79
    of least squares prediction equation, 89
    of mean, 39–40
Scatterplot, 73–75
Scheffé's test, 201–202
Significance, level of. See Level of signifi-
        cance.
Skewed distribution, 39
Spearman, Charles, 80
Standard deviation, in descriptive statis-
        tics, deviation formula, 48–49
    general properties, 47–48
    measure of distance, 49–50
    raw score formula, 50–51
    in inferential statistics, degrees of free-
        dom, 147–150
    deviation formula, 144
    raw score formula, 144
Standard error, of difference between
        means, dependent samples, 168–
        169
    independent samples, 156–157
    of mean, 111–112, 144
    of prediction, 88–89
Standard normal distribution. See Normal
        distribution.
Standard scores, general properties, 65
    other types, 66–67, 119–120
    T scores, 66
    z scores, 65–67
Statistics, descriptive, 1
    inferential, 1–2

Statistical significance, 162
Stem and leaf arrangements, 12–13
Sum of squares, 195–197, 212–213

Test of hypothesis. See Hypothesis test.
Ties in ranks, 42, 247–248
Treatment effect, 192, 210
t Test, and F test, 202
    and z test, 145
    assumptions, 147, 151, 161
    for correlation coefficient, 150–151
    for one population mean, 146–149
    for two population means, dependent
        samples, 169–171
    independent samples, 159–160
T Test (Wilcoxon), and ties, 247–248
    as replacement for t, 172, 242, 248
    calculation, 242–243
    decision rule, 243
    large sample approximation, 244–245
    tables, 248
Two-tailed and one-tailed tests, 124–126
Type I error, and effect of multiple tests,
        200–201
    compared with type II error, 140
    defined, 133
    probability of, 134
Type II error, and difference between true
        and hypothesized population means,
        134–136
    and sample size, 136–137
    compared with type I error, 140
    defined, 133
    probability of, 135–136

U Test (Mann-Whitney), and ties, 247–248
    as replacement for t, 161, 237, 248
    calculation, 237–239
    decision rule, 239
    large sample approximation, 241
    tables, 239

Variability, measures of, 46–53. See also
        Standard error.
    for qualitative data, 51–52
    for quantitative data, 46–51
    for ranked data, 52
    range, 46–47
    standard deviation, descriptive statis-
        tics, 47–49
    inferential statistics, 144
    variance, descriptive statistics, 51
    inferential statistics, 157, 194–195
Variance, defined, 51
    estimates of, in analysis of variance,
        194–195
    pooled, 157
Variance, homogeneity of, 156–157

Wilcoxon test. See *T Test.*

Yates' correction, 225–227

z Score, and correlation, 75–77
  and hypothesis tests, 119–121
  and normal distribution, 57–58

z Score, *(Continued)*
  and nonnormal distributions, 65–66
  and other standard scores, 66–67
  and prediction, 90
  critical, 122, 127
  defined, 56–57
z Test, compared with *t* test, 145
  for population mean, 119–121
  for two population means, 156
  tables of critical values for, 127